THE SUN AS A VARIABLE STAR:
SOLAR AND STELLAR IRRADIANCE VARIATIONS

.

Photograph of participants of IAU Colloquium No. 143 'The Sun as a Variable Star: Solar and Stellar Irradiance Variations', held in Boulder, Colorado, June 20–25, 1993. Nearly 200 scientists participated in this meeting.

THE SUN AS A VARIABLE STAR: SOLAR AND STELLAR IRRADIANCE VARIATIONS

*Proceedings of the 143rd Colloquium
of the International Astronomical Union
held in the Clarion Harvest House, Boulder,
Colorado, June 20–25, 1993*

Edited by

JUDIT M. PAP
*Jet Propulsion Laboratory, MS 171–400,
4800 Oak Grove Drive, Pasadena,
CA 91109–8099, U.S.A.*

CLAUS FRÖHLICH
*Physikalisch-Meteorologisches Observatorium Davos,
World Radiation Center, Davos Dorf, Switzerland*

HUGH S. HUDSON
*Institute for Astronomy, University of Hawaii,
Honolulu, Hawaii, U.S.A.*

and

W. KENT TOBISKA
*TELOS/JPL, MS 264-765,
4800 Oak Grove Drive,
Pasadena, CA 91109, U.S.A.*

Reprinted from Solar Physics, Volume 152, No. 1, 1994

SPRINGER-SCIENCE+BUSINESS MEDIA, B.V.

Library of Congress Cataloging-in-Publication Data

ISBN 978-94-010-4410-3   ISBN 978-94-011-0950-5 (eBook)
DOI 10.1007/978-94-011-0950-5

Printed on acid-free paper

TABLE OF CONTENTS

(The Sun as a Variable Star: Solar and Stellar Irradiance Variations)

Preface

The IAU Colloquium No. 143 "The Sun as a Variable Star: Solar and Stellar Irradiance Variations" was held on June 20 – 25, 1993 at the Clarion Harvest House, Boulder, Colorado, USA. The main objective of this Colloquium was to review the most recent results on the observations, theoretical interpretations, and empirical and physical models of the variations observed in solar and stellar irradiances. A special emphasis of the Colloquium was to discuss the results gained on the climatic impact of solar irradiance variability.

The study of changes in solar and stellar irradiances has been of high interest for a long time. Determining the absolute value of the luminosity of stars with different ages is a crucial question for the theory of stellar evolution and energy production of stellar interiors. Observations of the temporal changes of solar and stellar irradiances – in the entire spectral band and at different wavelengths – provide an additional tool for studying the physical processes below the photosphere and in the solar– stellar atmospheres. Since the Sun's radiative output is the main driver of the physical processes within the Earth's atmosphere, the study of irradiance changes is an extremely important issue for climatic studies as well. Climatic models show that small, but persistent changes in solar irradiance may influence the Earth's climate. Furthermore, to understand the human effect on global climatic change, the role of irradiance variations (as a significant source of natural climate changes) in terrestrial and climatic processes must be revealed.

The Colloquium was a historical meeting since this was the first time when a conference sponsored by the International Astronomical Union was entirely devoted to irradiance variations and their climatic impact. 200 scientists from 30 countries participated in this Colloquium. The Colloquium was divided into six sessions as defined by their key topics: (1) General Reviews on Observations of Solar and Stellar Irradiance Variability; (2) Observational Programs for Solar and Stellar Irradiance Variability; (3) Variability of Solar and Stellar Irradiance Related to the Network, Active Regions (Sunspots and Plages), and Large-Scale Magnetic Structures; (4) Empirical Models of Solar Total and Spectral Irradiance Variability; (5) Solar and Stellar Oscillations, Irradiance Variations and their Interpretation; and (6) The Response of the Earth's Atmosphere to Solar Irradiance Variations and Sun-Climate Connections. A special 1-day session of the "*Solar Electromagnetic Radiation Study for Solar Cycle 22*" (SOLERS22) was held on June 25, 1993, where the five working groups discussed their progress and future plans on measuring the absolute value of total solar and spectral irradiances and studying their temporal variations.

There were 36 invited talks and 110 contributed poster papers presented at the Colloquium. These papers have demonstrated that the solar energy output changes on different time scales: the short-term (from minutes to months) variations are related to surface modulations mainly caused by the evolution of active regions; the solar-cycle-related long-term variations are directly linked with the evolution of magnetic fields over the activity cycle; while the secular variations over centuries are associated with long-term internal modulations. It is now established that the terrestrial climate, radiative environment, and upper atmospheric chemistry are strongly influenced by the varying luminosity of the Sun. The human consequences are such that quantitative study over the solar magnetic cycle is imperative for societal planning. The solar variability together with the accumulation of anthropogenic pollutants determine the human milieu of the future.

Although considerable information exists on solar-stellar irradiance variations, their physical origin is not well-understood. The lack of adequate physical models of irradiance variations for predicting the solar- induced climatic changes led to extensive discussions of the consequences of the planned delay or even a possible termination of irradiance observations performed in space. Based on these discussions, the Scientific Organizing Committee released a resolution that addressed this issue to the leaders of the Space Agencies concerned. The resolution was also forwarded to the General Secretary of the International Astronomical Union by the President of IAU Commission 10. The following recommendations were implemented in the resolution:

- *NASA*

    (1) High precision observations of solar total and spectral irradiance from UARS and ATLAS are required for one full solar 11-year cycle to achieve progress in understanding key issues of climate change and ozone variation.

    (2) Future flight opportunities on EOS for continuos monitoring of the solar radiative outputs urgently need to be maintained, to study the longer-term trends of potential importance to global change. Failure to achive overlapping continuous measurement sets from successive satellites will reduce the usefulness of the excellent data already obtained (since 1980) addressing these issues.

- *ESA*

    Considering the results obtained from measurements of solar total and spectral irradiance of the Sun from the ultraviolet to the infrared, a second flight of EURECA would provide a unique opportunity to extend the measurements carried out by UARS and EURECA-1, to fill the gaps before and after the launch of SOHO. This proposal is in line with the general objectives of the European Community with respect to the global change program.

- *ISAS and NASDA*

    The contribution of *Yohkoh* to studies of solar global properties are recognized, and the participants of IAU Colloquium No. 143 recommend and encourage continued analysis of *Yohkoh* data, and of the development of future Japanese solar and applications satellites that can extend our research knowledge in these important directions.

The proceedings of the Colloquium have been printed in two volumes. This volume contains the contributed papers accepted for publication in Solar Physics. The invited papers have been published separately by Cambridge University Press.

Many people and many organizations helped in organizing the Colloquium and making the proceedings ready for publication. We are sincerely thankful to Dr. Jacqueline Bergeron, IAU General Secretary, and Dr. Immo Appenzeller, IAU Assistant General Secretary, for their help and assistance throughout the planning process of the meeting. We would like to take this opportunity to express our gratitude to Dr. Richard C. Willson (JPL, Solar Irradiance Monitoring Group) and Dr. Ernest Hildner (NOAA Space Environment Laboratory) for their continuing support. TELOS System Group provided support services which were greatly appreciated.

We also would like to express our appreciation and thanks to the members of the Scientific Organizing Committee for their support. The Scientific Organizing Committee consisted of Sallie Baliunas, Richard F. Donnelly, Peter Foukal, Claus Fröhlich (Co-Chair), Vic Gaizauskas, Ernest Gurtovenko, Tadashi Hirayama, Hugh S. Hudson (Co-Chair), Julius London, Judit M. Pap (Chair), Paul Simon, Henk Spruit, Béla Szeidl, Jean-Claude Vial, and Oran R. White. Here we are very sorry to report that Dr. Ernest Gurtovenko died on January 20, 1994. Dr. Gurtovenko headed the Solar Physics Department of the Main Astronomical Observatory of the Ukrainian Academy of Sciences for years and served on many national and international committees. His work gave a significant contribution to the models of solar radiation, and he built the space telescope *DIFOS* to be launched on *CORONAS* in 1994 for studying global oscillations of solar brightness. His work as a member of the Scientific Organizing Committee provided significant help to scientists from the former Soviet Union to attend the Colloquium.

We also would like to express our gratitude to the members of the Local Organizing Committee: Pat Bornmann, Peter Fox, Howard Garcia (Chair), Patrick McIntosh, Larry Puga, Gary Rottman, Andy Skumanich, Kent Tobiska (Co-Chair), and Tom Woods for their help and assistance. We are especially thankful to Pat McLane, Helga Mycroft, Liana The, Mary Eberle, Pam Bergstedt, Judy McGarvey, Gwen Dickenson, and Gregory Dickenson for all their time and hard work in handling the numerous details necessary for the planning and smooth running of the meeting. The staff of the Clarion Harvest House provided a pleasant environment for the meeting.

Financial support provided by the National Aeronautics and Space Administration, NASA Jet Propulsion Laboratory, California Institute of Tech-

nology, NOAA Space Environment Laboratory, and Ball Aerospace made it
possible to carry out this meeting. Travel support provided by the International Astronomical Union, National Science Foundation, International
Science Foundation, and the Department of Energy helped more than 30
scientists from different countries to attend the Colloquium. We are very
thankful for the financial assistance of the above organizations.

Judit M. Pap
Claus Fröhlich
Hugh S. Hudson
W. Kent Tobiska
For the Scientific Organizing Committee
March, 1994

CRYOGENIC SOLAR ABSOLUTE RADIOMETER - CSAR

J.E.MARTIN and N.P.FOX
Division of Quantum Metrology, National Physical Laboratory,
Teddington, Middlesex, UK. TW110LW.

ABSTRACT. The Cryogenic Solar Absolute Radiometer (CSAR) is a new primary radiometric standard for use in space traceable to the International System of Units (SI). CSAR will have the potential to measure the total irradiance of the Sun with an uncertainty of 0.01% and a resolution of 0.001%, more than a factor ten improvement over existing instruments. Since CSAR will be cooled by a mechanical cooler based upon the Stirling cycle its working lifetime is projected to be in excess of 10 years.

1. INTRODUCTION

Electrical substitution radiometers (ESRs) operating at ambient temperatures have been the traditional instruments for the precise measurement of black-body radiation. Unfortunately their performance has been limited by the thermal properties of materials at these temperatures ($\approx$ 293 K), and even with the most modern innovative design their absolute accuracy stubbornly remains in the range 0.1% to 0.3% (Gillham 1962, Boivin and Smith 1978, Willson 1979).

During the 1980's work at NPL showed that by cooling the ESR to liquid helium temperatures significant improvements could be achieved in resolution and accuracy. The first cryogenic radiometer constructed measured the power of black-body radiation from a black body held at the triple point of water and was used to determine the Stefan-Boltzmann constant with an uncertainty of 0.02% (Quinn and Martin 1985). The value obtained agreed with the theoretical value which can be calculated using other fundamental constants and is regarded as the best experimental determination to date (Cohen and Taylor 1986).

Following the success of the first cryogenic radiometer a second radiometer was built dedicated to the measurement of optical laser power with an uncertainty of 0.005% (Martin et al 1985) and to demonstrate the validity of this uncertainty the two radiometers were successfully compared in 1990 (Fox and Martin 1990). The second radiometer has become the Primary Standard radiometer (PS radiometer) upon which all UK optical radiation scales are based.

Thus the development of cryogenic radiometry has reduced the measurement uncertainty of optical radiation by at least a factor 10. This has led to most National Standards Laboratories adopting (or proposing to adopt) a cryogenic radiometer for use as their primary standard for optical radiation.

The current position in the uncertainty of the measurement of total solar irradiance is that which existed in the measurement of optical radiation before the advent of cryogenic radiometers.

Terrestrial solar irradiance measurements are based upon the World Radiometric Reference (WRR). The WRR is established from a group of ambient temperature ESRs, known as the World Standard Group (WSG). The individual ESRs in the WSG are periodically compared with one another using the Sun as a standard source; this leads to a reproducibility of the WRR of 0.02% but an absolute uncertainty of 0.3% (Fröhlich 1991).

Irradiance measurements made using three space radiometers of similar design to the WSG, the Solar Maximum Mission Active Cavity Irradiance Monitor (SMM-ACRIM), and the solar monitors aboard the Earth Radiation Budget Satellite (ERBS) and Nimbus-7 are shown in figure 1 (Lee et al 1991). The measurements agree within their uncertainties but

Solar Physics **152**: 1–8, 1994.

there is a maximum disagreement between the instruments of 0.4%.
    More recent data from ACRIM II on the Upper Atmosphere Research
Satellite (UARS-ACRIM) are 0.17% lower than the data from SMM-ACRIM
(just within the combined uncertainty of the instruments). This data
has been adjusted to the SMM-ACRIM data by comparing them with the data
from the NIMBUS-7 and ERBS experiments which have been taken in the
same time-frame as the SMM and UARS flights. Comparisons have also been
made with the shuttle ACRIM instrumentation flown on the ATLAS 1 and 2
missions (Willson 1993). The preliminary results from the Solar
Variability experiments (SOVA 1 and 2) aboard the European Retrievable
Carrier (EURECA) have been reported to fall within the overall spread
of results shown in figure 1 (Romero et al 1993).

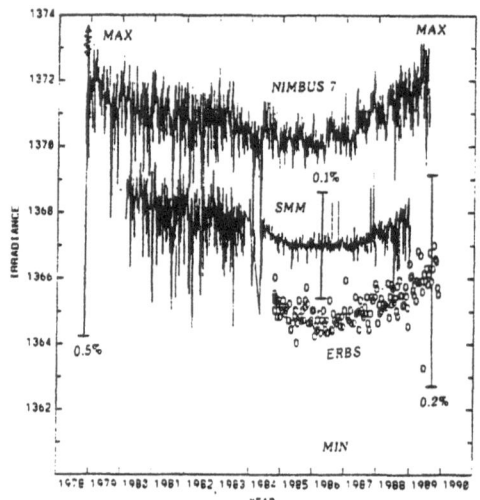

Figure 1. Comparison of solar irradiance values from ERBS, SMM and
Nimbus-7 experiments. The irradiance unit is Wm^{-2} and the error bar
denotes the instrument uncertainty.

    Although the data are the best available to date the
uncertainties in their absolute accuracies can lead to difficulties in
monitoring long term drifts in solar irradiance, for example, the 0.2%
difference between the maximum and the minimum of the eleven year solar
cycle. To overcome this problem the solar community have decided to
establish a continuous database relying on measurements from
overlapping satellite missions to normalise the data. This empirical
approach is not ideal since it is not based on an absolute radiometric
measurement standard and any break in the data chain means that data
recorded before and after the break cannot be compared with sufficient
accuracy. Thus this approach is very vunerable to instrument lifetime,
instrument failure, satellite malfunctions or simply delays in the
launch dates, and it is unlikely that a continuous database can be
generated to cover many solar cycles.
    The development of a Cryogenic Solar Absolute Radiometer, CSAR,
should provide the means of obtaining absolute data which negates the
need of a continuous database. Any break in the data chain can be
bridged by later measurements with a new absolute radiometer. In
addition, if flown in the near future it would link the present data
base to an absolute radiometric standard.
    The objective of CSAR is to measure the total solar irradiance
with at least a ten-fold improvement in the accuracy and resolution
over previous measurements. If this objective can be achieved then CSAR

will

    a. be capable of measuring the total solar irradiance with an uncertainty of 0.01% and a resolution of 0.001% and

    b. become a primary space radiometric standard traceable to a SI based radiometric standard (PS radiometer).

    CSAR also has the advantage of being a low noise instrument and thus it should be possible to improve the measurement of solar noise. Solar noise, that is, the departure of the total solar irradiance from that of a perfect quiet sun has four sources : granulation, mesogranulation, supergranulation and active regions. The total solar irradiance as measured by the present radiometers gives a power spectrum higher than the expected solar noise as derived from simulation (Anderson 1993). SMM-ACRIM measurements show in the range about 10 µHz a power level three to fives times higher than the simulated power (Fröhlich et al 1991). The significance of this result is difficult to assess since the experimental measurements contain a significant component of instrumental noise.
    CSAR could separate the solar activity phases as it seems that during the solar maximum the power is about ten times higher than during the solar minimum at 0.1 µHz, the difference being smaller for higher frequencies (Fröhlich et al 1991). At frequencies lower than 10 µHz the solar power spectrum is mainly due to active regions (sunspots and plages). Interpretation of the variations of the total solar irradiance in terms of the Photometric Sunspot Index and the Facular Index could be improved by measuring more precisely the solar noise. A distinction between the contribution of the active regions and the quiet Sun could be attempted.
    To achieve these objectives the technology and the experience gained with laboratory cryogenic radiometers will need to be transferred to a space-borne radiometer without a serious loss of performance. This paper draws attention to the areas where problems of degradation in accuracy may arise. However, CSAR is based on the state of the art technology and has the very considerable advantage of starting from a much lower uncertainty base than the conventional ambient temperature radiometer. Work already carried out by the authors and others suggests that CSAR can deliver the desired level of performance if it is designed with sufficient redundancy so that any degradation can be monitored in space after deployment.

2. ADVANTAGES OF CRYOGENIC RADIOMETRY

A cryogenic radiometer is an ESR operating at a low temperature (≈4 K). The principal advantages of operating at 4 K are:

i. the thermal diffusivity of copper (the material used for the absorbing cavity) at 4 K is about 1000 times greater than it is at 293 K. Therefore, a larger cavity can be constructed with a high absorptance, minimising the correction for the cavity absorptance that has to be applied for ambient temperature ESRs. The high diffusivity also gives the added advantage that the cavity has a relatively small time constant.

ii. superconducting wire can be used for the heater leads, ensuring all the electrical power is dissipated in the cavity heater and no corrections are required for power dissipated in the leads.

iii. the radiative heat loss from the cavity is small since the radiative coupling between the cavity and it's surroundings is negligible. Thus the effect of different temperature gradients in the cavity that are caused by electrical or radiant heating can be ignored. This is not the case for ambient temperature ESRs where the position of the heater is critical in order to reduce the size of the correction for this effect.

iv. the temperature of the cavity can be more easily monitored with miniature devices that have a high signal to noise ratio, and finally

v. the reference temperature heat sink can more easily be controlled.

Thus at low temperatures the over-riding principle of all ESRs, that is the equivalence of radiant and electrical power, can be clearly demonstrated with a minimum number of corrections combined with a low uncertainty.

3. CSAR

3.1. Description

The overall configuration of CSAR is shown in figure 2. The instrument requires 120 W of power, weighs approximately 50 Kg and fits within an envelope size of 280 x 480 x 750 mm.

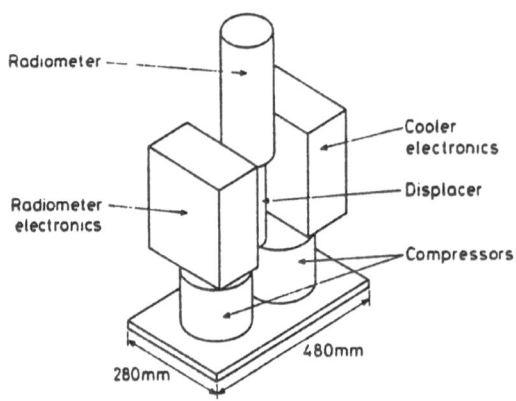

FIGURE 2: The overall configuration of CSAR.

To date all cryogenic radiometers have used liquid helium to provide the cooling. However, CSAR will be cooled by a mechanical cooler based on the Stirling cycle (Jones and Scull 1992). The coolers most recently developed have base temperatures approaching 20 K and are specifically designed for cooling space-borne instruments. There are only marginal differences in the performance of an ESR cooled to 20 K compared to 4 K. As part of the ESA development contract a reliability analysis was undertaken following the guidelines laid down in procedure documents, for example, ESA PSS-01-302. The ten year prediction for cooler reliability for the mechanical parts is 95.9%, the drive electronics 87.4% and the overall system 83.9% (ESA 1992). Bench life-testing of the coolers is in progress, a 80 K cooler has achieved 46,000 hours and a 20 K cooler 14,000 hours. A two stage 80 K cooler aboard ISAMS has achieved about 9000 hours in orbit.

Some of the advantages of cooling CSAR using a mechanical cooler compared to using liquid helium are,

a) the life-time of the instrument is considerably increased with a high probability of achieving in excess of 10 years,

b) the reduction in cost and size of an instrument with this long life-expectancy (it has been calculated that the volume of liquid helium to cool CSAR for ONE year is about 1000 litres) and

c) the cooler can be turned off allowing the instrument to warm and degradation caused from cryo-deposits to be assessed.

The schematic diagram of the radiometer head is shown in figure 3. It consists of three absorbing cavities any one of which can be irradiated by radiation from the Sun by opening a cold shutter. The radiation passes through a view limiting aperture and a defining aperture which are both attached to the cold radiation trap.

The cavities are electro-formed from copper; the external surfaces are gold plated and the internal surfaces are treated with a black coating formed using a plating technique (Kodama et al 1990). The reflectance of the black has been measured at NPL over the range 0.3 μm to 54 μm and is predominantly a diffuse reflector. The diffuse reflection below 1 μm is less than 0.1% rising to 1.3% at 5 μm and 9% at 20 μm. The cavity absorptance has been calculated using the series reflectivity method (Quinn 1990) for the range 0.3 μm to 20 μm (99.99% of the solar power is within this wavelength band). The measured reflectance data has been combined with the spectral power distribution of the sun and the calculated cavity absorptance is 0.99996 and with the 3 mm defining aperture in place 0.999997. The uncertainty in these calculations is 2 parts in 10^5 which stems from the uncertainty in the reflectance measurements. It will also be possible to verify these calculations by measuring the cavity absorptance at discrete laser wavelengths using an integrating sphere (Martin and Fox 1985).

The cavities (the time constant of each cavity is approximately 5 seconds) are attached to poor thermally conducting tubes soldered to a copper reference temperature heat sink which is connected to the second stage of the mechanical cooler via a stainless steel tube. The temperature of the heat sink can be controlled with a stability of 1 part in 10^5.

The cavities are surrounded by three thermal shields, one anchored to the second stage of the cooler, the second anchored to the first stage and the third forms part of the framework.

The electronic circuits for the devices used for the measurement of temperature and dissipating the electrical power have been designed using space qualified components and successfully evaluated using a SPICE computer software package.

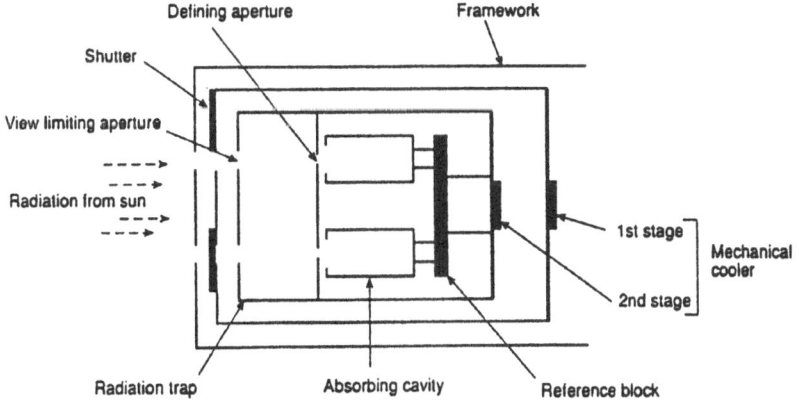

FIGURE 3: Schematic diagram of CSAR.

3.2. Operation

CSAR is designed to operate in a space environment but a complete testing of the instrument can be undertaken in a thermal vacuum chamber for space qualification.

After allowing sufficient time for outgassing the mechanical cooler is activated and the reference temperature heat sink controlled as CSAR approaches its working temperature. When temperature equilibrium is reached and with the shutter closed electrical power is dissipated in one of the cavity heaters in incremental steps and the thermometer response recorded for each step. Cryogenic radiometers are extremely stable instruments and it is anticipated that the thermometer response versus electrical power calibration will only have to be determined once every 12 hours at most. Calculations have shown that stability of the reference temperature heat sink will not be affected by perturbations of the temperature of the outside shield caused by day/night orbital thermal cycling provided they do not exceed 10 K.

With the electrical power off the shutter is opened and radiation from either the Sun or a laser source (used for space qualification testing) passes through the aperture system and is absorbed in the cavity. Since the area of the defining aperture has previously been determined in a separate experiment, the irradiance of the Sun or the laser source can be determined. The measurement system for the thermometer has been designed with an integration time of 10 seconds thus it will be possible to measure the irradiance of the Sun in the short and long term.

3.3. Accuracy of CSAR

Table 1 gives the uncertainties associated with CSAR that contribute to the absolute accuracy of the measured solar irradiance.

TABLE I: The uncertainties associated with CSAR
given as parts in 10^4 of the solar irradiance
at the one standard deviation level.

Electrical power measurement	0.5
Defining aperture area	1.0
Cavity absorptance	0.2
Diffraction of radiation	0.5
Noise	0.1
TOTAL (summed in quadrature)	1.3

The largest single uncertainty arises from the area of the defining aperture, A. It follows that the larger the area of A the smaller the uncertainty, however, other factors limit the size of A to be no larger than 3 mm in diameter. To achieve an overall accuracy of 0.01% the diameter of A must be determined with an uncertainty of 0.15 µm. This uncertainty has been achieved with a precision diamond-turned copper aperture measured on the NPL internal diameter measuring machine which is directly traceable to the National Standard of Length (Goodman et al 1988). The radiometric throughput of this aperture using a laser source has been compared to measurements made with other apertures and the results have been found to be consistent. It will also be possible to cool the aperture using a laboratory mechanical cooler and measure the throughput at low temperatures and comparing the values against those calculated using thermal expansion data for copper.

Finally it will be possible to verify the calculated uncertainties by comparing CSAR against the PS radiometer. This will be achieved using an integrating sphere source irradiated with an intensity stabilised laser beam through a vibrating optical fibre (Anderson et al 1992). The source is extremely stable and spatially uniform, of the order 0.01%, and can operate over the wavelength range 0.4 to 10 μm. Irradiance measurements using this source will be made using the PS radiometer and CSAR.

3.4. Degradation in space

It is recognised that the environment surrounding a satellite is not 'clean' and since CSAR is operating at 20 K there is a high risk that it will act as a 'getter' for the whole satellite. The parts of CSAR most likely to suffer from degradation are the defining aperture and the absorbing cavity.

The area of the defining aperture is the most critical part of CSAR. It is shown in section 3.3 that a 3 mm cold aperture can be measured to the required accuracy in the laboratory but a build up of cryo-deposits would obviously degrade this measurement. The position of the aperture is behind the view limiting aperture and enclosed by the radiation trap which are both at low temperatures, and only molecules etc. that pass directly along the optical axis of CSAR and through the view limiting aperture will impinge on the aperture. Provision has been made for a heater on the aperture mounting plate so that when the cooler is off the aperture can be warmed to above ambient and any cryo-deposits driven off.

Existing space-borne radiometers have used a specular black paint as the absorbing medium in their cavities. Specular absorbing cavities are susceptible to degradation caused by either surface damage to the black or simply the black paint not adhering to the substrate. For CSAR the black is formed using a plating technique directly onto the copper cavity and thus is chemically bonded. By choosing a diffuse reflector and designing a cavity with such a high absorptance the effects of surface damage are minimised and it is unlikely that the performance of CSAR will be significantly downgraded because of degradation of the cavity.

Finally, there is the possibility that in the presence of uv radiation some cryo-deposits may become permanently fixed. Periodically the irradiance as measured with the working cavity and its associated defining aperture will be compared to a measurement using one of the two other cavity/aperture combinations which are shielded when not in use and are unlikely to become degraded. If differences are found then the working cavity/aperture combination can be re-calibrated in space. Thus a check can be made for permament degradation of the working cavity and the defining aperture.

4. CONCLUSION

A long-term radiometric space standard for the 21st century comparable to the best laboratory standards and based on SI units is in the process of being developed. CSAR has the potential to transfer the state of the art technology used in National Standards Laboratories to realise optical radiation standards to the space environment. CSAR will place extra demands on satellite platform capabilities compared to the current ambient temperature radiometers but the rewards are absolute, accurate, low noise data of the total solar irradiance required by the solar physics community.

5. REFERENCES

Andersen B.: 1993, *Solar Phys.*, (to be published).
Anderson V.E., Fox N.P. and Nettleton D.H.: 1992, *Appl. Opt.* 31, 536.
Boivin L.P. and Smith T.C.: 1978, *Appl. Opt.* 17, 3067.

Cohen E.R. and Taylor B.N.: 1986, *Codata Bulletin* **63**, 11.
ESA Contract 9004/90/NL/PP, 1992. Doc. No. TP/MCC/A0163/BAe
Fox N.P. and Martin J.E.: 1990, *Appl. Opt.* **29**, 4686.
Fröhlich C.: 1991, *Metrologia* **28**, 111.
Fröhlich C., Foukal P.V., Hickey J.R., Hudson H.S. and Willson R.C.:
1991, *Sun in Time*, Univ of Arizona Press, Tucson.
Gillham, E.J.: 1962, *Proc. R. Soc. London, Ser.* **A 269**, 249.
Goodman T.M., Martin J.E., Shipp B.D. and Turner N.P.: 1988, *Inst.
Phys. Conf.* **92**, 121.
Jones B.G. and Scull S.R.: 1992, *Cryogenics* **32**, 850.
Kodama S., Horiuchi M., Kuni T. and Kuroda K.: 1990, *IEEE Trans* **39**, 230
Lee III R.B., Gibson M.A., Shivakumar N., Wilson R., Kyle H.L. and
Mecherikunnel A.T.: 1991, *Metrologia* **28**, 265.
Martin J.E., Fox N.P. and Key P.J.: 1985, *Metrologia* **21**, 147.
Quinn T.J. and Martin J.E.: 1985, *Philos. Trans. R. Soc. London, Ser.*
A 316, 85.
Quinn T.J.: 1990, *Temperature*, Academic Press, London.
Romero, J., Wehrli C., Fröhlich C., Crommelynck D., Fichot A. and
Penelle B.: 1993, *Solar Phys.*, (to be published).
Willson, R.C.: 1979, *Appl. Opt.* **18**, 179.
Willson, R.C.: 1993, *Proc. IAU Collquium No. 143*, (to be published).

Acknowledgement. The authors would like to thank Dr. J. Romero (WRC)
for a helpful discussion regarding solar noise.

A REVIEW OF THE NIMBUS-7 ERB SOLAR DATASET

H. L. KYLE

NASA/Goddard Space Flight Center, Greenbelt, MD 20771

D. V. HOYT

Research and Data Systems Corporation, Greenbelt, MD 20770

J. R. HICKEY

The Eppley Laboratory, Inc., Newport, RI 02840

ABSTRACT. Fourteen years (November 16, 1978 through January 24, 1993) of Nimbus-7 total solar irradiance measurements have been made. The measured mean annual solar energy just outside of the Earth's atmosphere was about 0.1% (1.4 W/m²) higher in the peak years of 1979 (cycle 21) and 1991 (cycle 22) than in the quiet Sun years of 1985/86. Comparison with shorter, independent solar measurement sets and with empirical models qualitatively confirms the Nimbus-7 results. But these comparisons also raise questions of detail for future studies: in which years did the peaks actually occur and just how accurate are the models and the measurements?

The total solar irradiance, at the mean Earth-to-Sun distance, varies in phase with the sunspot cycle (Figure 1). In solar cycle 21, the monthly mean irradiances varied by 3.46 Wm^{-2} with the maximum in May 1979 and the minimum in January 1984. To date, the range in cycle 22 has been 2.23 Wm^{-2} with the maximum in April 1991 and the minimum in April 1987. In the short-term (a few days or weeks), the irradiance drops sharply when large, dark sunspots face the Earth, but the irradiance increases for bright facular regions. Both faculae and sunspots follow roughly the same 11-year activity cycle. In the long-term, the faculae are dominant. Presently the solar cycles are about 10 years long. The Nimbus-7 measurements show a mean of 1371.92 Wm^{-2} for the 10-year period (November 16, 1978 through November 16, 1988). The Nimbus-7 solar measurements ended on January 24, 1993. The slow precession of the satellite's orbit moved the Sun out of the sensor's field-of-view on the next day. The Sun was also out of view from June 18 through September 1, 1992.

In Figure 2, the mean daily solar measurements from the SMM satellite (Willson and Hudson, 1991) and the ERBS spacecraft (Lee et al., 1991) are compared with the Nimbus-7 results. The vertical gaps between the three datasets are due to systematic errors in the absolute calibration. The uncertainties are reported as: Nimbus-7 (±0.5%), ERBS (±0.2%), and SMM (±0.1%). The Nimbus-7 measurements are shown for the period November 1978 through September 1992; the SMM values run from February 1980 to July 1989, and the ERBS from October 1984 to October 1992. Only the ERBS is still taking measurements; the SMM satellite ceased operating in late 1989. However, Willson has a solar sensor on the Upper Atmosphere Research Satellite (UARS) that started taking data in September 1991.

Solar Physics 152: 9–12, 1994.
© 1994 *Kluwer Academic Publishers.*

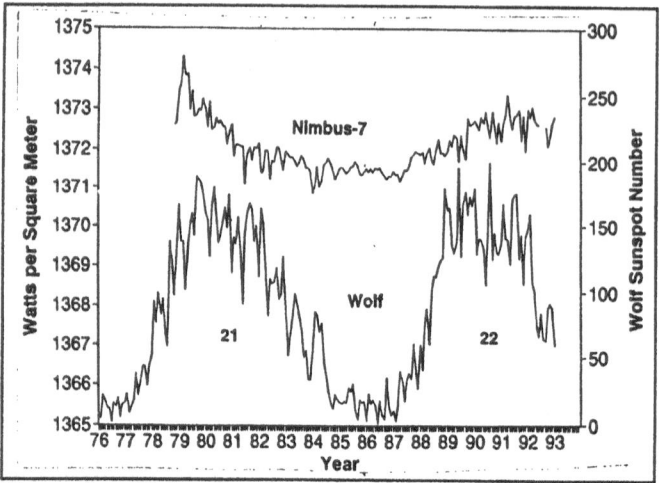

Figure 1. Nimbus-7 solar irradiances and Wolf sunspot numbers for solar cycles 21 and 22 (monthly means through January 1993). The Nimbus-7 measurements started in November 1978.

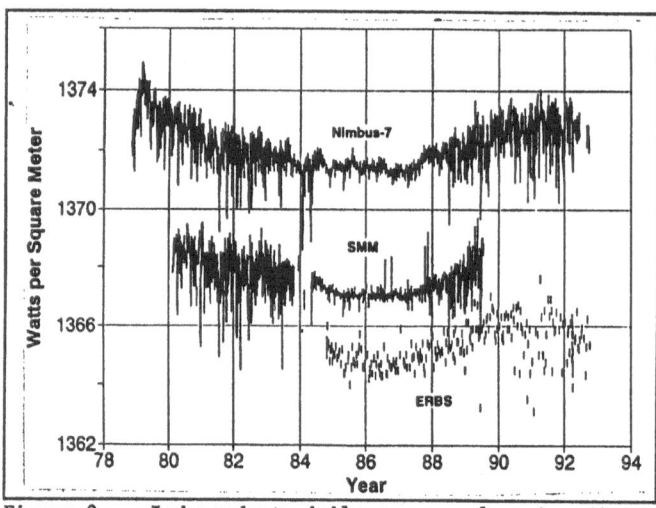

Figure 2. Independent daily mean solar irradiance measurements from the Nimbus-7, SMM, and ERBS Satellites.

The three sensors operated on different schedules. The Nimbus-7 sensor observed the Sun for a 3-minute period once per orbit (13.85 times per day). The SMM sensor normally observed the Sun for 65 seconds and then, for comparison, observed its own shutter for 65 seconds. In the Sun-tracking mode, the SMM could obtain up to 28 shutter cycles per orbit or several hundred in a day. The ERBS sensor was similar to the SMM but with a 32-second Sun observing period. Because of experimental constraints, the ERBS solar measurements usually consisted of two shutter cycles once every 2 weeks.

The mean stability (repeatability) of the measurements is an order of magnitude or so better than the reported absolute calibration. Hoyt et al. (1992) discussed the history and accuracy of the Nimbus-7 measurements. The worst case error in the long-term trend was estimated to be ± 0.5 Wm^{-2}, but comparison with SMM measurements suggests a better accuracy. The SMM was in its normal Sun tracking mode in the years 1980 and 1985 to 1988. For these years, the annual mean (Nimbus-7 minus SMM) difference is constant to within ± 0.07 Wm^{-2}. The SMM was in a spin-scan operating mode from December 1980 through March 1984. Willson applied a bias correction of 0.12 to the spin-scan measurements to bring them approximately into line with the Sun pointing results. If this correction had been 0.09% or 0.10%, the annual means of the two datasets would have agreed to ± 0.07 Wm^{-2} for all of the years. However, short-term perturbations of one or both of the sensors caused some daily measurements to differ by ± 1 Wm^{-2} or more while some monthly means differed by up to ± 0.3 Wm^{-2}. Thus, to maintain long-term continuity, successive measurement systems should overlap by about 1 year.

There is no known reason why the stability of the Nimbus-7 solar sensor should have differed during its lifetime. Nevertheless, critics have questioned the measurements at the peaks of both solar cycles 21 and 22. Figure 3 shows the daily mean difference (Nimbus-7 minus ERBS) for the period October 1984 to October 1992. The central line represents an 11-point running average to indicate trends. The sharp initial rise in the difference is attributed to the initial degradation of the ERBS sensor (R. B. Lee, private communication) but the remaining drifts are unaccounted for. From the fall of 1987 to the fall of 1989, the difference slowly drops by 0.5 Wm^{-2} and then fairly rapidly recovers. Then from late 1989 to the end of 1991, the difference slowly increases by another 0.5 Wm^{-2}. As a result, the ERBS measures the cycle 22 irradiance maximum in late 1989 as opposed to the April 1991 peak found by the Nimbus-7. In 1979, there were no competing solar measurements. However, Foukal and Lean (1990) used various solar indices to predict facular brightening and sunspot dimming and thus decipher past solar irradiance changes. Their model predicts that the cycle 21 irradiance peak should have occurred in the winter of 1980/81 and not in the spring of 1979 as measured by the Nimbus-7. This despite the fact that both the Nimbus-7 and SMM show a higher irradiance in early 1980 than in late 1980. Foukal and Lean assume unidentified sensor problems are present in both instruments, while Willson and Hudson (1991) counter that unidentified solar phenomena probably caused a real irradiance increase. Not all modelers agree with Foukal and Lean (Lean, 1991). Future studies may resolve some of these questions.

The ERB solar measurements were taken in a fair but not ideal satellite environment. The Nimbus-7 satellite carried eight separate experiments and in the early years, when most were active, they had to operate in a power-sharing mode. Thus, there were periods when the solar sensor was frequently turned on and off. Many observing opportunities were thus lost while, some measurements had to be rejected because the sensor was not in

thermal equilibrium. In 1992 and in January 1993, some data were lost
because the Sun was not in view. Due to these various problems, useful
measurements are available from only about 82% of the 71,753 orbits from
November 16, 1978 through January 1993. On a yearly basis, the useful
measurements varied from 63% in 1979 to 97% in 1988.

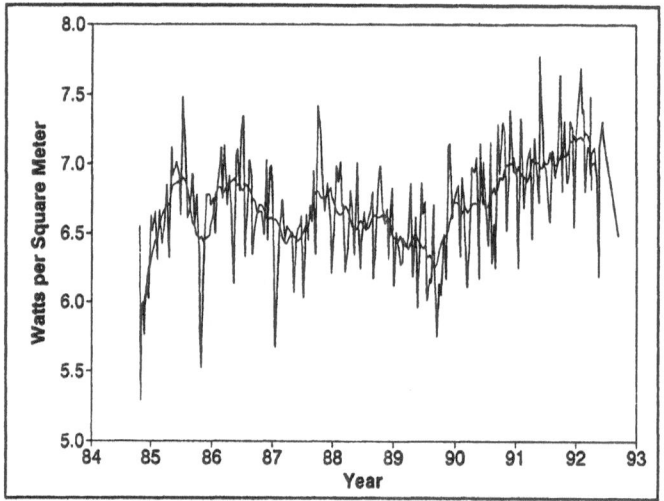

Figure 3. The difference (Nimbus-7 minus ERBS) in the
daily mean solar measurements. The central curve is an
11-point running mean.

The 14-year Nimbus-7 solar dataset extends from the peak of cycle 21 to
the descending phase of cycle 22. It clearly shows the Sun to be a low-
level variable star and reawakens interest in the possible past and future
influences of the Sun on the Earth's climate (Hoyt and Schatten, 1993).
However, comparison with independent solar measurements and total solar
irradiance models also raises questions. Can our understanding of our
present instruments be improved and can better instruments be made? How
well do we understand the causes of the solar variability and how well can
this variability be monitored by proxy indices? Clearly two or three
independent satellite solar monitors should be in simultaneous operation
to obtain an accurate long-term record. Can such a program be set up and
maintained? There is still much work to do.

References

Foukal, P. and Lean, J.: 1990, *Science* 247, 505.
Hoyt, D. V., Kyle, H. L., Hickey, J. R., and Maschhoff, R. H.: 1992, *J.*
    Geophys. Res. 97, 51.
Hoyt, D. V. and Schatten, K. H.: 1993, *J. Geophys. Res.* (in press).
Lean, J., 1991, *Rev. Geophys.* 29, 505.
Lee III, R. B., Gibson, M. A., Shivakumar, N., Wilson, R., Kyle, H. L.,
    and Mecherikunnel, A. T.: 1991, *Metrologia* 28, 265.
Willson, R. C. and Hudson, H. S.: 1991, *Nature* 351, 42.

LONG-TERM VARIATIONS IN TOTAL SOLAR IRRADIANCE

JUDIT M. PAP, RICHARD C. WILLSON
Jet Propulsion Laboratory
Pasadena, CA, U.S.A.

CLAUS FRÖHLICH
Physikalisch-Meteorologisches Observatorium Davos
World Radiation Center
Davos Dorf, Switzerland

and

RICHARD F. DONNELLY, LARRY PUGA
NOAA Space Environment Laboratory
Boulder, CO, U.S.A.

Abstract. For more than a decade total solar irradiance has been monitored simultaneously from space by different satellites. The detection of total solar irradiance variations
by satellite-based experiments during the past decade and a half has stimulated modeling efforts to help identify their causes and to provide estimates of irradiance data, using
'proxy' indicators of solar activity, for time intervals when no satellite observations exist.
In this paper total solar irradiance observed by the Nimbus-7/ERB, SMM/ACRIM I, and
UARS/ACRIM II radiometers is modeled with the Photometric Sunspot Index and the
Mg II core-to-wing ratio. Since the formation of the Mg II line is very similar to that of
the Ca II K line, the Mg core-to-wing ratio, derived from the irradiance observations of
the Nimbus-7 and NOAA9 satellites, is used as a proxy for the bright magnetic elements.
It is shown that the observed changes in total solar irradiance are underestimated by the
proxy models at the time of maximum and during the beginning of the declining portion
of solar cycle 22 similar to behavior just before the maximum of solar cycle 21. This disagreement between total irradiance observations and their model estimates is indicative
of the fact that the underlying physical mechanism of the changes observed in the solar
radiative output is not well-understood. Furthermore, the uncertainties in the proxy data
used for irradiance modeling and the resulting limitation of the models should be taken
into account, especially when the irradiance models are used for climatic studies.

1. Introduction

The precise and continuous determination of the value of the total solar
energy flux impinging on the Earth and its variation is one of the principal
concerns for both climatic and solar physics studies, since the solar radiative
output is the main driver of the physical processes within the Earth's atmosphere. Thus the monitoring and study of the changes in total irradiance
are extremely important. There are indications that changes in the solar
output influence the Earth's climate on time scales ranging from the Gleissberg cycle (Reid, 1987, Friis-Christensen & Lassen, 1991) up to the Maunder
Minimum type of climate anomalies (Lean et al., 1992, Ribes, 1993). Fur-

Solar Physics **152**: 13–21, 1994.
© 1994 *Kluwer Academic Publishers.*

ther, to understand the human effect in the global change of the climate, it is essential to reveal the role of the solar irradiance variations in the terrestrial and atmospheric processes.

For more than a decade total solar irradiance has been monitored from space by different satellites. These observations have revealed variations in total irradiance ranging from minutes to the 11-year solar cycle (Willson & Hudson, 1991). It has been shown that the very small, rapid irradiance fluctuations are due to solar oscillations (e.g. Fröhlich, 1992). The short-term variations (from days to months) are directly produced by the passage of active regions (Willson, 1982, Pap, 1985, Fröhlich & Pap, 1989) with dark sunspots and bright faculae (Chapman, 1987). The most important discovery of irradiance observations was the enhancement of solar luminosity during the high activity part of the solar cycle (Willson & Hudson, 1991, Hoyt et al., 1992) that is attributed to the enhanced emission of bright faculae and the magnetic network (Foukal & Lean, 1988).

Although considerable information exists about the variations in total solar irradiance, their physical origin is not well understood. Results of multivariate spectral analysis show that considerable variation in total irradiance remains unexplained after removing the effect of dark sunspots and bright magnetic elements, and the residual variability changes with the phase of the solar cycle (Pap & Fröhlich, 1992). Furthermore, empirical models of total solar irradiance, developed from various solar activity indices, such as the full disk equivalent width of the He-line at 1083 nm (HeI), various Fraunhofer lines, and the 10.7 cm radio flux (Foukal & Lean, 1988, Livingston et al., 1988, Pap et al., 1992, Fröhlich, 1993), disagree significantly with the irradiance observations at the maximum of solar cycle 21. The main purpose of this paper is to estimate total solar irradiance at the maximum of solar cycle 22 to clarify whether the disagreement between irradiance observations and the proxy estimates has an intrinsic solar or instrumental origin. The uncertainties in the proxy data used for irradiance modeling and the resulting limitation of the models are also discussed.

2. Observational Data

Measurements of total solar irradiance used in this study were performed by the Nimbus-7/ERB, SMM/ACRIM I, and UARS/ACRIM II radiometers. The irradiance observations on board the Nimbus-7 satellite began in November 1978 and data are available through November 30, 1992 (Hoyt et al., 1992). The high precision irradiance observations of the Solar Maximum Mission satellite started in February 1980 and ended with re-entry of the SMM spacecraft in November 1989 (Willson & Hudson, 1991). The daily ACRIM I irradiance data are available from February 1980 through June 1989. The second ACRIM radiometer began to operate on board the Upper

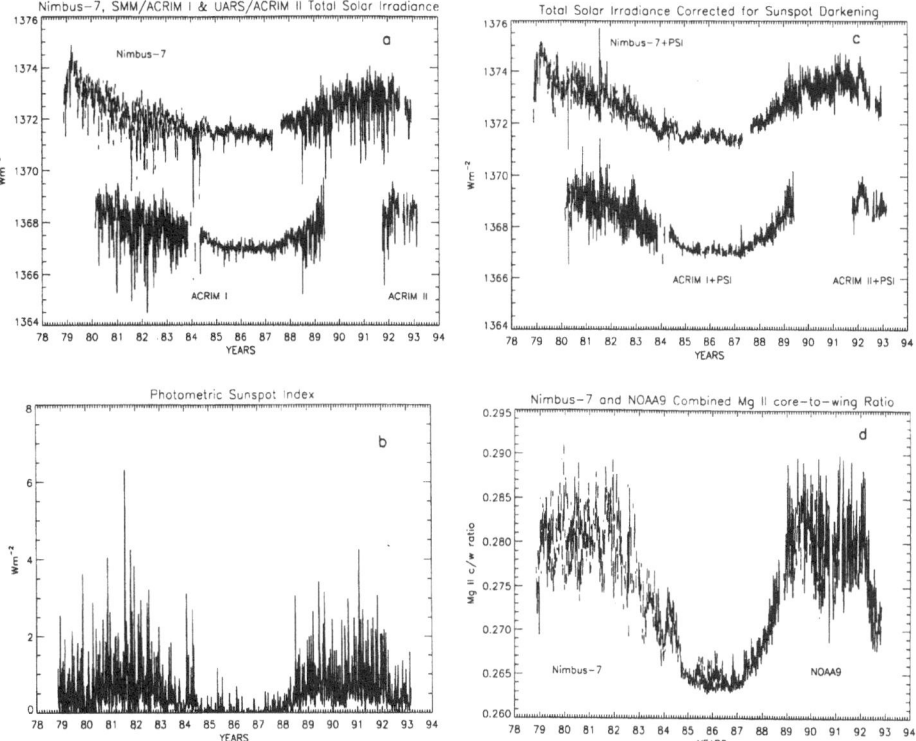

Fig. 1.   Time series of total solar irradiance measured by the Nimbus-7/ERB, SMM/ACRIM I and UARS/ACRIM II radiometers (a), the Photometric Sunspot Index (b), total solar irradiances corrected for sunspot darkening (c), and the combined Nimbus-7 and NOAA9 Mg c/w (d).

Atmospheric Research Satellite on October 4, 1991 and continues into the present. The UARS/ACRIM II irradiance data are available for this study from October 1991 through February 1993. The daily values of the three total irradiance data sets are presented in Fig. 1a. Note that the SMM/ACRIM I and UARS/ACRIM II data are adjusted to each other through their mutual intercomparisons with Nimbus-7/ERB. A detailed information on this topic is given by Willson, 1993.

The Photometric Sunspot Index (PSI) has been developed to study the effect of sunspots on total solar irradiance (e.g. Hudson et al., 1982). However, results of direct sunspot photometry discovered the dependence of the contrast of the sunspots on their area (Steinegger et al., 1990, Chapman et al., 1992). It has also been found that the former PSI models overestimate the sunspot effect on total irradiance by about 40% (Chapman et al., 1992). Therefore, a new PSI model has been developed by Fröhlich et al.,

1993 taking into account the area dependence of the sunspot contrast among other effects, such as screening, daily mean values and different limbdarkening functions. Mainly the screening for outliers improves significantly the homogeneity of the data set. The newly calculated PSI, used in this study, is shown in Fig. 1b. The daily values of total solar irradiance corrected for the sunspot darkening by adding PSI are presented in Fig. 1c.

The Mg II h & k core-to-wing ratio (Mg c/w) is used as proxy for the bright magnetic elements, including plages and the magnetic network (Fig. 1d). This quantity is calculated from the observations of the SBUV experiments on the Nimbus-7 and NOAA9 satellites (Heath & Schlesinger, 1986, Donnelly, 1991). The Nimbus-7 and NOAA9 Mg c/w were made consistent by the means of their overlapping observational time period in 1986, and also by the means of the full disk Ca II index. Since the formation of the Mg line is very similar to that of the Ca II K line, the Mg c/w provides a reasonable good index for the plage and network radiation.

3. Empirical Models of Total Solar Irradiance

Total solar irradiance corrected for sunspot darkening is estimated from the Mg c/w ratio using linear regression analysis. The estimated irradiance data are calculated from the equation $y = a + b \cdot x$, where a and b are the regression coefficients, x contains the daily mean values of the Mg c/w ratio and y gives the best-fit linear relationship (by minimizing χ^2) between total irradiance corrected for sunspot darkening and the bright magnetic elements represented by Mg c/w. Since the correlation between total irradiance and solar activity indices depends on the phase of the solar cycle (Pap & Fröhlich, 1992, Pap et al., 1992), the irradiance models were developed for the maximum, declining portion and minimum of solar cycle 21 and for the rising portion and maximum time of solar cycle 22, respectively. In a first step, the models are calculated for the period of the SMM/ACRIM I observations from February 1980 to June 1989. This period is divided in the following intervals:

－　(1) February 1980 to June 1984 (maximum and declining portion of solar cycle 21),
－　(2) July 1984 to August 1986 (minimum of solar cycle 21),
－　(3) September 1986 to June 1989 (rising portion of solar cycle 22).

To extend the irradiance models after June 1, 1989 the regression coefficients calculated for the time interval from February 1980 to June 1984 has been used. Fig. 2a shows the 81-day running means of the SMM/ACRIM I and UARS/ACRIM II total irradiance data after the correction by PSI. The solid line shows the 81-day running means and of the irradiance model estimates developed from the Mg c/w with linear regression analysis. The same model has been developed for the Nimbus-7/ERB total solar irradiance

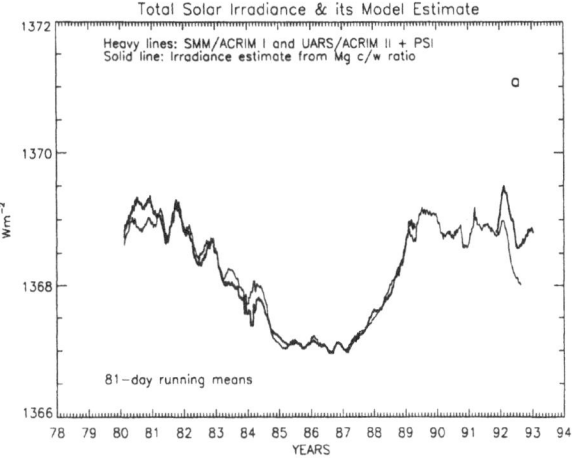

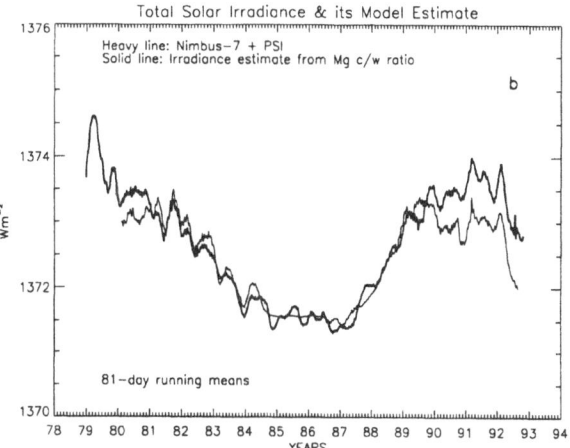

Fig. 2.   81-day running mean of the SMM/ACRIM I and UARS/ACRIM II total solar irradiance corrected by *PSI* (heavy line in a) and of the irradiance estimates calculated from the Mg c/w ratio (solid line in a). (b) same as (a) but for Nimbus-7/ERB irradiance data.

corrected by *PSI*, and is shown in Fig. 2b.

   As can be seen from Fig. 2, the estimated values of total solar irradiance corrected for sunspot darkening considerably underestimate the observed irradiance values at the maximum time and the beginning of the declining portion of solar cycle 22. Note that irradiance models calculated from the full disk He-line equivalent width at 1083 nm and 10.7 cm radio flux (Brandt et al., 1993, Fröhlich, 1993) show the similar discrepancy between the observations and model estimates. These results clearly show that the current irradiance models underestimate the real amplitude of the solar-cycle-related long-term changes in total solar irradiance and the models cannot reproduce the observed changes in total irradiance.

4. Uncertainties in the Models of Total Solar Irradiance

The breakdown between the observed total irradiance and its model estimates during solar maximum is one of the current outstanding problems in solar physics and is indicative of the fact that we do not understand the physical origin of the variability observed in the solar radiative output. The uncertainties in the proxy data used to construct these statistical irradiance models are surely part of the problem. The usefulness of these irradiance models for climate studies is limited by their inherent uncertainties.

One of the largest uncertainties in the irradiance models, especially on long time scales, originates from the lack of knowledge of the effect of faculae principally because of the lack of high quality synoptic data. While more than 90% of total solar irradiance is emitted from the photosphere, most of the irradiance variations related to the bright magnetic elements are modeled by chromospheric proxy data, such as the Ca II K plages, HeI, Mg c/w and 10.7 cm radio flux. However, the conversion factor between the area and intensity of the chromospheric plages and photospheric faculae is not known; the center-to-limb behavior of the contrast of white light faculae is quite different from that of Ca K plages (e.g. Chapman et al., 1992, Schatten & Mayr, 1992). The situation is far more difficult since the spatially resolved Ca K plage data observed at the Big Bear Solar Observatory does not include the remnants of plages and other network elements that give a significant contribution to the changes in total (and also in UV) solar irradiance (Pap et al., 1990). It should also be noted that the available resolved Ca K plage data are from three different sources:

-   (1) from December 1970 till September 1979 observed at the McMath Observatory,
-   (2) from October 1979 till August 1982 observed at the Mt. Wilson Observatory,
-   (3) from September 1982 till November 1987 observed at the Big Bear Solar Observatory.

This indicates that no homogeneous spatially resolved plage data set is available for studying irradiance variations, and after November 1987 no plage area and intensity data are measured and published on a routine basis (Marquette, 1992).

It has also been shown that the evolution of active regions plays an important role in the irradiance changes (e.g. Willson, 1982, Pap, 1985, Fröhlich &

Pap, 1989), but the evolutionary effects are not really included in the irradiance models. The additional sources of the Ca II K data (White et al., 1992), the Mg c/w ratio, the He-line equivalent width at 1083 nm, and the 10.7 cm radio flux are full disk indices, therefore, these proxies cannot provide information either on the evolutionary effects or the contribution of the plage and network radiation to the changes in solar irradiance. On the other hand, the variation of the He-line equivalent width and the 10.7 cm radio flux is rather complex and not well-understood (Harvey, 1984, Tapping, 1987). It has been shown that while most of the variations in the He-line equivalent width is related to the Ca K plages, a substantially large fraction of its variability is related to the filaments (Harvey & Livingston, 1993) that do not effect total solar irradiance. Furthermore, it is not clear whether the irradiance effect of the small network elements is included in the plage-related variation of the He-line equivalent width (Harvey & Livingston, 1993). The variability of the 10.7 cm radio flux is attributed to gyroresonant absorption and free-free (bremsstrahlung) emissions (Tapping, 1987). The latter is supposed to be related to weaker magnetic fields concentrated in plages and in the magnetic network, while the former one is related to the strong magnetic field of the sunspots. These results clearly show that the above proxies used in the current irradiance models give only a rough estimate for the behavior of the faculae and their effect on total solar irradiance.

The lack of the good synoptic data sets for sunspots has to be mentioned too. The publication of the high precision area and position of the sunspots in the Greenwich Catalogue ended in 1976. Since then the sunspot observations are reported in the NOAA-WDC Solar Geophysical Data catalogue. The current precision of the measurements of the area of sunspots is about 20-25% for large sunspots and it can be as high as 50% for small spots (Sofia et al., 1982). Fröhlich et al., 1993 show that the irregularities in the sunspot area reports cause the largest uncertainty in the PSI. Furthermore, no observational information exists on the area and intensity of the umbrae and penumbrae of sunspots. Part of the observed change of contrast with time (Fröhlich et al., 1993) could be due to changes in the umbra–penumbra intensity and/or area ratio. The results of sunspot photometry also show that the contrast changes with even during the passage over the visible disk and is normally overestimated in PSI calculations (Steinegger et al., 1990, Chapman et al., 1992). Moreover, the contrast of sunspots changes over the solar cycle (Maltby et al., 1986) and all these details are not incorporated in the irradiance models.

Another important aspect to be pointed out is that the current observations of total solar irradiance cover only 14 years, not much more than one solar cycle. From this short observing period it is hard to establish accurate irradiance models, mainly because it is difficult to judge the representativeness of the proxies, even if their quality is good. It must also be underscored that the current irradiance models are only empirical models developed with simple linear regression analysis and no adequate physical model of irradiance variations exists as yet. The large uncertainties in the irradiance proxies, the lack of adequate irradiance models, and the existing disagreements between the observed changes in total irradiance and its model estimates emphasize the need to provide continuous total solar irradiance monitoring while extending and expanding investigations aimed at better understanding the physical origin of irradiance changes.

5. Conclusions

A simple empirical model of total solar irradiance corrected for sunspot darkening has been developed from the Mg II h & k core-to-wing ratio with linear regression analysis. The newly calculated Photometric Sunspot Index (Fröhlich et al., 1993) was used for removing the effect of sunspots from total solar irradiance. It has been found that the models underestimate the observed irradiance values at the maximum time and the beginning of the declining phase of solar cycle 22, similar to solar cycle 21. These results show that the disagreement between irradiance models and observations, found first by Foukal & Lean, 1988 for the maximum of solar cycle 21, seems to be more of a problem with the models than with the instruments.

Part of the found discrepancies can be due to the uncertainties in the proxy data. The inherent limitation of these simple empirical models must have to be kept in mind when irradiance models are used. Considering the significance of irradiance variations as a causal mechanism for climate change, continuous observations of total irradiance from space are necessary to maintain a long-term, high precision irradiance data base for climatic studies. In parallel with the direct irradiance observations, advanced theoretical and statistical studies of the observed irradiance variations are required. The ultimate goal is to understand (1) why, (2) how, and (3) on what time scale the total solar irradiance varies in order to reconstruct and predict the solar induced climatic changes.

Acknowledgements

The research described in this paper was carried out by the Jet Propulsion Laboratory, California Institute of Technology under a contract with the National Aeronautics and Space Administration, by the Physikalisch-Meteorologisches Observatorium, partly funded by the Swiss National Science Foundation (C.F.), and by the NOAA Space Environment Laboratory (R.F.D. and L.P.). The authors wish also to thank Helen Coffey, NOAA, Boulder, Co., for providing sunspot data before publication.

References

Brandt, P.N., Stix, M., Weinhadrt, W.: 1993, *Solar Physics*, submitted.

Chapman, G.A.: 1987, *Ann. Rev. Astron. Astrophys* **25**, 633-667.

Chapman, G, Lawrence, J.K., and Hudson, H.S: 1992, in R.F. Donnelly, ed(s)., *Proc. of the Workshop on the Solar Electromagnetic Radiation Study for Solar Cycle 22*, NOAA ERL, Boulder, 135-136.

Donnelly, R.F.: 1991, *J. Geomagnet. Geoelect.* **43** Suppl., 835-843.

Foukal, P. and Lean J.: 1988, *ApJ.* **328**, 347–357.

Friis-Christensen, E. and Lassen, K.: 1991, *Science* **254** 698-700.

Fröhlich, C.: 1992, in Donnelly, R.F, ed(s)., *Proc. Solar Electromagnetic Radiation Study for Solar Cycle 22*, SEL NOAA ERL, Boulder, 1-11.

Fröhlich, C.: 1993, in J.M. Pap, C. Fröhlich, H.S. Hudson, S. Solanki, ed(s)., *Proc. The Sun as a Variable Star: Solar and Stellar Irradiance Variations*, Cambridge Univ. Press, in press.

Fröhlich, C. and Pap, J.M.: 1989, *Astron. Astrophys.* **220**, 272–280.

Fröhlich, C., Pap, J.M., and Hudson, H.S.: 1993, *Solar Physics*, submitted.

Harvey, J: 1984, in B.J.Labonte, G.A. Chapman, H.S. Hudson, and R.C. Willson, ed(s)., *Proc. Solar Irradiance Variations in Active Regions Time Scale*, NASA CP-2310, 197-211.

Harvey, J., and Livingston, W.: 1993, in in press, ed(s)., *Proceedings of IAU Symposium 154, Solar Infrared Astronomy*, , .

Heath, D. F. and Schlesinger, B.M.: 1986, *J. Geophys. Res* **91**, 8672-8682.

Hoyt, D.V., Kyle, H.L., Hickey, J.R., and Maschhoff, R.H.: 1992, in Donnelly, R.F, ed(s)., *Proc. Solar Electromagnetic Radiation Study for Solar Cycle 22*, SEL NOAA ERL, Boulder, 43-48.

Hudson, H.S., Silva, S., Woodard, M., and Willson, R.C.: 1982, *Sol.Phys.* **76**, 211-218.

Lean, J., Skumanich, A. and White, O.R.: 1992, *Geophysical Research Letter* **19**, 1591-1594.

Livingston, W., Wallace, L., and White, O.R.: 1988, *Science* **240**, 1765-1767.

Maltby, P., Avrett, E.H., Carlsson, M., Kjeldseth-Moe, Kurucz, R.L., and Loeser, R.: 1986, *Ap.J.* **306**, 284-303.

Pap, J.M.: 1985, *Solar Physics* **97**, 21-33.

Pap, J.M. and Fröhlich, C.: 1992, in Donnelly, R.F., ed(s)., *Proc.Solar Electromagnetic Radiation Study for Solar Cycle 22*, SEL NOAA ERL: Boulder, 62–75.

Pap, J.M., Marquette, W., and Donnelly, R.F.: 1990, *Adv. Space Res.* **11**, 271-275.

Pap. J.M., Willson, R.C., and Donnelly, R.F.: 1992, in K. Harvey, ed(s)., *Proc. The Solar Cycle Workshop*, 27, 491-502.

Reid, G.C.: 1987, *Nature* **329**, 142-143.

Ribes, E., Sokoloff, D., and Sadourney, R.: 1993, in Pap, J.M., Fröhlich, C., Hudson, H.S., Solanki, S., ed(s)., *The Sun as a Variable Star: Solar and Stellar Irradiance Variations*, Cambridge Univ. Press, in press.

Schatten K. and Mayr H.: 1992, in Donnelly, R.F., ed(s)., *Proc.Solar Electromagnetic Radiation Study for Solar Cycle 22*, SEL NOAA ERL: Boulder, 142-153.

Sofia S., Oster, L., and Schatten, K.: 1982, *Solar Physics* **80**, 87-98.

Steinegger, M., Brandt, P.N., Pap, J., and Schmidt, W.: 1990, *Astrophys.Space Sci.* **170**, 127-133.

Tapping, K.: 1987, *J. Geophys. Res.* **92** 829-838.

White, O.R., Livingston, W.C., and Keil, S.L: 1992, in Donnelly, R.F., ed(s)., *Proc.Solar Electromagnetic Radiation Study for Solar Cycle 22*, SEL NOAA ERL: Boulder, 160-165.

Willson, R.C.: 1982, *J. Geophys Res.* **87**, 4319-4326.

Willson, R.C.: 1993, in Pap, J., Fröhlich, C., Hudson, H.S. and Solanki, S., ed(s)., *Proc. The Sun as a Variable Star: Solar and Stellar Irradiance Variations*, Cambridge Univ. Press, in press.

Willson, R.C. and Hudson, H.S.: 1991, *Nature* **351**, 42-44.

SOLAR TOTAL IRRADIANCE VARIABILITY FROM SOVA 2
ON BOARD EURECA

J. ROMERO, CH. WEHRLI and C. FRÖHLICH

Physikalisch-Meteorologisches Observatorium Davos, World Radiation Center
Davos Dorf, Switzerland

Abstract. The solar total irradiance has been measured during the EURECA (EUropean Retrievable CArrier) mission by the radiometers PMO6 of the experiment SOVA 2 (SOlar VAriability, Experiment 2). The instruments are of the active cavity type with a sampling of 99 s. Their specification and behavior in space are described. The time series of total irradiance gathered with the radiometers covers 9 months, starting in August 1992. Solar variability on time scales from minutes to the mission duration except for the periods close to the orbit around the Earth is observed. The results are correlated with the Photometric Sunspot Index (*PSI*) and compared with results from other experiments.

1. Introduction

The SOVA (SOlar VAriability) investigation is the result of a collaboration between the Institut Royal Météorologique de Belgique (IRMB) in Brussels (Belgium), the Physikalisch-Meteorologisches Observatorium Davos, World Radiation Center (PMOD/WRC) in Davos (Switzerland) and the Space Science Department (SSD) of the European Space Agency (ESA) at ESTEC in Noordwijk (the Netherlands) with D.Crommelynck as Principle Investigator (Crommelynck et al., 1993). The experiment is composed of three parts, designed and manufactured by the IRMB for SOVA 1, the PMOD/WRC for SOVA 2 and the SSD for SOVA 3. SOVA 1 contains a CROM type radiometer, the results of which are described by Crommelynck and Domingo, 1993. SOVA 2 contains among other instruments a set of two redundant absolute radiometers of type PMO6, the results of which are presented in this paper. The data acquisition system is based on voltage-to-frequency converters with a resolution of 22 bits in 8.2s. Several house-keeping parameters are measured, such as temperatures, currents, voltages, necessary to monitor the health of the experiment and if needed to correct the acquired data. Moreover a sun fine pointing monitor is implemented in SOVA 2 to measure the attitude and relative pointing of the satellite with respect to the Sun's center. From this pointing information corrections for the measured irradiances are deduced. SOVA 3 contains the interface to the spacecraft's data handling and power system. These functions are performed with one of two redundant CPU in SOVA 3.

Solar Physics **152**: 23–29, 1994.
© 1994 *Kluwer Academic Publishers.*

2. The PMO6 radiometer and its operation

The continuously operated PMO6 radiometer is the 'active' instrument, while the other is the 'back–up' instrument serving to monitor a possible degradation of the 'active' radiometer. Until November 9, 1992, the 'back–up' instrument measured during one orbit per day. Afterwards, it measured during one orbit each 16 days. During special comparison campaigns (a few days in March and April 1993) both instruments were operated continuously.

The operation of the radiometers is briefly described in the following; more details can be found in Brusa and Fröhlich (1986). The solar radiation is absorbed in a cavity with an inverted cone and painted with a specular black paint and is coupled to a differential heat flow meter. The radiometer is electrically calibrated and operated in an active mode: the heat flux is held constant by controlling the electrical power fed to the heater in the cavity during shaded (reference) and irradiated (measuring) phases. The irradiated phase lasts 58 seconds and the shaded one 41 seconds yielding a value of the irradiance each 99 seconds. A value of the total solar irradiance S in Wm^{-2} is then obtained as the difference of the electrical power during both phases multiplied by the radiometric constant C_R:

$$S = C_R(P_c - P_o) \tag{1}$$

where P_c and P_o are the closed and open values of the electrical power calculated from the current and voltage across the heater.

The constant C_R is the inverse of the precision aperture area A times correction factors for the deviations from the ideal behavior:

$$C_R = \frac{(1+R)(1+SL)(1+LH)(1+D)(1+NE)}{A} \quad \text{in air}$$

$$C_R = \frac{(1+R)(1+SL)(1+LH)(1+D)}{A} \quad \text{in vacuum} \tag{2}$$

The value of R, SL, LH, D and NE and their meaning are given in Table I for both radiometers. The absolute accuracy of this type of radiometers, determined from the experimental uncertainties of the correction factors, is 0.15 % and the precision is 50 ppm.

During an orbit (60 minutes in the Sun, 30 minutes in eclipse) the temperature of the experiment is modulated by 10–15K at the front plate and by 2–3K at the heat sink of the radiometer. This has been accounted for by a model based on the geometry and temperature of the surfaces influencing the detector. The model includes also the infrared losses of the cavity to space. The magnitude of these corrections increased after the cooling loop of the satellite was switched off at the end of January 1993 and the temperature of the instrument rose by 18–20K to a mean experiment temperature of 323K.

TABLE I

Correction factors experimentally determined contributing to the radiometric constant of the two SOVA 2 PMO6 radiometers. The radiometric constant is calculated according to Eq.(2).

Factor	Initials	Active	Back–up	Units
Area	A	19.7169	19.5919	mm^2
Reflectivity	R	150	240	ppm
Stray-Light	SL	-150	-150	ppm
Lead-Heating	LH	370	320	ppm
Diffraction	D	-1000	-1000	ppm
Non-Equivalence	NE	3340	2700	ppm
Radiometric constant in air		50855.25	51149.16	m^{-2}
Radiometric constant in vacuum		50685.96	51011.43	m^{-2}

A further correction is applied due to the effect of the off-axis angle θ between the Sun and the optical axis by multiplying the irradiance by $1/cos\theta$. The mean value of θ was $0.2° \pm 0.5°$. Finally, in order to obtain the total solar irradiance at 1 AU, the measured irradiance was corrected for the satellite-Sun distance r and the Doppler effect due to the radial velocity towards the Sun v (Willson and Hudson, 1981). With the speed of light c the irradiance S_0 at 1AU becomes:

$$S_0 = \frac{Sr^2}{(1 + v/c)^2} \tag{3}$$

The orbital data around the Earth were provided by the European Space Operation Center and the Sun-Earth distance and velocity calculated by an algorithm given by Montenbruck & Pfleger (1989) accurate to $1 \cdot 10^{-6}$ AU.

3. Results and discussion

3.1. TOTAL SOLAR IRRADIANCE

The daily means of the total solar irradiance measured by the two radiometers of SOVA 2, 'active' and 'back–up', are shown in Fig. 1. Large dips around days 232, 295-305, 405-410 are due to the passage of sunspot groups. The irradiance blocked by sunspots during these days is of the order of 0.1%.

The exposure of the cavity to the solar radiation in the space environment could cause a degradation of the black paint in the cavity (see e.g. Willson and Hudson, 1991). The less often exposed 'back–up' radiometer monitors this effect. In Fig. 2 the relative difference of the 'active' and 'back–up' radiometers is plotted and no unambiguous trend can be deduced. Comparing the ratios at the beginning and at the end of the mission might suggest a

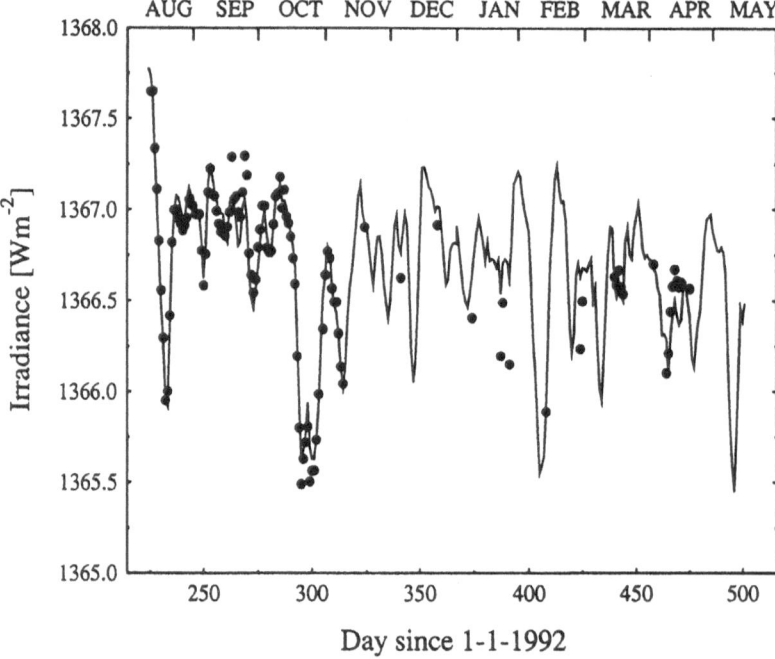

Fig. 1. Daily mean values of the total solar irradiance during the EURECA mission as measured by the 'active' (continuous line) and 'back–up' (dots) radiometers of SOVA 2.

loss of sensitivity of about 1 ppm/day similar to the value found by Willson and Hudson (1991) during the first year of operation of ACRIM. Interesting enough, the high ratios at the end of January 1993 occurred when the cooling system of the satellite was stopped and the temperature rose by about 20K and remained afterwards constantly high.

3.2. INFLUENCE OF SUNSPOTS

To account for the effect of the sunspot deficit, the Photometric Sunspot Index (PSI) calculated by Fröhlich et al. (1993) is used. This PSI takes into account the area dependence of the contrast of sunspots and calculates 'true' daily means for each observation using the latitude dependent surface rotation of the spots. Moreover, the observations are screened for outliers which improves the homogeneity of the data set substantially. A bivariate spectral analysis (see e.g. Koopmans, 1974) has been performed with the daily means of the total irradiance measured by PMO6 and the PSI for the period from 15 August 1992 till 15 February 1993. The spectral power density of the PMO6 irradiances is represented by the uppermost solid line of Fig. 3. The spectral power explained by PSI is the area between the

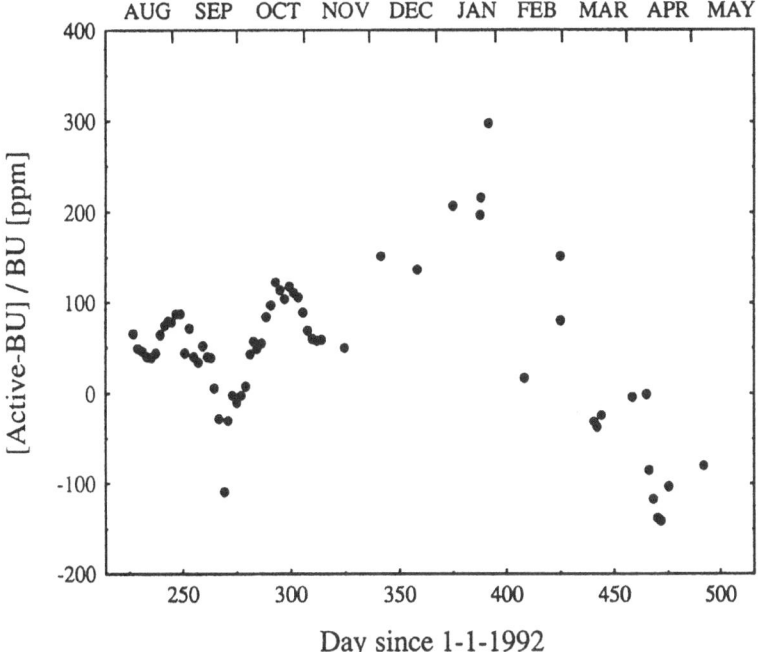

Fig. 2. Relative difference between the 'active' and 'back–up' radiometers during the EURECA mission.

upper and lower solid lines. The two dotted lines indicate the upper and lower 90% confidence interval of that area. Below 0.3μHz (35 day period) and between 0.6 and 1.25 μHz (19 and 9 day periods) more than 90% of the variance is explained by PSI. On the other hand, around the 27-day rotational period very little is due to sunspots ($<$60%).

The gain of the bivariate analysis is the factor by which the PSI has to be multiplied in order to get the observed irradiance change. A gain greater than 1 indicates that the contrast assumed in the PSI calculation is too small. The gain is shown in Fig. 4 with its 90% confidence intervals. The mean gain up to 1.1μHz weighted by the inverse of its uncertainty yields a value of 1.38 which is close to values found by Fröhlich et al. (1993) for the time period from October 1991 until November 1992, thus confirming the strongly time dependence of the contrast of sunspots. It is also interesting to note that the coherence[2] for the EURECA period is much higher (88%) than the one (72%) for the longer time interval including the sharp decrease of activity in spring 1992.

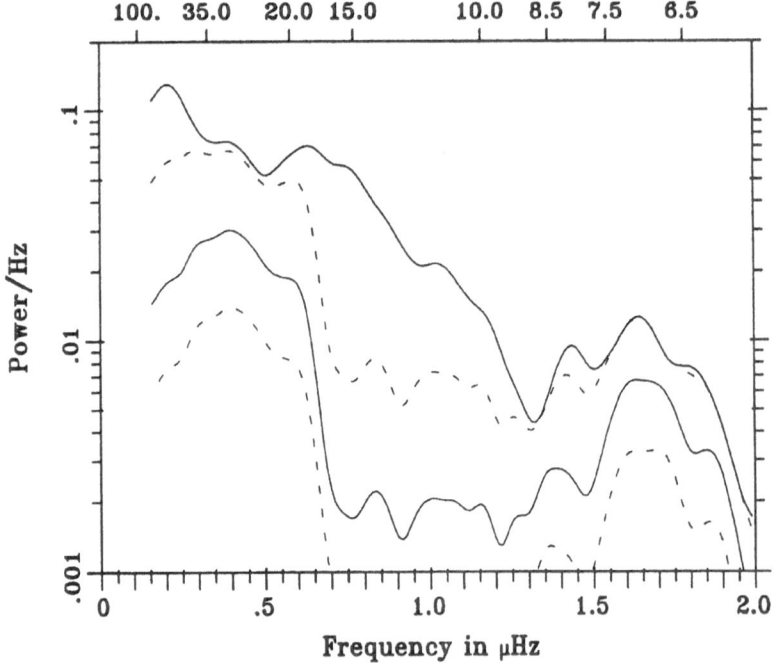

Fig. 3.   Spectral power density of the PMO6 irradiance (upper solid curve); spectral power density explained by the *PSI* (lower solid curve) with its 90% confidence intervals (dotted lines).

4. Conclusions

The results of these measurements demonstrate the ability of the PMO6 radiometers to monitor the total solar irradiance with high precision. The use of two independent instruments allows to assess possible degradation down to the level of 1ppm/day.

The variations of the total solar irradiance due to sunspot blocking is investigated by comparison with *PSI*. Results from similar analysis with data from other missions are confirmed by the present findings, mainly the time dependence of the contrast of sunspots, a still unexplained feature.

Acknowledgements

The support from the SOVA team at IRMB, PMOD/WRC and SSD/ESA is gratefully acknowledged as well as the continuous financial support of the Swiss National Science Foundation.

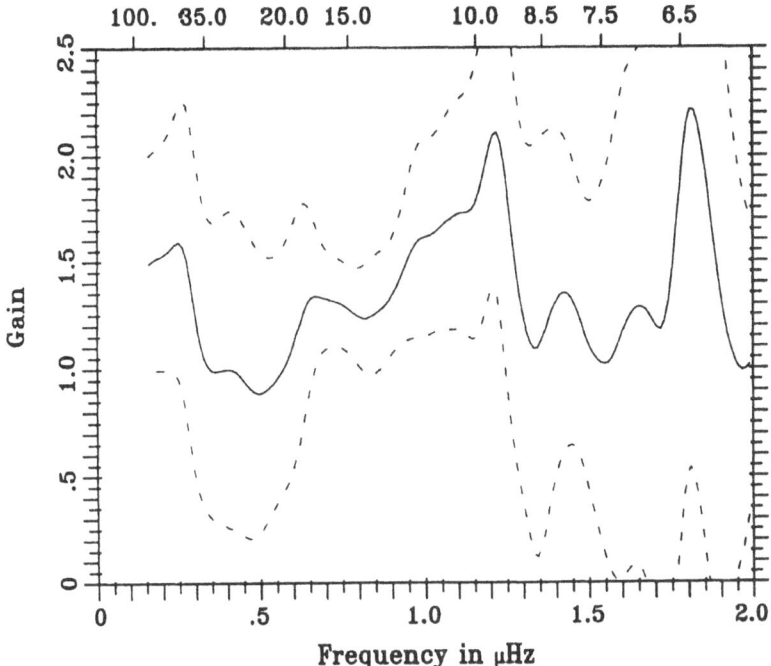

Fig. 4.   Gain of the bivariate analysis of PMO6 with *PSI* (solid line) with its 90% confidence intervals (dotted lines). The gain is the factor by which the *PSI* has to be multiplied in order to get the observed irradiance change. A gain greater than 1 indicates that the contrast assumed in the *PSI* calculation is too small.

References

Brusa, R.W. and Fröhlich, C.: 1986, *Appl. Optics* **25**, 4173.

Crommelynck,D., Domingo, V.: 1993, *Sol. Phys.* , submitted.

Crommelynck,D., Domingo, V.,Fichot, A.,Fröhlich, C., Penelle, B.,Romero J., and Wehrli Ch.: 1993, *Metrologia* **30**, in press.

Fröhlich, C., Pap, J., Hudson H.S.: 1993, *Sol.Phys.* , submitted.

Koopmas, L.H.: 1974, *The Spectral Analysis of Time Series*, Academic Press: New York.

Montenbruck, O., Pfleger, T.: 1989, *Astronomie mit dem Personal Computer*, Springer, Heidelberg.

Willson, R.C. and Hudson, H.S.: 1981, *Adv. Space Res.* 1, 285.

Willson, R.C. and Hudson, H.S.: 1991, *Nature* **351**, 42.

THE PERIODICITY OF SOLAR ACTIVITY CYCLES

Yu.S.ROMANOV, N.S.ZGONYAIKO
Astronomical Observatory, Park Shevchenko, Odessa 270014, Ukraine

ABSTRACT. On the basis of published Wolf Numbers and Schove Row, solar activity cycles in the interval of 11 to hundreds of years have been investigated. In this case the method of investigation of pulsating stars showing the Blazhko effect was applied. The elements of cycles and O-C were calculated and compared with results of solar activity parameters determined by classical methods.

1. INTRODUCTION

This work analyzes the characteristics of solar activity from many years' data. Such investigations have been carried out by many other authors too (see, for instance, Vitinsky, 1973; Vasiliev, 1970). However, our approach to the analysis of variations of solar activity features is different: we consider variations in solar activity by methods used in variable star investigations. Therefore, the appropriate terminology will be used below.

2. WOLF NUMBER ANALYSIS

The series of average annual Wolf numbers, W, from 1749 througt 1971 (Vitinsky, 1973) and international average annual Wolf number W for 1972 – 1990 have been used. Fourier analysis of W(t) has been made. In Figure 1 spectral density is plotted as a function of frequency, $F = 1/P$. A peak of spectral density is distinctly shown at the frequency of $f_1 = 0.09049$ that corresponds to the period $P_1 = 11.051$ years. However, one more peak is observed nearby which value is comparable at the frequency of $F_2 = 0.1003$, i.e. with the period $P_2 = 9.97$ years. If one considers the values of the periods to be significant, then the beats of processes occuring for the periods should lead to an exhibition of period Π (determined from the ratio $1/P_1 - 1/P_2 = 1/\Pi$). In the given case, Π equals $P_3 = 101.92$, which corresponds to the peak at the frequency $f_3 = 0.00981$, i.e. to the period of 101.94 years. Thus, a frequency analysis of Wolf numbers from 1749 through 1990 shows the existence of two processes of solar activity with periods 9.97 and 11,05 years which brings about beating with a period of 101.92 years.

Moreover, the frequency analysis of Wolf numbers indicates that the presence of processes with characteristic time duration of the order of 294.1 years is possible.

Besides that, the information contained in Wolf numbers was treated by another method.

From these data we have determined moments of maxima and minima of the 11-year solar cycle and made comparison of these with the results of other autors' determinations.

Elements M_0 and P were estimated by the least-squares method. These elements of 11-year solar cycles are involved in the well-known ratio $M_i = M_0 + PE_i$, where M_i is the moment of its activity extremum, P is the period and E_i is the epoch extremum. It should be noted that such a determination was made thrice:

a) from maxima of solar activity;
b) from minima;
c) jointly from maxima and minima, with an average time interval between maximum and nearby minimum.

In case a), the following values are found:

$M_{0\ max}$ = 1748.57 ± 0.88 (in years)

P =11.137 ± 0.075 (in years)

In case b):

$M_{0\ min}$ = 1743.82 ± 0.65

P =11.1483 ± 0.0519;

In case c):

$M_{0\ min}$ = 1744.42 ± 0.32

$M_{0\ max}$ = 1749.31 ± 0.32

P = 11,1349 ± 0.0223.

With these elements, $(O-C)_i$ values have been calculated for moments of maxima and minima obtained from Wolf numbers. A frequency analysis of this relation has been made. In the power spectrum (Figure 2) peaks are distinctly seen correspondingly to periods P_1 = 196, P_2 = 88.8 and P_3 = 59.52 years.

3. THE INVESTIGATION OF SCHOVE SERIES

We decided to analyze the Schove series from 653 B.C.E. to 1990 by a similar way. Unlike the work of Zhukov, Mouzalevsky (1969) we used the typical methods of investigation of variable stars. Initial moments of maxima and minima and an average period were determined by the least-square method:

$M_{0\ min}$ = 1744.809 (in years)

$M_{0\ max}$ = 1749.851 (in years)

P = 11.1066 (in years).

Then from the above elements the (O-C) values were calculated for all the maxima and minima of solar activity. From the given relation a Fourier analysis was made. The results of the Fourier analysis are presented in Figure 3. Four periods are seen: P_1 = 2296.7; P_2 = 200.56; P_3 = 133.34; P_4 = 78.39 years. (The 80-year solar cycle was often discussed in the literature, but in spectral density plots from Zhukov and Mouzalevsky, 1969 it is practically not visible).

Thus, frequency analysis of more long-term series features of solar activity from Schove series shows the existence of processes with a period of 200 years which is a multiple with respect to a 100-year cycle (though there may be noticeable cycles with other values of periods).

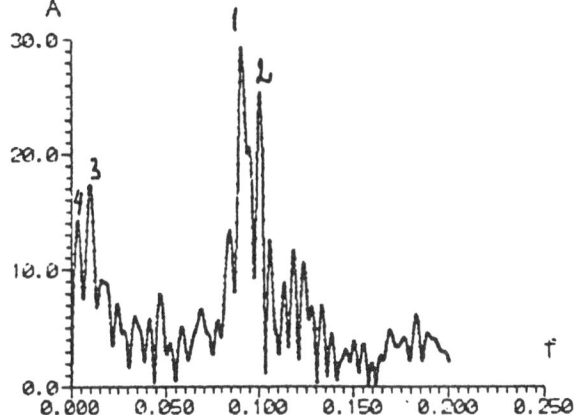

Fig.1   Power spectrum of the dependance W(T)

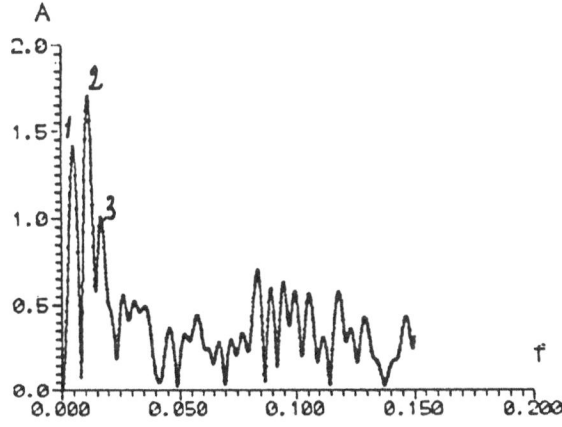

Fig.2   The (O-C) power spectrum of Wolf numbers maxima and
                              minima

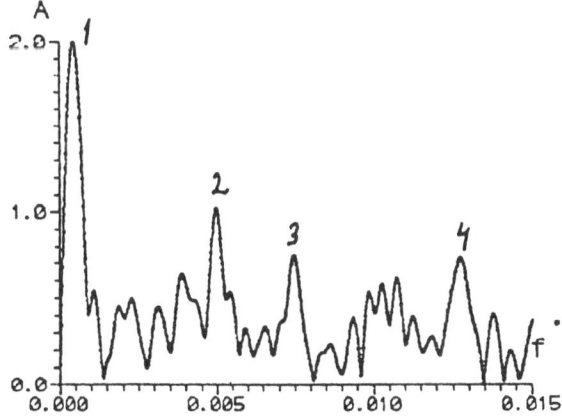

Fig.3   Power spectrum of (O-C) Schove series

4. CONCLUSION

Solar activity is a complicated process with cycles of the order of 11 and 10 years leading to beats with a cycle of the order of 100 years, and possibly, 200 years. The main cycle of solar activity from determinations of maximum and minimum of Wolf numbers can be described by the elemehts:

M_{max} = 1744.42 (± 0.32) + E 11.135 (± 0.022)

M_{min} = 1749.31 (± 0.32) + E 11.135 (± 0.022)

With that, we have to take into account that there are processes with a period of the order of 9.97 years that result in the exhibition of cycles in solar activity with a characteristic time duration of the order of 100 and more years. For reliable determinations of elements of long-term cycles, unfortunately, there are not enough observational data for the present.

REFERENCES

1.Ju.I.Vitinsky, The Cyclic recurrence and prognoses of Solar activity, 1973, Leningrad.
2.O.V.Vasiliev, Frequency-Time Spectrum of the Zurich Series of Wolf's Numbers (1701 - 1964), 1970, The Solar Data, 1.

3.M.S.Eigensson, M.N.Gnevishev, A.I.Ole, V.M.Rubashev, The Solar activity and its earth manifestation, 1948, Moscow, Leningrad.
4.L.V.Zhucov, Ju.S.Mouzalevsky, Correlativ-spectral analysis of the periodicitics of Solar activity, 1969, Astron.J. 46, part 3.

MAXIMUM AND MINIMUM TEMPERATURES AT ARMAGH OBSERVATORY, 1844-1992, AND THE LENGTH OF THE SUNSPOT CYCLE

C.J. BUTLER

Armagh Observatory, College Hill, Armagh, BT61 9DG, N. Ireland

Abstract.
The question of whether or not the Earth's climate is influenced by solar activity has received considerable attention since the mid-nineteenth century. Most investigations have adopted the sunspot number as the parameter of solar activity. Recently, however, it has been shown by Friis-Christensen and Lassen (1991) that the mean northern hemisphere temperature, from 1861-1990, follows a strikingly similar trend to the *length* of the sunspot cycle, suggesting that the recent global warming could, at least in part, arise from changes in solar activity. In view of the importance of this result, we have examined a set of continuous meteorological records, maintained at Armagh Observatory since 1844, to assess, first, whether data from a single site can give meaningful information on global trends, and second, whether the data from this particular site for the period 1844-1866 can be used to extend the baseline of the comparison with solar activity. We find that both are indeed the case and that there is a strong correlation between the solar cycle length and the mean temperature at Armagh over the past 149 years.

1. Introduction

In studies of climatic variability, and in particular global warming, it has been common practice to use the combined data from many sites, spread geographically as widely as possible. Whilst this gives a valid global picture in the modern era, when accurate instruments and systematic methods are commonplace, it becomes more difficult to move backwards in time, particularly before 1880 when there were far fewer meteorological stations and records were not made in a standardized way. Changes in measurement techniques, in the siting of instruments and in the immediate environment of the meteorological station, can all be the source of small differences that could be confused with real climatic changes. An average of the results from a number of stations may be expected to reduce the impact of the deficiencies in any one of them and the loss of individual stations and their replacement by others becomes a less serious problem. However, in some cases, where data have been gathered over a long period by careful and assiduous observers, where there has been little change in the environment, due, say, to urban encroachment, the data from a single site may be of particular value in assessing long-term trends in the Earth's climate. Armagh Observatory (see Bennett, 1990) where meteorological observations have been carried out since 1795, is believed to be one such site.

The Armagh meteorological station is situated close to the centre of the Observatory grounds on the top of a drumlin at 61m above mean sea level.

Solar Physics **152**: 35–42, 1994.
© 1994 *Kluwer Academic Publishers.*

Throughout much of the nineteenth century, the third director at Armagh Observatory, Thomas Romney Robinson, who is principally remembered for his cup anemometer, maintained a keen interest in meteorology. It was he who established the current series of meteorological measurements with daily pressure and temperature from 1833, rainfall from 1836 and hourly anemometer readings from 1846. Maximum and minimum temperatures, which form the basis of this study, were started in 1843. Though a considerable amount of earlier data survives, it is unfortunately not continuous (see Butler and Hoskin, 1987).

TABLE I

Maximum and Minimum Thermometers in use at Armagh Observatory, 1843-1899

Type	Period	Thermometer	Correction
Maximum:	Dec. 1843 - May 1860	Newman	
	Dec. 1860 - Sep. 1882	Casella	-1.7°C (-3.0°F)
	Oct. 1882 - Oct. 1892	Negretti 3404	-0.3°C (-0.5°F)
	Nov. 1892 - Oct. 1899	ditto	-0.6°C (-1.0°F)
Minimum:	Dec. 1843 - Sep. 1882	Newman	-0.3°C (-0.6°F)
	Oct. 1882 - Nov. 1899	Casella 427	0.0°C

2. Corrections for Thermometers

One of the principal difficulties associated with this type of study is to ascertain the reliability of the thermometers which were used, their calibrations, and any changes in measuring techniques or exposure which might introduce systematic errors. After 1900, the thermometers in use at Armagh were generally accurate to within 0.06°C. In Table 1, we list the details of the maximum and minimum thermometers in use at Armagh Observatory, up until 1900. Only two minimum thermometers were employed in the period 1844-1899 and, because of the extended period of their use, they were more reliably calibrated. On the other hand, three maximum thermometers were used and for the earliest of these, the Newman thermometer, we have not found any calibration data. Therefore, for the maximum temperatures, 1844-1860, we have assumed a correction of zero. This thermometer was unfortunately broken in May 1860 and replaced later that year with a Casella thermometer which had an unusually large error of $-1.7°C$ ($-3.0°F$). This correction was determined by reference to a standard thermometer from

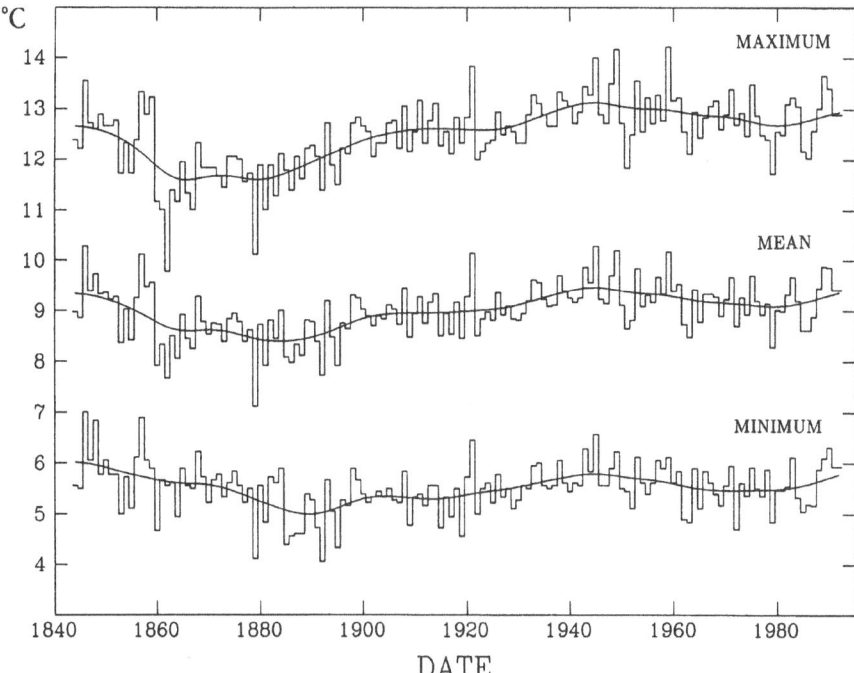

Fig. 1. The annual mean maximum, annual mean minimum and annual mean temperature at Armagh Observatory: 1844-1992

Kew on several occasions in the period 1870-1882 and found to be stable.

Initially, the maximum and minimum thermometers at Armagh were placed in a ventilated metal box suspended 3.5m above ground level on the north side of the east tower (still there in 1993). The adjacent building which housed the Mural Circle and Transit Instrument was not heated at this time. From January 1885 the thermometers were placed in a Stevenson Screen, 43m south of the earlier site. During the period 1868-1883, the British Board of Trade maintained an automatic weather station at Armagh Observatory, one of seven established in the British Isles, from which hourly measures of temperature, pressure, rainfall and wind-speed were published by Scott (1870-1883). The thermograph, which was set up in a north wall screen, 6.1m southeast of the original thermometer site, allowed Robinson, Dreyer and subsequently ourselves to confirm the calibrations of the maximum and minimum thermometers in use at that time. *

---

* Also it has revealed a discrepancy in published data for maximum and minimum temperatures by Scott (1884-1910) which were given without correction for thermometer error, contrary to statements in those publications.

3. Maximum, Minimum and Mean Temperature at Armagh Observatory

In Figure 1 we show the mean annual maximum and minimum temperature at Armagh Observatory for the years 1844-1992 with the data corrected for the thermometer errors listed in Table 1. Superimposed on the data is a running mean with a gaussian shaped filter which has a half-width of five years. With reference to Figure 1, we note the following: (1) The amplitude of the variation in minimum temperature is less than that in maximum temperature. (2) Years with higher or lower than average maximum tend to have higher or lower than average minimum - but this is not invariably the case, as for instance in 1862 which had the lowest mean maximum (2°C below normal) on record, but had a mean minimum close to average. This was also an exceptionally wet year with cloudy summer months. (3) There was an exceptionally warm period in the 1840s when both maximum and minimum temperatures were high; minima noticeably so, with higher values than at any other time during the 149 year coverage. The Great Famine in Ireland occurred at this time as a result of the devastation of the potato crop by fungal attack (blight). (4) The long-term variations in maximum and minimum temperatures have the same general behaviour, with a colder than average period in the second half of the 19th century, a significantly warmer period around 1950, a fall in temperature in the 1960s and 1970s and subsequent rise in the 1980s. This behaviour has been noted in many climatic studies, for example Lamb (1977). We should also note that, due to the lack of any calibration data for the first maximum thermometer, prior to 1860, the maximum curve could be less reliable than the minimum curve.

4. Relationship between Armagh and Northern Hemisphere Mean Temperature

Climatological data from single sites, though of interest, becomes substantially more important if they can be used to indicate global trends. In particular, if we can establish a clear relationship between them, then we can use the longer baseline data for a single site to predict the global trend at times in the early 19th and late 18th centuries, when the geographical distribution of meteorological stations was insufficient to determine the global picture. In Figure 2 we show the mean temperature at Armagh, for the years 1844-1992, together with the deviation from the northern hemisphere mean for 1880-1985 given by Hansen (1987). There is general agreement in the behaviour of the two curves, over the range in common, though the amplitude of the change is greater at Armagh than in the NH mean. It is noticeable in Figure 2 that the rise in the NH mean has already begun in the 1970s, whereas this does not occur in the Armagh data until the 1980s. It is now well known, (see Lamb, 1977 and Hansen, 1987), that the long-term variations in temperature are more extreme at higher latitudes (>64°N) than at low latitudes (<44°N). Therefore, it is not surprising that the variation of the mean temperature at Armagh (latitude 54.3°) is greater than the average for the northern hemisphere.

In Figure 3 we show the mean temperature at Armagh (from the smoothed curve in Figure 1) plotted against the mean NH temperature anomaly extended to include 1861-1880 and 1985-1990, as given by Friis-Christensen and Lassen (1991). We note a good correlation, though the two most recent points for 1982 and 1985 come below the general relation defined by the other points. This is presumably due to the effect, noted above, of the delay in the most recent rise in temperature at Armagh compared with the average for the northern hemisphere. We may also note that the three earliest points, for 1866, 1873 and 1878, which required a large instrumental correction for the maximum temperature (−1.7°C), lie well within the band defined by the remaining points. This confirms that the correction that has been applied to the maximum temperatures in the period 1860-1882 is reasonable.

5. The Relationship between Mean Temperature at Armagh and the Length of the Solar Cycle

It has recently been shown by Friis-Christensen and Lassen (1991) and Burroughs (1992) that the behaviour of the mean NH temperature anomaly, over the period 1865-1990, follows closely the variation of the sunspot cycle length. However, this conclusion is based on similarity in behaviour over two cool periods (circa 1890 and 1970) and one warm period (circa 1950) followed by a rise to current levels. The correlation would be strengthened if it could be shown that the same agreement continues back into the early 19th century. The length of the sunspot cycle, which is known with reasonable precision back to circa 1750, shows a peak around 1840, a minimum near 1805 and a further peak around 1770 (see Figure 3 of Friis-Christensen and Lassen, 1991). Our temperature data allow us to explore the connection with the solar cycle length back to 1844, close to the next peak back in time before the 1861 limit of the mean NH data. It is therefore important to establish whether or not the mean temperature was falling with advancing time, in the period 1840-1865. Our data clearly suggest that this was the case. In this connection we may note that the long Central England Temperature Series, given by Lamb (1977), which is based on observations from a number of sites linked together, does not show a clear fall in temperature at this time. This is surprising as the Central England Series covers a region only 400 kms distant from Armagh. However, a pronounced drop in temperature at this time is evident in data from some other northern hemisphere stations. Possibly the discrepancy can be explained by the greater influence

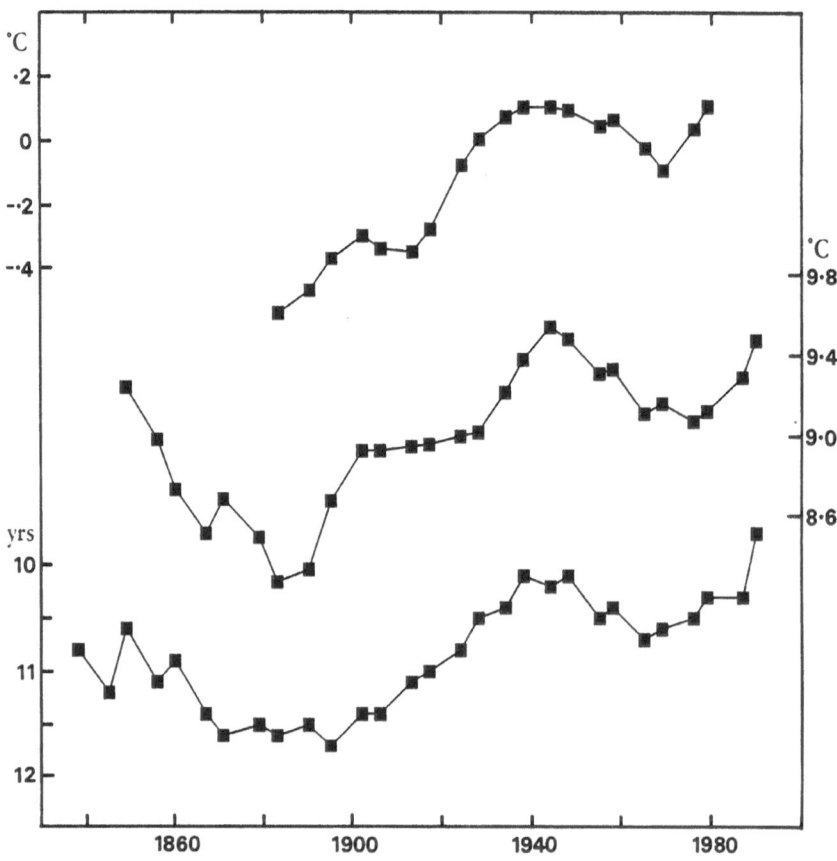

Fig. 2. *top* - Deviation of NH mean derived by Hansen (1987), *middle* - Mean temperature at Armagh 1844-1892, *bottom* - Length of Sunspot Cycle. Points on the top and middle curves are means for 11 year intervals centred on years of sunspot maximum and minimum. Points on the bottom curve are smoothed sunspot cycle lengths derived by Lassen and Friis-Christensen (1992).

of the Atlantic Ocean on the temperature at Armagh which has a strongly maritime climate and predominantly westerly winds.

In the lower panel of Figure 2 we show the length of the sunspot cycle as determined from both the minimum and maximum of the sunspot number by Lassen and Friis-Christensen (1992). The excellent agreement with the behaviour of the maximum, minimum and mean temperature recorded at Armagh over the period 1844-1992 is striking and confirms the results of the Copenhagen group determined over a shorter interval. In Figure 4, the mean

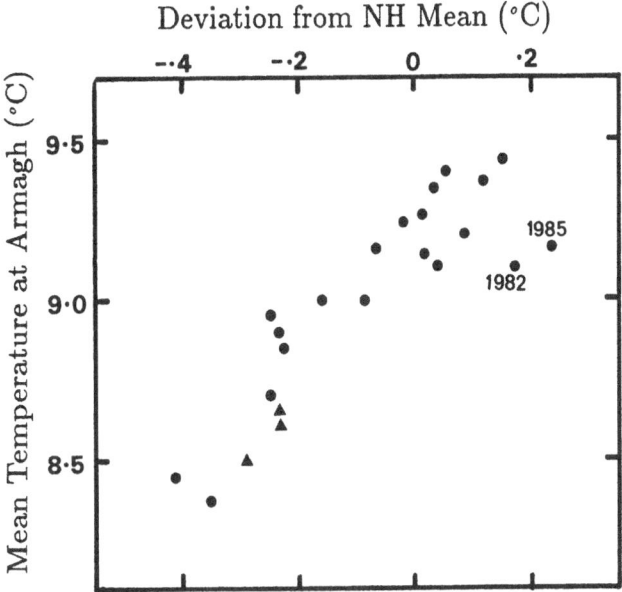

Fig. 3. The mean temperature at Armagh plotted against the NH mean (extended). The triangles represent values for the intervals centred on 1866, 1873 and 1878. The two most recent points, centred on 1982 and 1985 are indicated.

temperature at Armagh, over 11 year intervals centred on the maximum and minimum of the solar cycle, is plotted directly against the solar cycle length. This diagram gives further convincing evidence for a strong correlation between mean temperature and the sunspot cycle length, indicating that solar activity, or something closely related to it, has been a dominant influence on the temperature of the lower atmosphere in the northern hemisphere over the past 149 years.

6. Acknowledgements

The author thanks all members of staff at Armagh Observatory who have made meteorological observations during the past 150 years and in particular Mr Robert Scott, BEM who died tragically in 1992. Research at Armagh Observatory is grant-aided by the Department of Education for Northern Ireland.

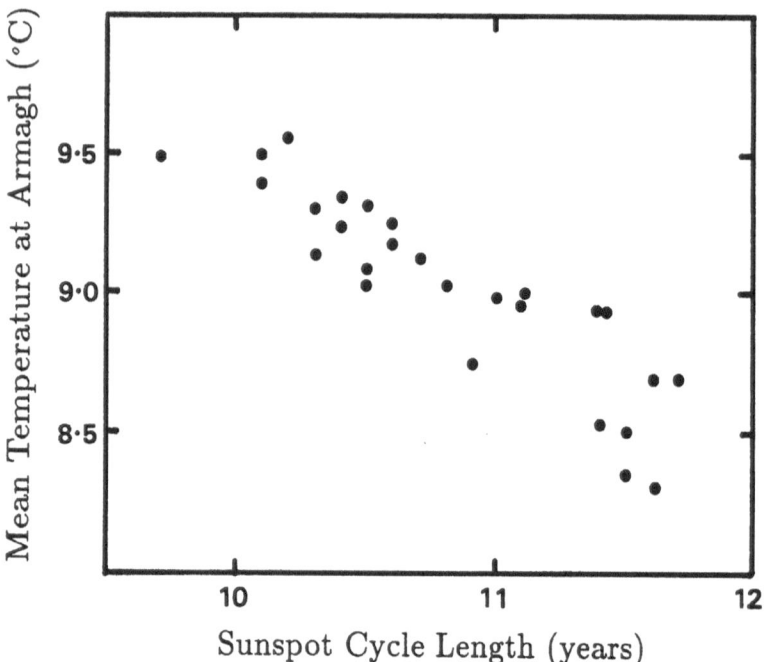

Fig. 4.   The mean temperature at Armagh for 11-year intervals, centred on years of
sunspot maximum and minimum, plotted against the sunspot cycle length.

References

Bennett, J.A.: 1990, *Church, State and Astronomy in Ireland - 200 years of Armagh
    Observatory*.
Burroughs, W.J.: 1992 *Weather Cycles - Real or Imaginary*, Cambridge, p132.
Butler, C.J. and Hoskin, M.A.: 1987, *J. Hist. Astron.* **18**, 295.
Friis-Christensen, E. and Lassen, K.: 1991, *Science*, **254**, 698.
Hansen, J.: 1987, *J. Geophys. Res.*, **92**, 13,345.
Lamb, H.H.: 1977, *Climate: Present, Past and Future - Vol 2: Climatic History and the
    Future*, Methuen, London.
Lassen, K. and Friis-Christensen, E.: 1992, *Danish Met. Inst. Technical Report 92-8*,
    Copenhagen.
Scott, R.H.: *1870-1883*, Hourly Readings from the Self-Recording Instruments at the Seven
    Observatories, in connection with the Meteorological Office, *H.M. Stationary Office*,
    London.
Scott, R.H.: *1884-1910*, Meterological Observations at Stations of the Second Order, *H.M.
    Stationary Office, London*.

PHOTOMETER "DIFOS" FOR THE STUDY OF SOLAR BRIGHTNESS VARIATIONS

E.A. GURTOVENKO, I.G. KESEL'MAN, R.I. KOSTYK AND S.N. OSIPOV
Main Astronomical Observatory, Ukrainian Academy of Sciences, 252127, Kiev, Ukraine.

N.I. LEBEDEV, I.M. KOPAYEV, V.N. ORAJEVSKY, AND YU.D. ZHUGZHDA
IZMIRAN, Russian Academy of Sciences, 142092, Troitsk, Russia

ABSTRACT. A photometer has been designed for measuring solar irradiance within three wide spectral bands with a relative error 0.00001 and time resolution of 16 sec. It is elaborated according to the international space project KORONAS and is planned to be launched at the begining of 1994. A description of its layout and operation is given briefly.

1. LAYOUT OF THE PHOTOMETER

The device consists of three identical channels whose optical axes are parallel and placed at equal (25 mm) distances from each other. Spectral transmittances of the channels are given in the following table:

Channel number	Filters	Wavelengths of maximum of transmittance	Range of transmittance, nm
1	neutral	-	-
2	green	526	494-600
3	red	710	690-790

A section of the photometer through a plane containing any pair of channel axes is displayed in Figure 1. When designing the photometer, besides the limitation of its weight and dimensions, we also had to take into account the following factors.

1. During the experiment the receivers' photo-sensitive surface (a, diameter is 2.5 mm, Figure 1) must be illuminated by radiation from the whole solar disk. At the same time the angle of vision, α, from each point of the photosensitive surface (a) in the direction of the photometer optical axis should be minimum, in order to prevent the unpredicted parasitic illumination from the Earth surface, upper atmospheric layers and apparatus mounted on the satellite platform, and also to diminish heating of the photodiode frame.

2. Estimation of the angle, α, should be made while accounting for the accuracy of centering of the entrance photometer apertures (b) and photodiode apertures (a) on the optical axis (the inaccuracy in both cases is 0.2 mm) as well as the error of photometer mounting (several minutes of arc) on the satellite platform.

3. The photosensitive photodiode area should not be illuminated by light passing through entrance apertures of both neighboring channels.

4. Transmittance of filters may be different at different parts of their area. Under limited accuracy (10 min. of arc) of satellite axis pointing to the Sun this may cause false effects in the output signal.

Accounting for the photometer dimensions (Figure 1) and the factors noted above in points 1 and 2, the angle of vision

$$\alpha = 2 \arctan (b - a)/2 \ c$$

Solar Physics **152**: 43–46, 1994.

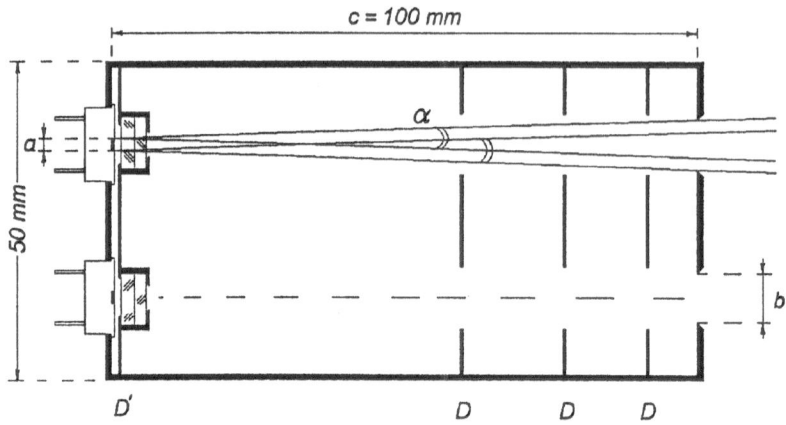

Fig. 1. Layout of the photometer DIFOS (see text)

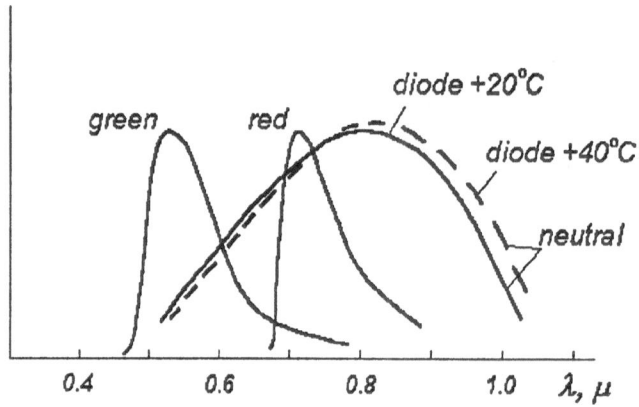

Fig. 2. Spectral transmittance of the photometer channels

is put equal to 200″ and the size of the entrance aperture is $b = 8$ mm. Diaphragms D meet the demands noted in point 3. They also act as traps for scattered light. In order to avoid the effects noted in point 4, the filters are put onto the photodiode frame. Diaphragms D' behind filters are designed in such a way that their apertures are always illuminated by radiation from the whole solar disk, even if the maximum rocking of the satellite axis takes place.

2. OPTIMUM CONDITIONS FOR OPERATION OF RECEIVERS

Silicon photodiodes FD-293 with the band of spectral sensitivity 440-950 nm are used in all three channels. The spectral transmittance of photometer channels is displayed in Figure 2.

The optimum operating regime of the photodiodes is with photocurrent between 80-160 µA. To meet this demand it was necessary, first - to equalize the fluxes in all channels

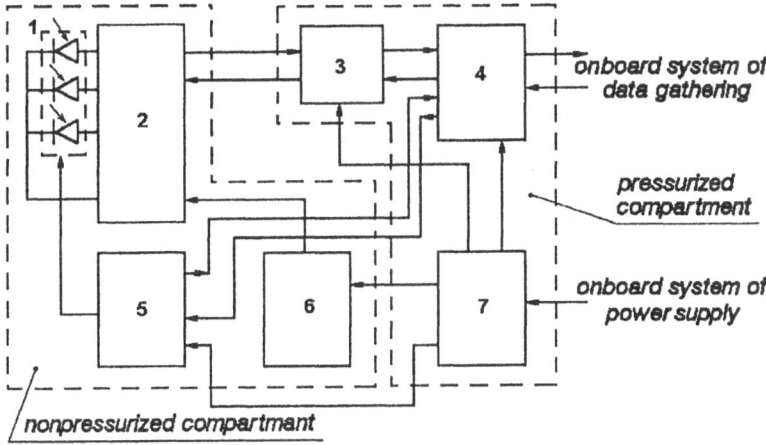

Fig. 3. Structural scheme of the photometer:
1 - photodetectors; 2 - analogue-digital converter (ADC) (analogue part); 3 - controller of ADC; 4 - adapter of interface of onboard data system gathering; 5 - thermoregulator; 6 - source of ADC power supply.

(using an appropriate neutral filter) and secondly - to equalize them in such a manner that the required photocurrent (80-160 µA) would be obtained under working conditons in space. Thus, the photometer was installed before the main mirror of solar telescope ATsU-5 in the parallel solar light beam, and was oriented along the telescope optical axis. The photometer's recorded photocurrent is decreased (in comparision with the conditions during the space experiment) due to the reflectivity of the coelostat mirrors and to atmospheric extinction. Ectinction by the coelostat mirrors is small, and can be estimated easily. In order to account for the atmospheric extinction we have measured the photocurrent over a whole day, and then by using "cosecant method" we extrapolated it to its value in space. Thereupon the neutral filters were sorted out in all channels so as to make the extrapolated photocurent approximately (within 10 percent) equal to the optimum value.

Adjustment of the photometer and its testing were performed with the help of a specially designed stable light source, whose spectral radiance corresponds to the solar one. A halogen battery filament lamp KGM-24-150 (24V, 150W) was used as a light source.

3. OPERATION OF THE APPARATUS

Detectors are mounted on the autonomous platform which provides a high equality of their temperatures as well as perfect thermoisolation from the photometer body.

To eliminate the influence of temperature variations upon receiver sensitivity a system of active thermoregulation was designed. Because the surrounding temperature (T_s) can not be predicted reliably, we have utilized the principle of adaptive setting of thermoregulation temperature T_{th}: preliminary measured temperature of the surroundings, T_s, establish the temperature of thermoregulation, T_{th}. A diode serves as the temperature sensor and is mounted on the receiver's platform with perfect thermal contact with it.

The thermoregulator represents a system of proportional regulation. It consists of the following elements: transducer which transforms the temperature of the crystal diode into voltage, digital-analoque converter which converts the input current (corresponding to the temperature of thermoregulation) into voltage, amplifiers of the signal mismatching and of the power gain, a heater which is fastened to the platform of receivers, and a controller.

Conversion of photocurrent into its equivalent digital code is performed with the help of a three-channel parallel analogue-digital converter (ADC) of integrating type. Choice of this ADC is conditioned by the demands of high resolving power (about 16 bits) and of high noise stability.

As the photometer is destined for measuring the relative solar variations the ADC is designed as a differential converter. Thus, an additional standard compensation current is inserted into the system. It corresponds to the mean value of the photocurrent, and during observations the deviation of signal from its average value is recorded. Two ranges of recording are forseen - fine and rough. On the rough range the photocurrent is diminished by two times.

Utilization of all techniques mentioned above enables us to attain a relative resolving power equal to 0.00001 under 16 sec storge time and with the unit signal-to-noise ratio (theoretical resolving power is 0.000002). The analogue part of the ADC is placed into the electromagnetic shield which is made of an alloy having a high magnetic permeability. Electric power for the ADC is supplied by the autonomous stabilized voltage converter which performs also the galvanic isolation of the ADC from the internal network of the photometer power supply. Control bus is optoisolated from the analogue part of the ADC.

Photometer is fed from the board net of electric power supply via autonomous voltage converter which performs the galvanic isolation of the photometer electrical circuits from the board network and also forms the required stabilized voltages.

Structural scheme of the apparatus is shown in Figure 3.

Solar irradiance variations will be measured on the background of the solar flux trend caused by the change of the Sun-Earth distance. The trend may also serve as a control for the stability of the photometer sensitivity.

Orbit of the satellite allows to perform continuous observations of the Sun during 20 days, periodically.

Spacecraft KORONAS is planned to be launched at the begining of 1994.

THE BRIGHTNESS OF THE SOLAR DISK IN THE CONTINUUM
IN THE REGION 1.0-2.4 μm

E.A. Makarova, E.M.Roshchina, and A.P.Sarychev

Sternberg State Astronomical Institute
Moscow State University

Universitetsky prospect, 13, Moscow 119899, RUSSIA

ABSTRACT. From the results of a critical analysis of some absolute solar radiance measurements, we determine two of the most reliable series for the range 1.0-2.4 μm. The absolute scale for both series are corrected using modern experimental data. The resulting data show the presence of an unknown source of absorption in the region 2.0-2.4 μm in the photosphere.

1. INTRODUCTION

The energy distribution in the solar spectrum contains information on the wavelength dependence of the absorption coefficient in the solar atmosphere. On the basis of the elementary theory of radiative transfer one can show that the absorption coefficient, k_λ, at the optical depth τ_λ is connected with the source function, B_λ, by the equation:

$$k_\lambda(\tau_\lambda) = \gamma\lambda(1 - \exp(\frac{-c_2}{\lambda T}))^{-1} B_\lambda(\tau_\lambda) / \frac{dB_\lambda(\tau_\lambda)}{d\tau_\lambda}, \qquad (1)$$

where the parameter γ depends on the geometrical depth of the atmospheric layer, but does not depend on the wavelength, λ. T is the temperature as a function of depth τ_λ; c_2 is the second radiation constant. To obtain Equation 1 it is necessary to assume that the source function is equal to Planck's function:

$$B_\lambda(\tau_\lambda) = B_\lambda(T(\tau_\lambda)), \qquad (2)$$

Using the Equation 1, one can calculate the dependence of the absorption coefficient on wavelength at a geometrical level with a temperature T if the value of the derivative $dB_\lambda/d\tau_\lambda$ is known. The absorption coefficient in that case is calculated with the precision of the constant γ. The necessary dependence of the source function on the optical depth is found from solutions of the inverse problem of transfer theory. To carry it out, it is necessary to derive the function $B_\lambda(\tau_\lambda)$, knowing the measured radiation intensity of the solar disk. If one makes the source function equal to the Planck value, one can calculate the τ_λ value at a level with a temperature T and the corresponding value of the derivative $dB_\lambda/d\tau_\lambda$. This

Solar Physics **152**: 47–52, 1994.
© 1994 *Kluwer Academic Publishers.*

procedure is known as the method of calculation of the "observed" absorption coefficient in the solar photosphere (for example, Pierce and Waddel, 1961).

2. THE CALCULATION OF THE "OBSERVED" OPACITY AND THE COMPARISON WITH PHYSICAL DATA.

An approximate solution of the inverse problem consists in a preliminary definition of the dependence of $B_\lambda(\tau_\lambda)$ as a function with several free parameters determined from the distribution of intensity, I_λ, across the solar disk. We have determined the approximate dependence of $B_\lambda(\tau_\lambda)$ by the exponential integral function $E_n(x)$ (Kourganoff,1949):

$$B_\lambda(\tau) = I_\lambda(0)\left[a_\lambda + b_\lambda\tau_\lambda + c_\lambda E_2(\tau_\lambda) + d_\lambda E_3(\tau_\lambda)\right], \qquad (3)$$

The limb darkening defined by this function must fit observations in the continuum. This was examined in the continuum between 0.3-2.4 μm using the measurements by Pierce and Slaughter (1977), Pierce et al.(1977), and also by results of the average of experimental data of many authors published by Makarova et al., 1991a. In both cases the errors in the approximation were less than the errors of the limb darkening measurements (Makarova et al., 1991b). If we let $d=0$, then the approximation of the limb darkening will be unreliable in the visible and near UV regions of the spectrum. We emphasize that the difference between measurements by the different experimenters greatly exceeds the inner repeatability of the data from each source. This was shown by the transfer to standard wavelength, $\lambda=500$ nm, for five series of measurements made in the spectral interval $(\lambda \pm 10)$ nm (Saryshev et al., 1992).

John (1989,1991) showed that the wavelength dependence of the observed absorption agreed with that of the negative hydrogen ion to within limits of 10% in the region 0.45-2.5 μm. In these articles, the data on brightness of the solar disk came from work by Neckel and Labs (1984) for $\lambda < 1.25$ μm, Labs and Neckel (1968) for $\lambda > 1.25$ μm; and the parameters of limb darkening a_λ ,b_λ ,c_λ for functions like Equation 3 with $d=0$ from the article by Pierce and Waddel (1961). Here we proceed to define the observed absorption more precisely with the help of new experimental data. We use the value of the brightness of the disk center from Table IX of the monograph by Makarova et al. (1991a) and the measurements of the limb darkening from the articles by Pierce and Slaughter (1977) and Pierce et al. (1977), which are analyzed critically in the work of Makarova et al.(1990). The spectral dependence of the observed absorption we obtain is close to that calculated for the H$^-$ ion in the region 0.45-1.8 μm, but as the wavelength increases, these dependences diverge. The observed absorption at λ >1.8 μm exceeds the absorption by the H$^-$ ion, and the difference arises both from the change in wavelength and the depth of the photospheric layer. This difference for 2.4 μm and at $\tau_{0.5} \approx 0.8$ reaches 50%. Since these results contradict the above-mentioned results from John (1989,1991), our study addresses the cause. According to our calculations, the main reason lies in the values taken for the brightness of the disk center.

3. ANALYSIS OF THE ABSOLUTE MEASUREMENTS OF THE SOLAR SPECTRUM

Let us consider the data of Labs and Neckel (1968) in the region 1.25-2.5 μm. This article states that the data were obtained by the transformation of relative intensities from Pierce (1954) to absolute values of the brightness. It is easy to prove that the measured distribution of brightness over the spectrum was distorted by this transformation. Let us calculate the ratio of I_λ (disk center) taken from work of Labs and Neckel (1968) to the values given by Pierce (1954). If these two sets of data are consistent, this ratio will not depend on wavelength. However, this ratio systematically changes along the spectrum, as is shown in Figure 1. We conclude that the Labs and Neckel (1968) data are not suitable as observed values for our purposes.

We critically analyze the available data given by Pierce (1954), Peyturaux (1952), Arvesen et al.(1969) and Thekaekara and Drummond (1971) in order to define the energy distribution more precisely in the near IR spectrum of the Sun. On occasion, the absolute scale for each series of measurements was connected to modern experimental data. For example, the measurements by Pierce (1954) were compared with absolute values of the brightness of the continuous spectrum from Neckel and Labs (1984) in the spectrum range 1.0-1.25 μm which is common for both measurement series. Using a least squares method, we find the multiplier 5.774 10^{12} erg/(cm^2 s ster cm) to transform the relative intensity data given by Pierce (1954) to the absolute values of the brightness was found by the least squares method. The "absolutized" measurements by Pierce (1954) are shown in Figure 2 in the form of the brightness temperature of the disk center. The measurements by Peyturaux are connected to the measurements by Pierce (1954) by the same method in the common spectral range 1.0-2.3 μm, and are also shown in Figure 2. Here we can see that the internal repeatability of the measurements by Peyturaux (1952) is worse then those by Pierce (1954), and they are not in agreement with the new measurements by Peyturaux (1978) also shown in Figure 2. We will not use the Peyturaux (1952) measurements since they have low precision.

One can evaluate the reliability of spectral measurements of the solar radiation flux by comparison with experimental results of the photometry of solar-type stars (Makarova et al., 1989). Using this criterion, the data from Thekaekara and Drummod (1971) in the region $\lambda < 1$ μm were recognized as unreliable. Makarova et al.(1972) showed that the data in the region 2.5-5.0 μm are wrong . It is difficult to accept that the data given by Thekaekara and Drummod (1971) in the band 1.0-2.5 μm are more reliable than outside; therefore, we exclude these measurements from the analysis.

According to the article by Makarova et al., (1989) the measurements of the solar irradiance by Arvesen et al.(1969) are acceptable by the stellar photometric criterion in the region $\lambda < 1$ μm. We improve the absolute scale of the Arvesen

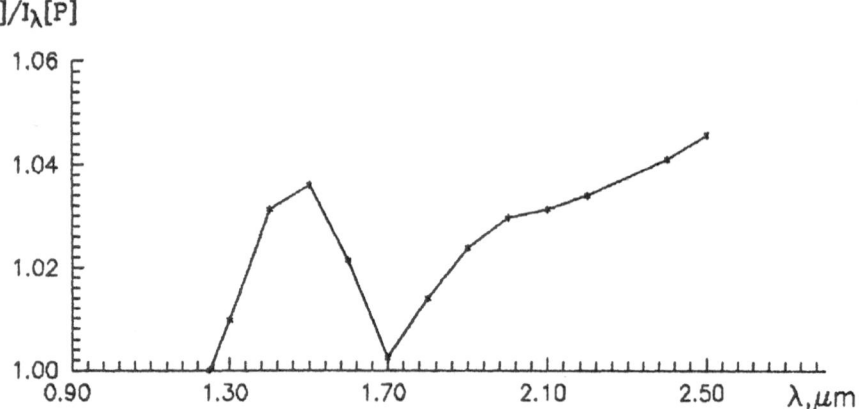

Fig.1. The ratio of the intensity on the solar disk center in the continuum (Labs, Neckel, 1968)-[LN], and (Pierce, 1954)-[P]: the ratio is chosen equal to unity for l=1.25μm.

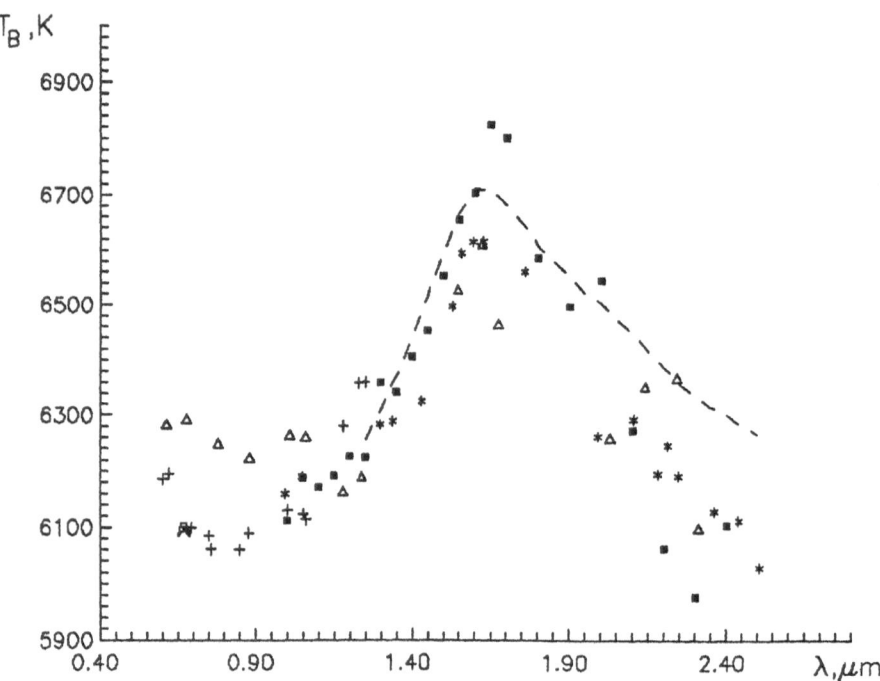

Fig.2. The brightness temperature of the disk center in the continuum from the data of different authors: Δ - (Peyturaux, 1952), * - (Pierce, 1954),) ■ - (Arvesen et al., 1969), ⋈ - (Peyturaux, 1978), --- - (Labs and Neckel, 1968), + -(Neckel and Labs, 1984).

measurements using modern measurements of the total solar irradiance with a precision better then a few tenths of a percent. In this case, the absolute measurements are used as relative measurements, and new absolute measurements are calculated by integrating between 0.3 and 2.4 μm. The value of the integral obtained is compared with the corresponding part of the total irradiance, S_0. Only the value 4.8% of S_0 remains outside the spectral interval, as is shown in Table XII of the book by Makarova et al. (1991). Therefore, even a 10% error in this part of S_0 cannot change its value more then 0.5%. Thus, the normalization of relative measurements of the solar constant is reliable to within 1% or less. We find that the irradiance of the Sun outside the Earth's atmosphere given by Arvesen et al. (1969) must be decreased by 1.5% in order to agree with the modern value of the solar constant 1369 W/m^2. In correcting the value of irradiance, we accounted for details of the line blanketing from Table XXVI of the monograph by Makarova and Kharitonov (1972) and the limb darkening functions from Makarova et al.(1990).

4.CONCLUSION

Thus, a reliable value of the brightness of the solar disk center in the spectral region 1.0-2.4 μm may only be obtained on the bassis of two measurement series: Pierce (1954) and Arvesen et al. (1969). We think that the most reliable experimental data at the present time are the mean weighted values of these two absolutized measurements. In addition, double weight must be given to the measurements by Pierce (1954), since they do not require additional correction for blanketing effects and limb darkening. The mean weighted values are presented in Table I.

Table I: The Mean Weighted Value of the Brightness of Solar I_λ (0) at Disk Center in the Units of 10^{13} erg/(cm^2 s ster cm).

λ,μm	$I_\lambda(0)$	λ,μm	$I_\lambda(0)$	λ,μm	$I_\lambda(0)$	λ,μm	$I_\lambda(0)$
1.00	12.68	1.3	6.712	1.60	3.950	2.10	1.473
1.05	11.45	1.35	6.014	1.65	3.625	2.20	1.227
1.10	10.21	1.40	5.457	1.70	3.274	2.30	1.033
1.15	9.155	1.45	4.976	1.80	2.635	2.40	0.899
1.20	8.231	1.50	4.605	1.90	2.146		
1.25	7.374	1.55	4.326	2.00	2.788		

These values are close to the data the monograph by Makarova and Kharitonov (1972). More reliable data I_λ (0) das not fit the absorption by the H$^-$ ion in the spectrum range near 2.0-2.4 μm. Thus these observations data show the availability of existence an unknown source (sources) of the absorbtion in the region $\lambda > 2$ μm in the photosphere. However, the reality of the unknown source of absorption requires the additional confirmation. For this purpose it is necessary to carry out the new absolute measurements in the near IR spectrum of the Sun with the precision better then 1%- 2%. Such accuracy is quite possible now.

REFERENCES

Arvesen, J.C., Griffin, R.N., and Pearson, B.D.:1969, *Appl.Optics* **8**, 2215

John,T.L.: 1989, *Mon. Not. Roy. Astron. Soc.* **240**, 1.

John,T.L.: 1991, *Astron. Astrophys.* **244**, 511.

Kourganoff, V.: 1949, *Contrib. Inst. Astroph. Paris* **228**, 2011.

Labs, D. and Neckel,H.: 1968, *Z.f.Astrophys.* **69**, 1.

Makarova, E.A. and Kharitonov, A.V.: 1972, *Distribution of Energy in the Solar Spectrum and the Solar Constant.*, Nauka, Moscow.

Makarova, E.A., Knyazeva, L.N., and Kharitonov, A.V.:1989, *Astron. Zh.* **66**,583.

Makarova,E.A., Roshchina, E.M., and Sarychev, A.P.:1990, *Kinematika and Phys. Celest. Bod.* **6**, 21.

Makarova, E.A., Kharitonov, A.V., and Kazachevskaya, T.V.: 1991a, *Solar Radiation Flux*, Nauka, Moscow.

Makarova, E.A., Roshchina, E.M., and Sarychev, A.P.:1991b, *Astron.Zh.* **68**, 855.

Neckel, H. and Labs, D.:1984, *Solar Phys.* **90**, 205.

Peyturaux, R.: 1952, *Ann. Astrophys.* **15**, 302.

Peyturaux, R.: 1978, *Astron. Astrophys.* **69**, 305.

Pierce, A.K.: 1954, *Astrophys. J.* **119**, 312.

Pierce, A.K. and Waddel, J.H.: 1961, *Mem. Roy. Astron. Soc.* **68**, 89.

Pierce, A.K. and Slaughter, C.D.: 1977, *Solar Phys.* **51**, 25.

Pierce, A.K., Slaughter, C.D., and Weinberger, D.: 1977, *Solar Phys.* **52**, 179

Sarychev, A.P., Roshchina, E.M., and Makarova, E.A.: 1992, *Astron. Zh.* **69**, 1322.

Thekaekara, M.P. and Drummond, A.J.: 1971, *Nature* **229**, 6.

IRRADIANCE OBSERVATIONS OF THE 1-8 Å SOLAR SOFT X-RAY FLUX FROM GOES*

MARKUS J. ASCHWANDEN *

University of Maryland, Astronomy Department, College Park, MD 20742

Abstract. The solar 0.5-8 Å soft X-ray flux was monitored by the NOAA Geostationary Operational Environmental Satellites (GOES) from 1974 to the present, providing a continuous record over two solar activity cycles. Attempts have been made to determine a soft X-ray (SXR) background flux by subtracting out solar flares (using the daily lowest flux level). The SXR background flux represents the quiescent SXR flux from heated plasma in active regions, and reflects similar (intermediate-term) variability and periodicities (e.g. 155-day period) as the SXR or hard X-ray (HXR) flare rate, although it is determined in non-flaring time intervals. The SXR background flux peaks late in Solar Cycle 21 (2-3 years after the sunspot maximum), similar to the flare rate measured in SXR, HXR, or gamma rays, possibly due the increasing complexity of coronal magnetic structures in the decay phase of the solar cycle. The SXR background flux appears to be dominated by postflare emission from the dominant active regions, while the contributions from the quiet Sun are appreciable in the Solar Minimum only (A1-level). Comparisons with full-disk integrated images from YOHKOH suggest that the presence of coronal holes can decrease the quietest SXR irradiance level by an additional order of magnitude, but only in the rare case of absence of active regions.

Key words: Solar irradiance – Soft X-rays – GOES spacecraft

1. The GOES satellites

The solar 0.5-8 Å soft X-ray flux is monitored by the full-disk soft X-rays sensors (XRS) onboard the *Geostationary Operational Environmental Satellites* (GOES), operated by NOAA, since 1974. Earlier versions of the soft X-ray sensors (XRS) were flown on the Solar Radiation (SOLRAD) satellite (Kreplin et al. 1977), starting in 1964, and then on the NASA Synchronous Metereological Satellite (SMS) series, in 1974. The currently operational spacecrafts are GOES-6 and GOES-7, simultaneously operating in a geostationary EAST and WEST position. Because the GOES spacecrafts are in geostationary orbits they have almost continous coverage of the Sun. The present GOES are spinning platforms, so that the XRS flux is modulated by solar and nonsolar signals. The main source of nonsolar signals is local particle contamination, detectable on a typical level of $\approx 10^{-8}$ W m^{-2}. Two bands of X-rays (0.5-4 Å, 1-8 Å) are measured, in 3-second intervals, by two gas-filled ion chambers. Instrumental descriptions and details on the calibration are given in Grubb (1975), Unzicker & Donnelly (1974), Donnelly et al. (1977), and Garcia (1993). The calibration of the XRS sensors is checked by intercomparison between different GOES spacecrafts, or with SOLRAD.

* Presented at IAU Colloquium No. 143, "The Sun as a Variable Star: Solar and Stellar Irradiance Variations", Boulder, CO, June 20-25, 1993

No trend for a long-term drift in the sensor calibration was found (Bouwer 1983). However, below about 10^{-7} W m^{-2} the relative error can increase to about 50% because of low-flux instrumental problems (Bouwer et al. 1982).

The effective temperature T and emission measure EM of an isothermal plasma can be determined from the ratio of the two SXR energy channels measured by GOES, using the analytical expressions which have been derived by Thomas, Starr & Crannell (1985), by folding the theoretical SXR spectra with the GOES detector transfer function. The evolution of the flare properties T and EM provide also constraints on the evaluation of the flare-unrelated background flux (Bornmann, 1990). Intercomparisons of measurements of T and EM between GOES, BCS/SMM, HINOTORI, and PROGNOZ agree within < 20% (Antonucci et al. 1984; Tanaka 1986; Garcia 1993).

2. Soft X-ray background flux measurements

The following reasons were brought forward to use the 1-8 Å flux to monitor solar irradiance instead of using other solar indices (Bouwer 1983): (1) the source of SXR flux is confined to the solar corona and has no chromospheric quiet Sun contribution (opposed to the 10.7 cm flux), (2) no center-to-limb darkening effects of the SXR flux and higher sensitivity at and behind the limb (compared with the 10.7 cm flux), (3) high dynamic range between solar minimum and maximum (about 3 orders of magnitude), and (4) importance of ionization in ionospheric D region during high solar activity. The long-term temporal variations of GOES X-rays is important not only to the D-region of the ionosphere, but also for modeling the temporal variations of coronal EUV emissions, which are important to the E and F regions of the ionosphere and to the thermosphere. The solar 1-8 Å flux above $\approx 10^{-6}$ W m^{-2} rivals cosmic rays and Lyman α as a source of ionization and excitation in the D region.

The daily background flux should be an indicator of the quiescent X-ray flux from active regions, where variable emission from flares and coronal mass ejections is largely subtracted out. The Soft X-ray flux in the 1-8 Å band is of the order 10^{-6} W/m^2, which is about a factor of 10^9 smaller than the white light solar constant of 1368 W m^{-2}. Methods to remove the effects of solar flares from a background flux are described in Bouwer et al. (1982) and Wagner (1988): The daily background flux is defined by the minimum hourly value, either taken in the middle 8 hours of the day, or interpolated from the other 16 hours to the middle of the day.

Wagner (1988) determined the daily background X-ray fluxes in the form of monthly averages and annually-smoothed (13-month) values of the 1-8 Å flux for 1974 to 1988 (Solar Cycle 21). He found that intermediate-term variations (on the scale of months) of the 1-8 Å flux roughly mimic those of other chromospheric and coronal indices such as Ca, K, He 10830 Å , and 10 cm radio flux. The annually-smoothed daily background X-ray flux was found to peak late in the solar cycle, and is best matched by photospheric white-light facular areas. A factor of 85 was found for the SXR background flux between the Solar Maximum in Sept 1981 and the last Minimum in Oct 1986, which is a much smaller variation than found between the most powerful (class X30.0; $3.0 \cdot 10^{-3}$ W m^{-2}) and the smallest detectable flares (class $< A1.0$; 10^{-8} W m^{-2}), differing by a factor of $> 3 \cdot 10^5$. Despite this efficient method of subtracting out the principal flare component, the residual SXR background flux and its intermediate-term variability seems still to reflect the flare-associated (post-flare) emission of heated plasma from the flaring active regions. A comparison of the monthly averaged GOES background 1-8 Å flux (Fig.1, middle) with the monthly flare rate counted either in SXR (Fig.1, second row: GOES $> M1$ class flares) or in hard X-rays (Fig.1, first row: ISEE-3, HXRBS/SMM and BATSE/CGRO > 25 keV flare rate) shows a high degree of correlation between these 3 solar activity indicators, which all show the 155-day variability equally well. Thus, the so-called "SXR background flux" is still dominated by flare-related SXR emission.

3. Long-term variability and periodicities

The nonflare temporal varations of solar activity indicators are usually characterized by 3 time scales: (1) short-term variations (over several weeks, caused by solar rotation and the evolution of active regions), (2) intermediate term variations over several months (caused by episodes of major activity or long-lived active regions), and (3) long-term solar cycle variations (Donnelly et al. 1986).

Bouwer (1983) performed an anharmonic frequency analysis of the 1-8 Å SXR background flux for the data from 1977-1981 and found that the frequencies most closely corresponding to the synodic solar rotation period (27.28 days) are concentrated at 22, 25, and 34 days. Donnelly, Hinteregger & Heath (1986) compared the temporal variations of the 1-8 Å flux with the EUV variability and found that the SXR flux exhibits day-to-day variabitlity related to flare activity rather than to EUV emission, and that the short-term variations in SXR are not well-correlated with the 10 cm flux or EUV, sometimes even out of phase (e.g. the 13-day periodicity), or anti-correlated.

The long-term variations of the SXR background flux is different from most of the other standard solar activity indicators: the SXR background flux tends to *peak 2-3 years later in the cycle* than the sunspot number (Wagner 1988; reproduced in Donnelly 1989; 1990) for cycle 21.

This delay in the peak is also manifest in other solar activity indicators, such as in the coronal green-line (Fe XIV at 5303 Å) emission (in the Slovak

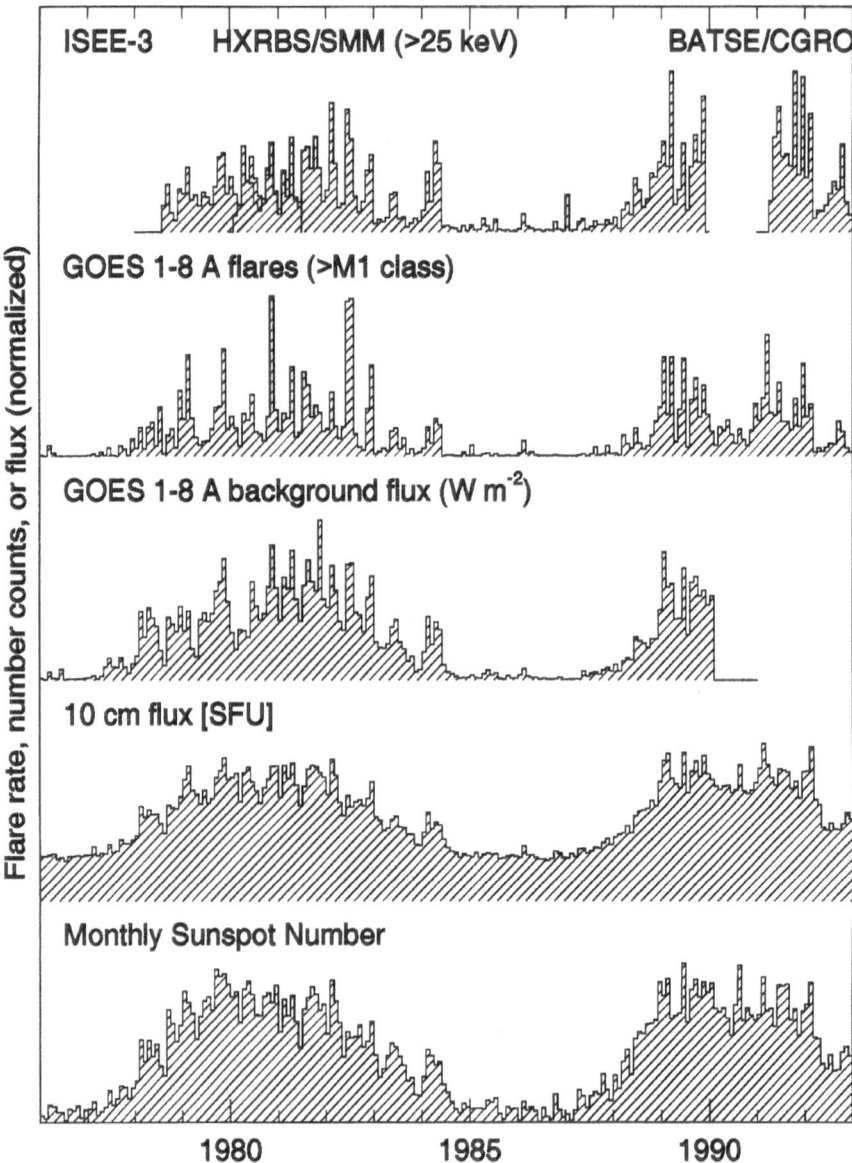

Fig. 1.   The GOES 1-8 Å background flux (monthly averages) in Solar Cycle 21 & 22 (middle) compared with solar flare rates in hard X-rays (top: ISEE-3, HXRBS/SMM, and BATSE/CGRO) and soft X-rays (second row: GOES >M1 class flares), 10-cm flux (forth row), and monthly sunspot number (adapted from Aschwanden & Dennis, 1993).

data; Ribansky et al. 1988), or in the hard X-ray and gamma ray flare rate (Aschwanden & Dennis 1993). The variability of the hard X-ray flare rate is not only a function of the number and size of active regions, but also a function of the magnetic complexity in active regions. Large, complex active regions with highly sheared (nonpotential) magnetic fields are more likely to produce X-ray flares. The fact that the peak amplitude of the hard X-ray flare rate steadily increases over 3 years after the first sunspot maximum may be interpreted in terms of increasing complexitiy in coronal structures during the decay phase of the solar cycle. Consequently, solar activity indicators of coronal origin, such as the SXR flare rate or the SXR background flux are expected to mimic the same behavior, and may, therefore, deviate from photospheric and chromospheric activity indicators.

4. The origin of the solar soft X-ray flux

The spectrum from 1-8 Å is a combination of continuum and line contributions, produced by free-free bremsstrahlung, free-bound recombination, and two-photon emission (Mewe 1972; Kato 1976; Mewe & Gronenschild 1981). Generally, the 1-8 Å flux is dominated by continuum emission, and the contribution of line emission varies from 18%, at 30 MK, to 54%, at 6 MK (Thomas, Starr & Crannell, 1985). The excitation of SXR line emission strongly depends on the temperature of the plasma and varies drastically from quiescent active region conditions to flare conditions. In the 1-8 Å range, the SXR flux is dominated by contributions from the hotter corona ($> 10^6$ K), while the contributions from the cooler parts of the corona are negligible. Thus, the sensitive temperature range of the 1-8 Å SXR flux is complementary to that of the EUV flux, which is sensitive to cool ($< 10^6$ K) plasma in the chromosphere and transition region. The SXR flux is optically thin and does not show a center-limb variation (Mosher 1979) as the 10 cm flux does (Donnelly 1982).

The solar 1-8 Å flux is believed to originate from heated plasma confined in closed magnetic loops in active regions. In earlier models, the disk-integrated SXR flux was simply modeled by summing the SXR emission measure from a variable number n of active region loops (with cross-section A_l and length L_l), neglecting contributions from the quiet sun outside active regions:

$$\Phi \approx a(T) \int N_e^2 dV \approx a \sum_{Active\ Regions} [nA_lL_lN_e^2\Phi(\alpha)/\Phi(0)]$$

where the center-to-limb dependence $\Phi(\alpha)/\Phi(0)$ as function of the aspect angle is constant for optically thin SXR, and has a FWHM of $\approx 205^0$ due to the coronal altitude of the SXR-emitting plasma (White 1964; Donnelly et al. 1982). Imaging observations are imperative to disentangle the origin of the various components contributing to the full-disk integrated SXR flux. The analysis of a solar eclipse observation showed that 98% of the disk-integrated SXR flux came from 4 active regions (Bornmann & Matheson

1990). Soft X-ray images from SKYLAB or YOHKOH confirm that the soft X-ray flux outside active regions, especially in coronal holes, is drastically decreased. From YOHKOH observations, a typical temperature of 5.7 MK and an emission measure of $5 \cdot 10^{28}$ cm^{-5} was determined in a bright loop of an active region, while the corresponding values for the quiet corona were found to be 2.7 MK and $1.3 \cdot 10^{26}$ cm^{-5} (Hara et al. 1992). Thus, this active region loop contributes a factor of ≈ 300 more to the SXR flux than an equal area from the quiet corona.

YOHKOH observations allow us to estimate the solar irradiance level of the quiet Sun in SXR by subtracting out the dominant contributions from active regions. The GOES background level amounts roughly to the C1-class level (10^{-6} W m^{-2}) during the solar maximum, and drops to the A1-class level (10^{-8} W m^{-2}) during the solar minimum. Although the contribution from active regions in these SXR background fluxes is minimized, there are additional local fluctuations of the quiescent SXR emission level due to remnants of old active regions, X-ray bright points (supposedly tiny bipolar magnetic regions), arcades of quiet Sun loops, filament channels, emerging flux regions, polar plumes, and coronal holes (zones with open magnetic field lines). K.Strong (private communication) found that X-ray bright points typically have a SXR flux (SXT/YOHKOH count rate per pixel) exceeding the quiet Sun flux by an order of magnitude, while coronal holes typically exhibit a flux an order of magnitude below the quiet Sun flux. Because of the softer response of YOHKOH (5-45 Å) compared with GOES (0.5-8 Å), a scaling law of $F_{YOHKOH} \propto F_{GOES}^{1/3}$ was found for the full-disk integrated SXR flux measured by both instruments, based on statistics during the period of 1993 May 1-20, in the range of GOES B- to M-class levels (K.Strong, private communication). Correcting for their area compared with the full Sun disk, X-ray bright points add little to the full-disk integrated SXR flux, while the presence of coronal holes can reduce the full-disk integrated SXR flux seen by YOHKOH by a few 10%. Given the nonlinear scaling between GOES and YOHKOH full-disk flux, coronal holes are expected to reduce the total quiet Sun flux seen by GOES by up to an order of magnitude, that corresponds to the Z1-class level (10^{-9} W m^{-2}) during the solar minimum. This low irradiance level cannot reliably be measured by GOES (because of low count statistics and contamination from magnetospheric particles), while SXT/YOHKOH still records a (disk-integrated) count rate of $\approx 10^6$ c/s.

5. Concluding remarks

The solar soft X-ray flux exhibits a variability by a factor of more than 10^6 between the largest flares and quietest periods. Even by subtracting out solar flares, the so-called SXR background flux still varies a factor of $\approx 10^2 - 10^3$

(depending on the time scale of averaging) between solar maximum and minimum, completely governed by free-free bremstrahlung and line emission of heated flare plasma confined in active region loops (mainly post-flare emission). Even the quiet component of the corona (without active regions) is estimated to vary by an order of magnitude due to the varying size of coronal holes and quiet Sun loops. Given this variability, there is no such thing like a "solar constant in soft X-ray irradiance". The understanding of the variability of the full-disk integrated SXR flux crucially depends on the evolution and confinement of heated plasma in flaring active regions.

References

Antonucci, E., Gabriel, A.H., and Dennis, B.R. : 1984, *Astrophys.J.* **287**, 917.

Aschwanden, M.J. and Dennis, B.R. : 1993, in preparation.

Bornmann, P.L.: 1990, *Astrophys.J.* **356**, 733.

Bornmann, P.L. and Matheson L.D.: 1990, *Astron.Astrophys* **231**, 525.

Bouwer, S.D., Donnelly, R.F., Falcon, J., Quintana, A., and Caldwell, G. : 1982, *NOAA Techn.Memo.*, ERL SEL-62.

Bouwer, S.D.: 1983, *J.Geophys.Res.* **88**, No.A10, 7823.

Donnelly, R.F.: 1982, *J.Geophys.Res.* **87**, No.A8, 6631.

Donnelly, R.F., Heath, D.F., and Lean, J.L.: 1982, *J.Geophys.Res.* **87**, No.A12, 10318.

Donnelly, R.F.: 1989, in J.Lastovicka (ed.), *Proc. of the IAGA Symposium for Solar Activity Forcing of the Middle Atmosphere*, MAP Handbook Vol. 29, p. 1-8.

Donnelly, R.F.: 1990, "Temporal variations of solar UV, EUV and X-ray fluxes, sunspot number and F10", in *Solar-Terrestrial Predictions*, Vol.1, NOAA/ERL, Boulder, Colorado, p.381.

Donnelly, R.F., Grubb, R.N., and Cowley, F.C.: 1977, NOAA Techn.Memo. ERL/SEL-48.

Donnelly, R.F., Hinteregger, H.E. and Heath, D.F.: 1986, *J.Geophys.Res.* **91**, No.A5, 5567.

Garcia, H.A.: 1993, '', *Solar Phys.* , submitted.

Grubb, R.N.: 1975, "The SMS/GOES Space Environment Monitor Subsystem", NOAA Techn.Memo. ERL SEL-42, NOAA/ERL/SEL, Boulder Colorado.

Hara, H., Tsuneta, S., Lemen, J.R., Acton, L.W., and McTiernan, J.M.: 1992, *Publ. Astron.Soc.Japan* **44**, L135.

Kato, T.: 1976, *Astrophys.J.Suppl.Ser.* **30**, 397.

Kreplin, R.W., Dere, K.P., Horan, D.M., and Meekins, J.F. 1977, in "The Solar Output and its Variations", ed. White, O.R., Colorado Assoc.Univ.Press, Boulder, p.287.

Mewe, R.: 1972, *Solar Phys.* **22**, 459.

Mewe, R. and Gronenschild, E.H.B.M.: 1981, *Astron.Astrophys.Suppl.Ser.* **45**, 11.

Mosher, J.M.: 1979, *Solar Phys.* **64**, 109.

Ribansky,M., Rusin, V., and Dzifcakova, E.: 1988, *Bull.Astron.Inst.Czech.* **39**, 106.

Tanaka, K.: 1986, *Publ.Astron.Soc.Japan* **38**, 225.

Thomas, R.J., Starr, R. and Crannell, C.J.: 1985, *Solar Phys.* **95**, 323.

Unzicker, A. and Donnelly, R.F.: 1974, NOAA Techn.Rep. ERL 310-SEL, p.31.

Wagner, W.J.: 1988, *Adv.Space.Res.* **8**, No.7, 67.

White, W.A.: 1964, AAS/NASA-SP-50, p.131.

COMPARISONS OF THE MG II INDEX PRODUCTS FROM THE NOAA-9 AND NOAA-11 SBUV/2 INSTRUMENTS

M. T. DeLAND and R. P. CEBULA

Hughes STX Corporation, Greenbelt, MD 20770 USA

ABSTRACT. The Mg II index is a proxy indicator of solar UV activity which is produced from measurements of the chromospheric Mg II absorption line at 280 nm. Mg II index data sets have been derived from the NOAA-9 and NOAA-11 SBUV/2 irradiance data sets using both discrete scan measurements about the Mg II line and continuous scan (sweep) measurements over the UV spectrum from 160-400 nm. This paper will discuss the rationale behind the creation of the different Mg II index products, and make a quantitative assessment of the differences between these products. Recommendations for future use of the Mg II index will also be presented.

INTRODUCTION

Solar ultraviolet variability in the 200-350 nm wavelength region is the primary driver for ozone variations in the upper stratosphere. In order to fully understand the contribution of solar activity to ozone variations, knowledge of long-term solar change to an accuracy of 1% is required. This goal has proven to be difficult to achieve with absolute irradiance measurements because of the significant changes in instrument response exhibited by satellite instruments (*e.g.* Schlesinger and Cebula, 1992; Cebula and DeLand, 1992; Brueckner *et al.*, 1993). In lieu of direct measurements of solar UV variability, proxy indexes are used to represent these changes. One such index is the Mg II index, first derived by Heath and Schlesinger (1986) from Nimbus-7 SBUV measurements using the Mg II absorption line at 280 nm. The irradiance in the core of the unresolved Mg II doublet is representative of chromospheric activity, while the irradiance in the wings of the line approximates the local continuum. Using the ratio of these quantities removes most instrumental change effects, and provides a good indicator of solar UV variability on both rotational and solar cycle time scales (Heath and Schlesinger, 1986). Scale factors have also been derived to estimate solar irradiance variations in the 170-400 nm region from the Mg II index variations (Cebula *et al.*, 1992; DeLand and Cebula, 1993). Mg II index products have been derived from NOAA-9 and NOAA-11 SBUV/2 irradiance measurements using slightly different source data and algorithms (*e.g.* Cebula *et al.*, 1992; Donnelly, 1988, 1991). This paper presents a comparison of these SBUV/2 Mg II index products, and gives recommendations for further use of the data.

DATA SETS

The "classical" Mg II index proposed by Heath and Schlesinger (1986), and employed by Cebula *et al.* (1992) and DeLand and Cebula (1993), uses a total of 7 wavelength positions about the Mg II absorption feature (Figure 1). The average irradiance from 3 core wavelengths comprises the numerator of the ratio, while the irradiances from 2 pairs of wave-

Solar Physics **152**: 61–68, 1994.

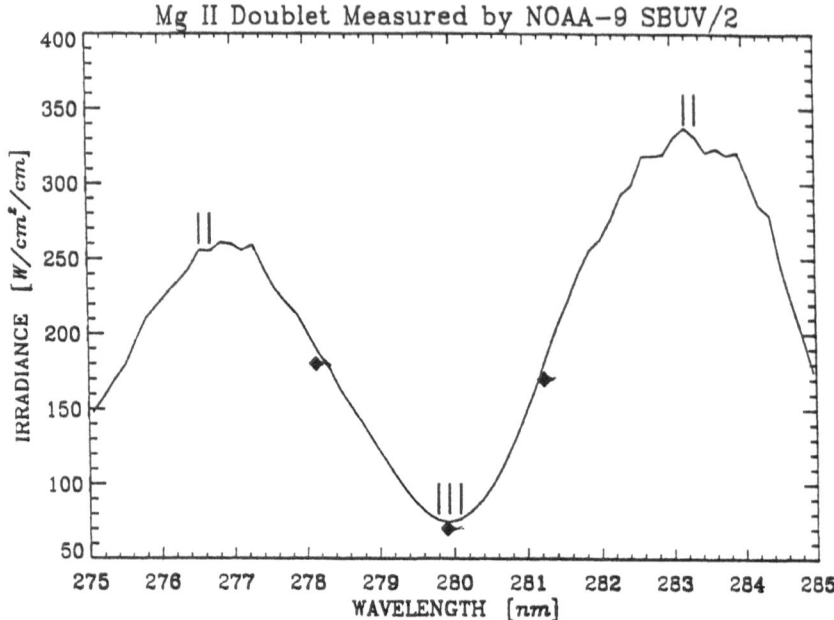

Fig. 1. The Mg II doublet at 280 nm as observed in the NOAA-9 SBUV/2 "Day 1" sweep mode spectrum. Positions of the 7 wavelengths used in the "classical" Mg II index are indicated by heavy lines, and the 3 discrete mode wavelengths used in the "modified" Mg II product are marked with diamonds. Adapted from DeLand and Cebula (1993) with permission.

lengths in the wings of the line are combined to give the denominator of the ratio. The Nimbus-7 Mg II index used irradiances taken from the daily average of three continuous scan measurements over the 160-400 nm region ("sweep mode") to construct the Mg II index. These measurements have been continued beginning in March 1985 for NOAA-9 SBUV/2 and December 1988 for NOAA-11 SBUV/2, with only 2 sweep mode scans per day available. The Mg II index products derived from these data are shown in Figures 2(a) and 3(a) respectively, and are only plotted through October 1992 for consistency with the "modified" Mg II index described below. Solar rotational modulations of up to 7% can be seen, and the solar cycle amplitude of the "classical" Mg II index is approximately 9-10% (Cebula et al., 1992). No corrections have been made to the data for long-term drift of the nominal wavelength scale calibration. All data were processed with an algorithm which incorporates a full treatment of the instrument characterization (goniometry, PMT temperature dependence, interrange ratio time dependence). The interrange ratio time dependence was determined from in-flight Earth radiance measurements (e.g. Laamann and Cebula, 1993).

Beginning in May 1986 for NOAA-9 and in February 1989 for NOAA-11, additional irradiance measurements of the Mg II absorption feature were begun using step scan measurements at 12 wavelength positions ("discrete mode"), which included the 7 "classical" wave-

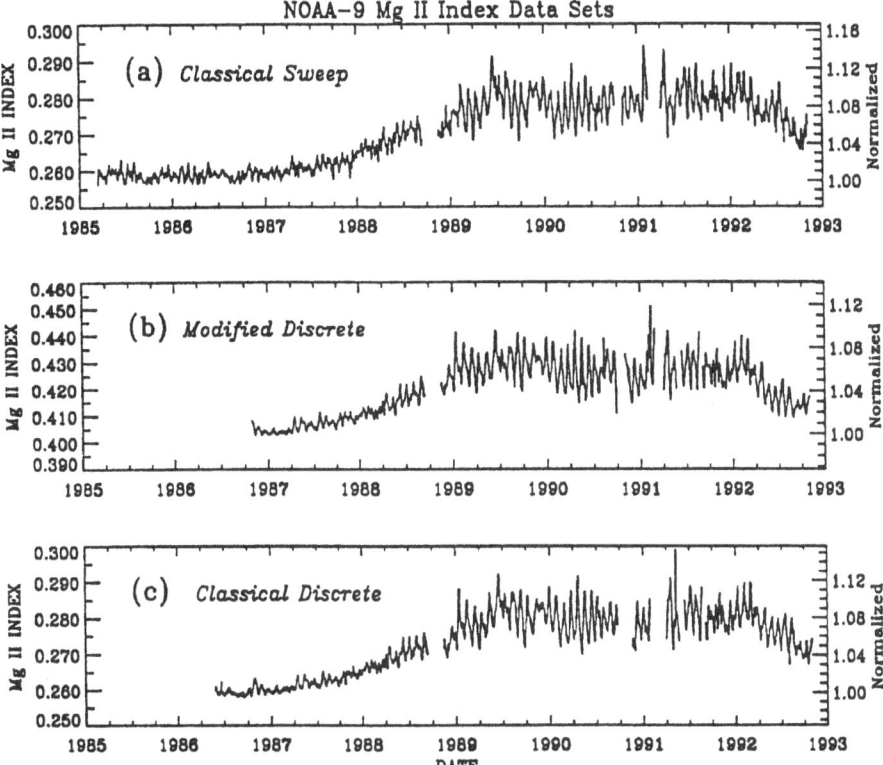

Fig. 2. Time series of NOAA-9 Mg II index products: (a) "classical" sweep mode; (b) "modi-fied" discrete mode; (c) "classical" discrete mode. The sweep mode Mg II time series has been smoothed with a 5-day binomial average. The right-hand Y-axis shows the Mg II data normalized to the average of 8-12 November 1986 for each data set.

lengths and 5 more positions along the sides of the line (see Figure 1). The discrete mode irradiance measurements are inherently more precise than the sweep mode data because the sample integration time is increased by a factor of 12.5, the instrument steps and locks into each wavelength position prior to taking data, and 8-9 scans are available to construct daily average values.

Donnelly (1988, 1991) has used the NOAA-9 and NOAA-11 discrete mode measurements of the Mg II line to construct a "modified" Mg II index product using a total of three wave-lengths (1 position in the core of the line, and 1 position on each side to represent the continuum). The wavelengths chosen to represent the continuum irradiance lie closer to the core of the line than the "classical" wing wavelengths in order to avoid the transition between electronic gain ranges described by Cebula *et al.* (1992), which introduces noise into the sweep mode Mg II index. The "modified" Mg II index derived by Donnelly (1988, 1991) also uses a limited instrument characterization, including a simplified correction for

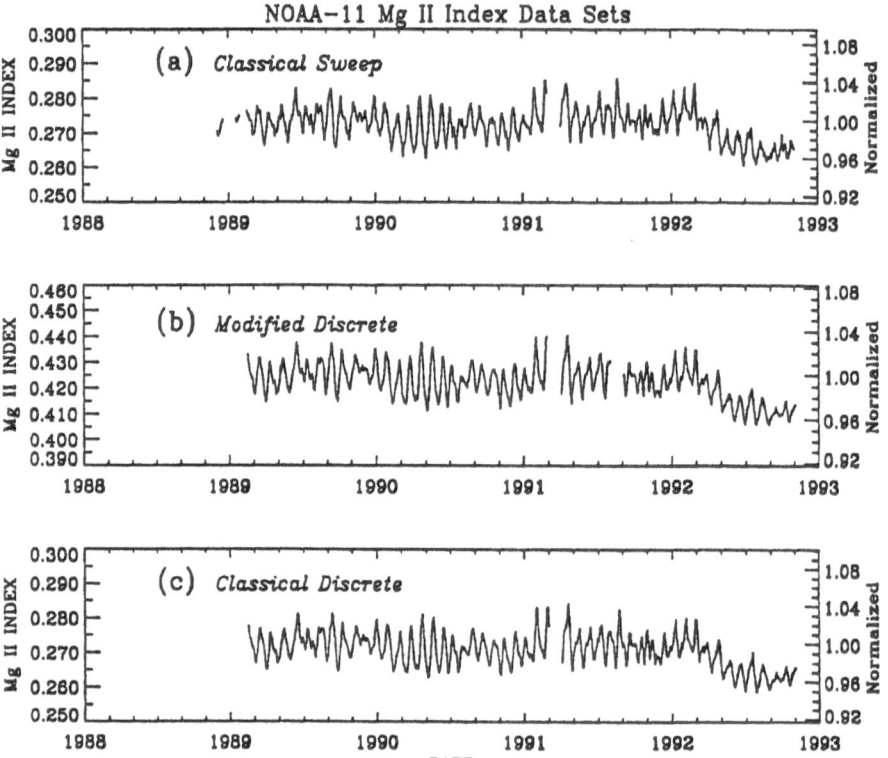

Fig. 3. Time series of NOAA-11 Mg II index products: (a) "classical" sweep mode; (b) "modified" discrete mode; (c) "classical" discrete mode. The sweep mode Mg II time series has been smoothed with a 5-day binomial average. The right-hand Y-axis shows the Mg II data normalized to the average of July 1989 for each data set.

goniometry (L. Puga, private communication). The "modified" Mg II index products for NOAA-9 (October 1986 to October 1992) and NOAA-11 (February 1989 to October 1992), kindly provided by R. F. Donnelly and L. C. Puga, are shown in Figures 2(b) and 3(b) respectively. The data after April 1990 are preliminary, and require further corrections before quantitative use. No correction for wavelength scale drift has been applied. The magnitude of the day-to-day noise is considerably reduced from the "classical" sweep mode Mg II index due to the improved signal-to-noise ratio in the irradiance data. The absolute magnitude of the "modified" Mg II index is greater than the "classical" Mg II index because the wing wavelengths chosen lie closer to the Mg II line core. This selection also reduces the amplitude of the solar variability signal observed by the "modified" Mg II index, because the wing wavelengths are now more responsive to chromospheric variations.

A "classical discrete" Mg II index product can also be constructed from the NOAA-9 and NOAA-11 discrete mode solar irradiance measurements, using the same algorithm and wavelengths chosen for the "classical sweep" Mg II index. These data have not been corrected for wavelength scale drift. The "classical discrete" data sets are shown in Figures 2(c) and 3(c), also plotted through October 1992 only, and have the same magnitude and response to solar variability as the "classical sweep" Mg II index to within 0.5%. However, as discussed previously, the statistical noise in the "classical discrete" product is reduced by a factor of 7 relative to the "classical sweep" product due to changes in signal integration time, wavelength selection repeatability, and number of daily scans. This gives a substantial improvement in the representation of solar variability during periods of low solar activity (compare 1986-1987 in Figures 2(a) and 2(c)).

MG II INDEX COMPARISONS

The difference in response between the "classical" and "modified" Mg II indexes to solar activity variations in the core of the Mg II line can be estimated by using the solar variability scale factors for each wavelength of DeLand and Cebula (1993). These scale factors give the solar irradiance change at each wavelength relative to the "classical" Mg II index change. Thus, if the 0.2 nm gridded scale factors of DeLand and Cebula (1993) are interpolated to the exact wavelengths used in the "classical" Mg II index, the difference between the Mg II core and Mg II wing scale factors would ideally be 1.00. The calculated result for the "classical" Mg II index is 1.03, which is within the combined 1 σ uncertainties of the scale factors. The difference between the interpolated Mg II core and wing scale factors for the "modified" Mg II index is 0.93. The decrease is caused by the larger scale factors for the wing wavelengths of the "modified" Mg II index, reflecting the increased contribution of chromospheric activity at those wavelengths relative to the "classical" wing wavelengths. This result suggests that the "classical" Mg II index response to solar chromospheric activity should be approximately 10% larger in magnitude than the "modified" Mg II index response for both short-term and long-term solar variations.

The slope of a linear regression fit between the "classical" and "modified" data sets, normalizing both to a common date to remove the effects of the absolute offset, should indicate the relative magnitude of the long-term solar variation response. The ratio of the "classical" and "modified" scale factor results corresponds to a slope of 1.11. Regression fits between the "modified" and "classical sweep" Mg II data sets give slopes of 1.13 ($R = 0.864$) for NOAA-9 and 1.04 ($R = 0.919$) for NOAA-11, while regression fits with the less noisy "classical discrete" Mg II data give slopes of 1.14 ($R = 0.951$) and 1.07 ($R = 0.996$) respectively. These values are consistent with the scale factor analysis. For the observed rise of 9-10% in the "classical" NOAA-9 Mg II index during solar cycle 21, the regression fit results suggest that the increase in the "modified" Mg II index should be approximately 1.0-1.5% less, or approximately 7.5-9% for solar cycle 21.

Comparisons have also been made between the "classical discrete" and "classical sweep" Mg II index data sets for NOAA-9 and NOAA-11. The absolute values of these two products agree to within 0.5% for both instruments, as expected due to their identical design. An indication of the increased sensitivity of the "classical discrete" Mg II index is its detection of differences of 0.3-0.6% in rotational modulation strength between the "classical discrete" and "modified" Mg II indexes for 1989-1991 NOAA-11 data, when 27-day solar variability was strong and persistent at the 3-6% level. The magnitude of the difference in short-term response, approximately 10% of the total amplitude, is consistent with the estimate derived from the scale factor analysis.

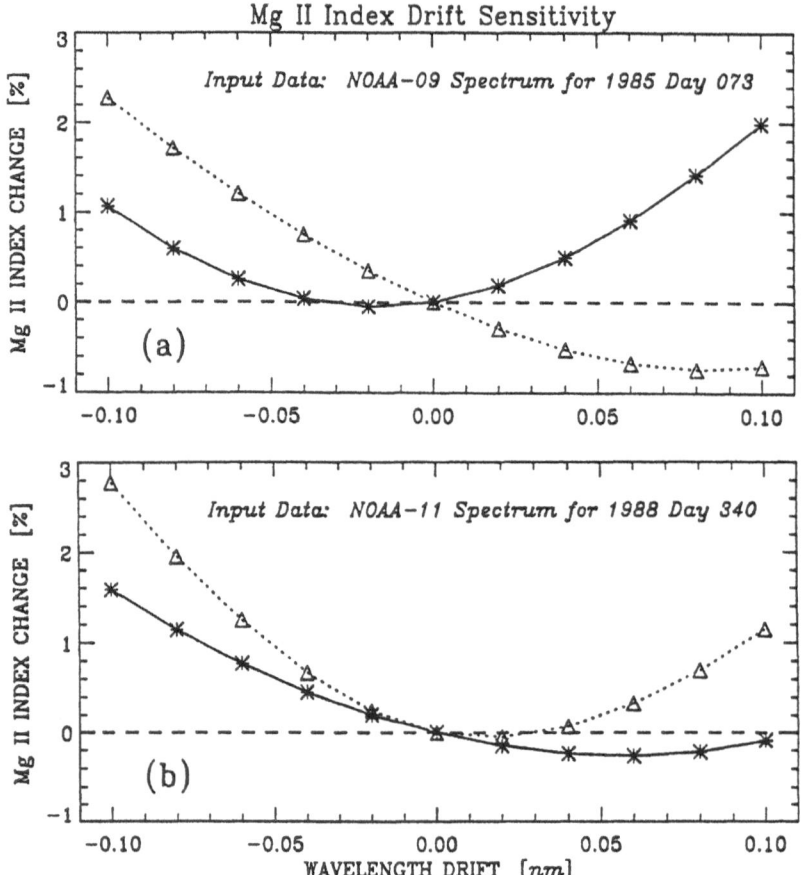

Fig. 4. (a) Calculated changes in the NOAA-9 Mg II index value as a function of wavelength scale drift. The stars and solid line represent the "classical" Mg II index results, the triangles and dotted line indicate the "modified" Mg II index results. (b) Calculated changes in the NOAA-11 Mg II index value as a function of wavelength scale drift. Identifications are as in part (a).

WAVELENGTH SCALE DRIFT

When a time series of the difference between the NOAA-9 "classical" and "modified" Mg II indexes is constructed, the magnitude is found to be approximately 3% by 1992, rather than the 1-1.5% predicted above. Some of this discrepancy is probably caused by the complicated thermal history of the NOAA-9 spacecraft induced by its drifting orbit (Cebula and DeLand, 1992). However, there are also changes in the Mg II index values due to wavelength scale drift with time. The wavelength scale calibration of the SBUV/2 instrument is monitored approximately bi-weekly during flight by tracking the measured positions of emission lines from an on-board mercury lamp. A similar analysis is performed with solar absorption lines using sweep mode data. Further details are given in DeLand et al. (1992). SBUV/2 instrument wavelength scale changes are largest for sweep mode measurements, where the data at each wavelength are integrated from a continuous scan without stopping at predetermined positions. The NOAA-9 sweep mode wavelength scale drift is estimated to be approximately $\Delta\lambda_{swp} \approx +0.10$ nm over 6 years (DeLand et al., 1992), which corresponds to a change of approximately +2% in the "classical sweep" Mg II index based on the reference wavelength positions (Figure 4(a)).

The magnitude of SBUV/2 discrete mode wavelength scale drift is much smaller than for sweep mode data, because the instrument "steps" to each wavelength position and locks before taking data. The estimated NOAA-9 discrete mode wavelength scale change over the same period is approximately $\Delta\lambda_{dis} \approx +0.03$ nm (DeLand et al., 1992). Calculations with a reference irradiance spectrum show that the "modified" and "classical" Mg II index products have similar sensitivities to wavelength scale drift for small drifts, but the "modified" Mg II index tends to be more sensitive for larger changes (Figure 4). This difference is primarily due to the use of wing wavelengths for the "modified" Mg II index located in spectral regions where the irradiance changes rapidly and unequally between the short wavelength and long wavelength sides of the Mg II line (see Figure 1). The impact of this difference in response to wavelength scale drift is approximately ΔMgII = 1% for the NOAA-9 discrete mode Mg II index products using Figure 4(a), proportioned roughly equally between an increase in the "classical discrete" product and a decrease in the "modified" product. For NOAA-11, the estimated wavelength scale drift of $\Delta\lambda_{dis} \leq 0.01$ nm through mid-1992 from DeLand et al. (1992) gives a change in Mg II index values of ΔMgII < 0.3% for both products (Figure 4(b)). However, the continuing drift of the NOAA-11 orbit may lead to a greater effect in the future. Because of the non-negligible impact of wavelength scale drift on the long-term response of the Mg II index, it is clear that each current Mg II product must be corrected for this effect.

CONCLUSIONS

The Mg II index has been shown to provide a good proxy of solar UV variability on both short and long time scales. Two versions of the Mg II index have been constructed and published from the NOAA-9 and NOAA-11 SBUV/2 data. The "classical" Mg II index constructed from sweep mode data is consistent with the first Mg II index product from Nimbus-7 SBUV, but is significantly affected by noise as a result of the SBUV/2 instrument design. The "modified" Mg II index reduces the noise through the use of discrete mode data and different wavelength selection. This choice of wavelengths also reduces the sensitivity of the "modified" Mg II index to solar variations, and increases the potential impact of long-term wavelength scale drift. Both current Mg II products must be corrected for the effects of wavelength scale drift to improve their value as a long-term solar UV proxy.

The most appropriate SBUV/2 Mg II index product for future use would seem to be a combination of these two products, namely the "classical discrete" Mg II index as shown in Figures 2(c) and 3(c). We feel that our experience with characterization of the SBUV/2 instruments allows us to understand and correct for effects such as time-dependent gain range ratio and wavelength scale changes. The "classical discrete" Mg II index combines the improved quality of the discrete-mode data with the historical precedent and lesser sensitivity to wavelength scale drift of the "classical" set of wavelengths. We suggest that future "composite" Mg II index products which link together Mg II index data sets from different instruments (*e.g.* Donnelly (1991), DeLand and Cebula (1993)) use the "classical discrete" Mg II index product for SBUV/2 data.

ACKNOWLEDGEMENTS

The continued assistance of W. G. Planet, J. H. Lienesch, and H. D. Bowman of NOAA/NES-DIS in providing the NOAA-9 and NOAA-11 SBUV/2 data is greatly appreciated. We thank R. F. Donnelly and L. C. Puga for valuable discussions. This work was supported by NASA contract NAS5-31755.

REFERENCES

Brueckner, G. E., Edlow, K. L., Floyd IV, L. E., Lean, J. L., and VanHoosier, M. E., 1993: *J. Geophys. Res.* **98**, 10,695.

Cebula, R. P., and DeLand, M. T.: 1992, in R. F. Donnelly (ed.), *Proceedings of the Workshop of the Solar Electromagnetic Radiation Study for Solar Cycle 22*, NOAA ERL, p. 239.

Cebula, R. P., DeLand, M. T., and Schlesinger, B. M.: 1992, *J. Geophys. Res.* **97**, 11,613.

DeLand, M. T., Weiss, H., Cebula, R. P., and Laamann, K.: 1992, *Rep. HSTX-3036-112-MD-92-013*, Hughes STX Corporation, Lanham, MD.

DeLand, M. T., and Cebula, R. P.: 1993, *J. Geophys. Res.* **98**, 12,809.

Donnelly, R. F.: 1988, *Adv. Space Res.* **8**, (7)77.

Donnelly, R. F.: 1991, *J. Geomagn. Geoelectr.*, **43**, suppl., 835.

Heath, D. F., and Schlesinger, B. M.: 1986, *J. Geophys. Res.* **91**, 8672.

Laamann, K., and Cebula, R. P.: 1993, *Rep. HSTX-3036-112-KL-92-023*, Hughes STX Corporation, Lanham, MD.

THE SOLAR Ca II K INDEX AND THE Mg II CORE-TO-WING RATIO

R. F. DONNELLY

Space Environment Lab., NOAA ERL, Boulder, CO 80303, U.S.A.

O. R. WHITE

High Altitude Observatory, NCAR, Boulder, CO 80307, U.S.A.

W. C. LIVINGSTON

National Solar Observatory, NOAO, Tucson, AZ 85726, U.S.A.

ABSTRACT. The 1 Å index of the solar Ca II K line is compared with the core-to-wing ratio of satellite measurements of the Mg II h and k lines. The correlation coefficient r = 0.976 for the Nimbus-7 Mg II ratio during solar cycle 21 and *r* = 0.99 for the NOAA9 Mg II ratio in cycle 22. Linear regression analysis for the full dynamic range of both data sets is used to combine the Nimbus-7 and NOAA9 Mg II data. These relations permit the ground-based Ca K index to estimate the solar UV flux.

1. Introduction

Satellite measurements of the core-to-wing ratio of the Mg II h and k lines (R_{Mg}) are important because of their long-term precision and their use in estimating solar UV flux temporal variations as a function of wavelength in the range important to the stratosphere and ozone layer (Heath and Schlesinger, 1986). The relation between R_{Mg} and the Ca K 1 Å index is important because it aids in developing the relation between the time series of R_{Mg} data from one satellite with those from a later satellite and because it allows the ground-based Ca K 1 Å index to be used to estimate solar UV flux variations. So what is the relation?

Fredga (1971) and Lemaire (1984) showed that the Mg k emission core intensity correlated highly with that for Ca K as a function of spatial position on the Sun. Since the increase or decrease in area and brightness of spatial features causes the temporal changes of full-disk fluxes, the high correlation of brightness as a function of spatial position implies that the temporal variations of the full-disk fluxes for the emission cores of these Mg II and Ca II lines should be highly correlated. Avrett's (1992) theoretical modeling of the upper photosphere, chromosphere, transition region, and lower corona for a spherically stratified, average quiet Sun shows that the emission core of the Mg II k line originates from slightly higher but similar altitudes in the chromosphere than does the emission core of the Ca K line. Therefore, the Mg II and Ca K activity brightenings should involve essentially the same source regions, which supports the expectation that the temporal variations of the full-disk fluxes for the emission cores of the Mg II h and k lines and the Ca II K line should be highly correlated.

2. The Solar Ca K and Mg II h and k Line Profiles and Indices

The Ca K line near 393 nm, shown in Figure 1, consists of the strong absorption feature that produces the overall large V shape with two small emission peaks on either side of the cen-

DONNELLY ET AL.

ter of the line (the zero location on the relative wavelength scale). The Ca K emission peaks, which originate in the chromosphere, vary strongly with solar activity. The K index is simply the full-disk intensity integrated as a function of wavelength across the 1 Å interval centered on the core of the Ca K line (between the two vertical lines in Figure 1) and then divided by a measure of the solar continuum intensity per Å at two reference wavelengths near 4020 Å and 3875 Å (White and Livingston, 1981).

Figure 1 also shows a fine-wavelength-resolution observation of the Mg II h and k lines (Allen et al., 1978). The emission cores, labeled h and k, are so close together that the short-wavelength absorption wing of the h line overlaps with the long-wavelength wing of the k line to produce the interwing maximum near 280 nm. Most of the variation with solar activity occurs in the chromospheric emission cores and not in the photospheric wings or weak lines. The lower left part of Figure 1 shows the Mg II h and k lines seen with the broad bandwidth of the Nimbus-7 measurements from the Solar Backscatter UV (SBUV) experiment (Heath and Schlesinger, 1986), which are similar to the NOAA9 SBUV2 monitoring measurements. The h and k emission cores, the broad absorption wings, and all the fine structure from weak absorption lines are smoothed into one broad absorption feature. Strong changes in the emission cores produce small variations near the minimum of the unresolved lines.

Heath and Schlesinger (1986) defined the center-to-wing ratio for the Mg II h and k solar absorption lines for solar flux measurements made by the SBUV experiment aboard the Nimbus-7 satellite as follows:

$$R_{Mg}(t) = \frac{4[F(279.8\ nm,t) + F(280.0\ nm,t) + F(280.2\ nm,t)]}{3[F(276.6,t) + F(276.8,t) + F(283.2,t) + F(283.4,t)]} . \tag{1}$$

$F(\lambda,t)$ is the measured solar flux at wavelength λ and time t. The wavelengths in the numerator were selected to have a strong signal of solar variability from the large percentage variations of the h and k emission cores. The far-wing measurements in the denominator were selected to be close in wavelength to the core of the line but to have very weak solar signals. Consequently, the ratio has a strong solar signal while being insensitive to drifts in instrumentation throughput that are weak functions of wavelength over the range involved in equation (1).

The arrows in Figure 1 mark the approximate locations of the wavelengths involved in equation (1). There are two problems evident in this figure that illustrate the difficulty in comparing R_{Mg} results from measurements made by two different instruments. The center wavelength of the three core wavelengths used by Nimbus-7 does not appear to line up with the minimum of the solid or dashed curves, yet the Nimbus-7 flux values for these three wavelengths indicate that the center one is close to the minimum. This is a consequence of a difference in the wavelength scales used for these two instruments. Secondly, the maximum in the short-wavelength wing of Hall and Anderson's (1988) balloon flight data is too low for R_{Mg}, derived from their data, to ever be low enough to agree with Heath and Schlesinger's (1986) results. This may result from the effect of the ozone layer on the balloon measurements being incompletely corrected. Other problems in comparing R_{Mg} val-

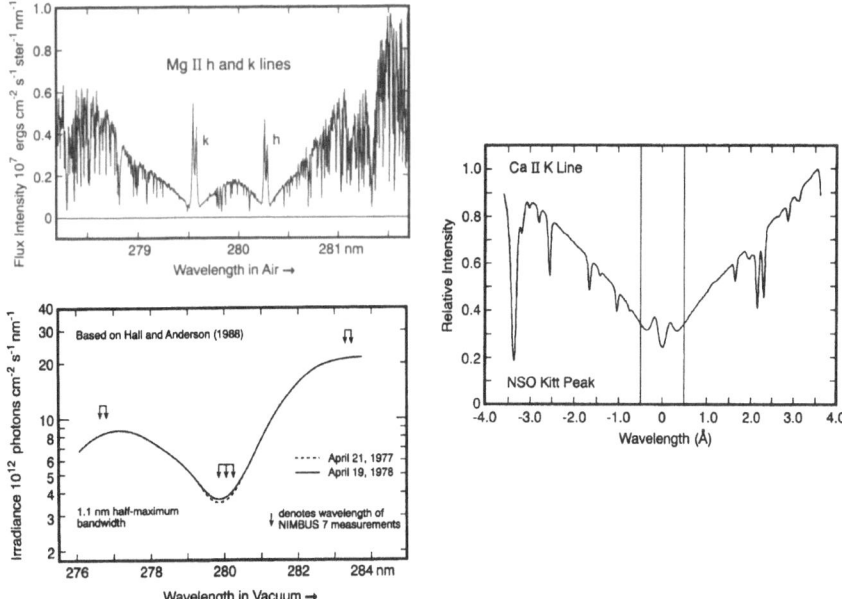

Fig. 1. Wavelength dependence of the solar Ca K line (top right) and the Mg h and k lines with fine wavelength resolution (top left) and with the low wavelength resolution of the SBUV monitoring measurements (bottom left). The many narrow dips in the top two graphs are weak absorption lines that vary little with solar activity.

ues from different instruments involve differences in scattered light (the data in the top left frame of Figure 1 has a high level of scattered light) and differences in bandwidths or wavelength step sizes. Nevertheless, R_{Mg} results from different instruments should have relative temporal variations that are linearly related.

3. Time Series of Ca K 1 Å Index and the Mg II Index

Figure 2 shows the Ca K 1Å index as a function of time from late 1974 through late 1990. All of solar cycle 21 and the rise and maximum of cycle 22 are included. The data recording rate of about 3 to 4 consecutive days each month samples the solar cycle or long-term variation very well, samples the intermediate-term variations (4–8 months) quite well, and provides a subsample of the short-term or day-to-day and week-to-week variations.

Figure 3 illustrates the Mg II index based on the research and data of Heath and Schlesinger (1986). These data start with the last part of the rise phase of solar cycle 21. They also include the entire peak phase and decline of solar cycle 21 and the minimum between cycles 21 and 22 in September 1986. Solar UV measurements were made by the Solar Backscatter Ultraviolet (SBUV) experiment aboard Nimbus 7, typically on three consecutive days and with the instrument turned off on the fourth day. So both Nimbus-7 data and

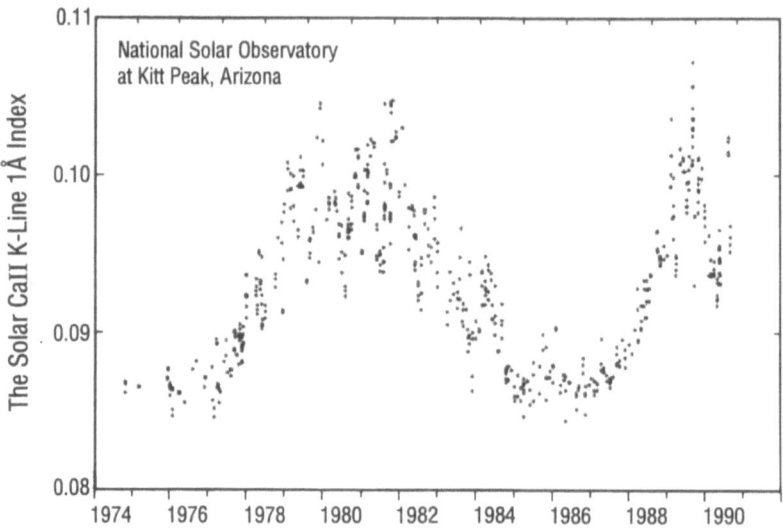

Fig. 2. Time series of the Ca K 1 Å index.

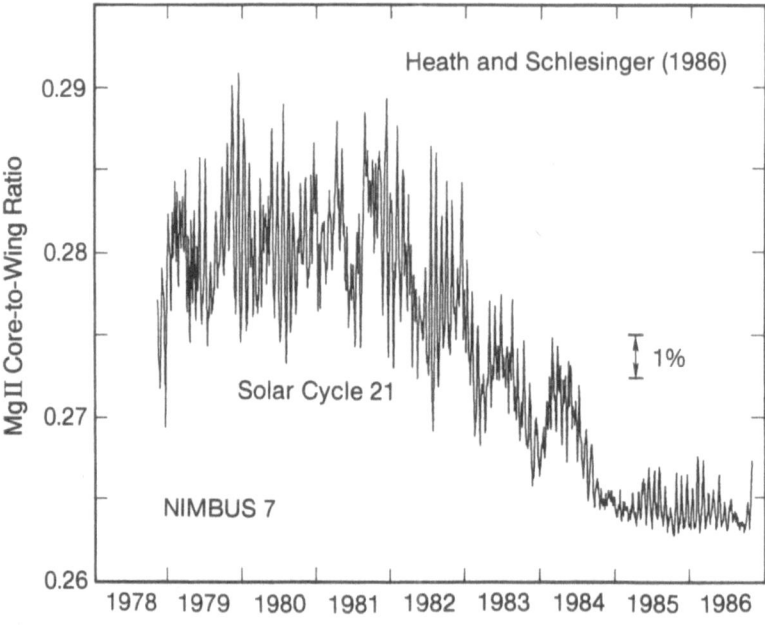

Fig. 3. The Nimbus-7 core-to-wing ratio of the Mg II h and k lines.

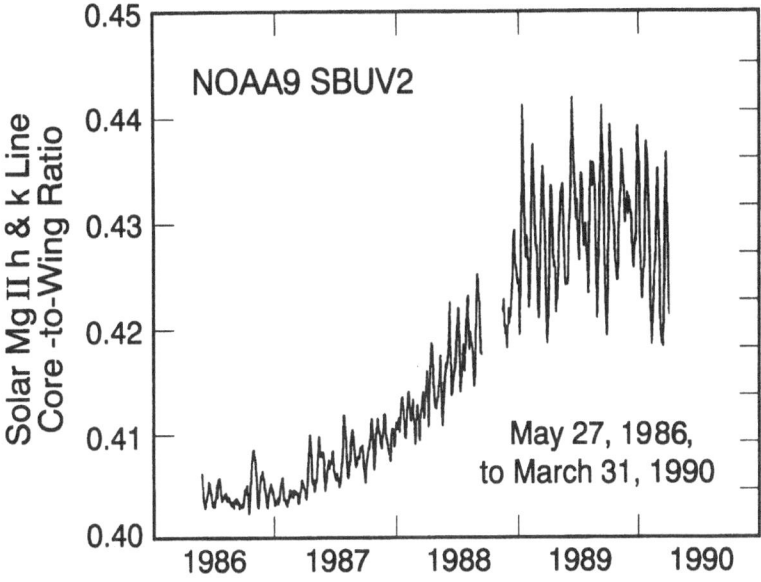

Fig. 4. The NOAA9 modified core-to-wing ratio from solar cycle minimum through the maximum of cycle 22.

the Ca K index were usually available on the same day for at least 2 days per month over an 8-year period.

Figure 4 presents the NOAA9 SBUV2 data for the solar Mg II index. These values differ from those in Figure 3 because a modified ratio is used for NOAA9. The output dynamic range for NOAA9 is covered by three overlapping ranges, each range having its own electronic system and digital telemetry output. Drifts have occurred in the relation between the strong signal output range and the medium signal range. To avoid these problems, the wing wavelengths used for the NOAA9 R_{Mg} were moved to the sides of the absorption line walls; both the core measurements and these new "wing" measurements involved in the ratio can then use outputs from the same (medium) intensity range, thereby avoiding the range-to-range drift problems. Because the wing measurements are now on the steeper part of the line profile with respect to wavelength, this modified ratio requires very accurate wavelength repeatability, or low wavelength jitter, which has been successfully achieved in the "discrete-wavelength" mode data used here. One undesirable consequence is that the modified long-wavelength wing measurement includes a weak solar signal from the emission cores. The first-order effect of this is to linearly reduce the relative amplitude of the modified ratio's solar variability; this is also reduced by the lower amplitude "wing" measurements, causing the average value of the ratio to be higher. These changes are all taken into account by using the linear regression relations described below to convert the modified ratios to equivalent Nimbus-7 values. (A very small but not negligible second-order nonlinear effect has not yet been corrected.) The modified core-to-wing ratio for the Mg II h and k lines used for the discrete-wavelength mode of the NOAA9 SBUV2 measurements is given by the following:

$$R_{Mg,NOAA9} = \frac{2\ F(279.92\ nm,\ t)}{[F(278.14,\ t) + F(281.24,\ t)]} \cdot \tag{2}$$

The average daily values derived from (2) are presented in Figure 4. These data show the minimum between solar cycles 21 and 22 in September 1986, the rise of solar cycle 22, and the main peak of the cycle. These observations were interrupted in the fall of 1988 when the Sun was unexpectedly occulted by part of the satellite.

4. Linear Regression Analysis

Linear regression analysis was computed for the Nimbus-7 Mg II core-to-wing ratio as a function of the Ca K index for the same-day pairs (177 matched pairs) of the data shown in Figures 2 and 3. The best-fit linear equation is given by

$$R_{Mg,Nimbus7} = 0.14391 + 1.3888\ [\text{Ca K index}]\ , \tag{3}$$

with the linear correlation coefficient $r = 0.976$. This means that about 5% of the variance is not explained by the linear relation with respect to the Ca K index. Some of that 5% is caused by both the Mg II and Ca K indices being based on daily samples over several minutes of the day where the samples are not taken at the same time of day for the two observing programs. Given R_{Mg}, Heath and Schlesinger's model allows one to estimate the solar UV flux as a function of wavelength and time in the 160–400 nm wavelength range. Equation (3) therefore lets one use the ground-based measurements of the Ca K index to compute the solar UV flux. Although the high correlations found above provide accurate conversions between the Nimbus-7 and NOAA9 Mg II ratios and the Ca K 1 Å index, the errors in the Heath and Schlesinger model are much larger than those in R_{Mg} and are also a function of wavelength.

Figure 5 shows the scatter diagram for the same-day pairs of Ca K index and the Nimbus-7 Mg II core-to-wing ratio. The straight line with the higher Ca K values in the upper right corner is that given in equation (3). The reverse relation for estimating the Ca K index from the Nimbus-7 Mg II index is also of interest; it is given by

$$[\text{Ca K index}] = -0.09422 + 0.68577\ R_{Mg,Nimbus7}\ . \tag{4}$$

Equation (4) allows one to estimate the Ca K index for most of the days each month when Ca K is not measured. The second line shown in Figure 5, the one with lower Ca K values in the upper right corner, is given by (4). These two lines are nearly the same because these data are highly correlated. Figure 5 also presents a scatter diagram for 109 matched pairs of daily samples for the Ca K index and the NOAA9 modified Mg II ratio. The equation for the line is

$$[\text{Ca K index}] = -0.10925 + 0.48468\ R_{Mg,NOAA9}\ . \tag{5}$$

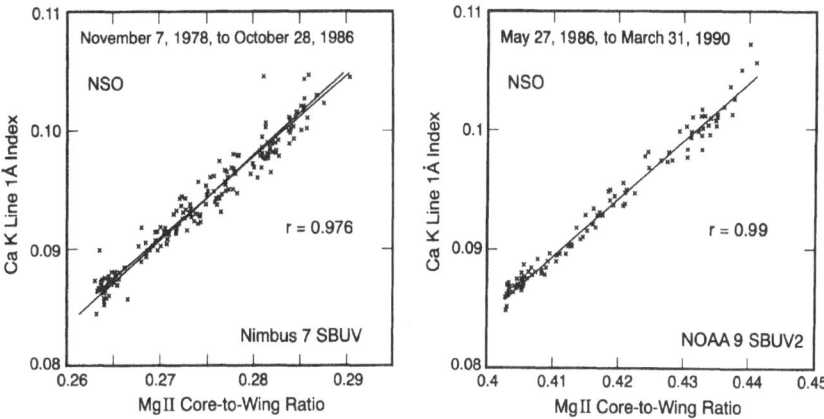

Fig. 5. Scatter diagrams for same-day pairs of the Ca K index versus Nimbus-7 values of the Mg II
ratio (left) and versus NOAA9 values of the modified MG II ratio.

Equation (5) allows one to estimate the Ca K 1 Å index from NOAA9 data. No second or
reverse relation line is shown in this case because the correlation is so high (r = 0.99) that
the two lines are almost the same.

Substituting (5) into (3) provides estimates of equivalent Nimbus-7 values of R_{Mg} from
the modified-ratio NOAA9 results, namely:

$$R_{Mg,\text{Nimbus7}} = -0.007814 + 0.67313\, R_{Mg,\text{NOAA9}} \cdot \qquad (6)$$

The advantages of converting the NOAA9 data to equivalent Nimbus-7 values through
comparisons with the Ca K 1 Å index are the following: (1) the comparisons with Ca K
involve very high correlations, in the high 0.9s; (2) the full dynamic range of the solar cycle
is included; (3) several years of overlapping data are included (8 years for Nimbus 7 and
over 3 years for NOAA9); and (4) short-term variations are included in the samples but do
not highly dominate the comparisons as they would in direct comparisons of data from two
satellites with an overlap period of the order of a year.

Figure 6 shows the results of applying equation (6) to the NOAA9 data shown in
Figure 4 and then combining those results with the Nimbus-7 values in Figure 3. These
observed and *equivalent* Nimbus-7 results can be used with Heath and Schlesinger's (1986)
wavelength scaling function to estimate solar UV flux variations in the 170–290 nm range.

5. Discussion

Ground-based measurements of the Ca K index and satellite measurements of the core-to-
wing ratio (R_{Mg}) of the Mg II h and k lines are very highly correlated. Comparisons of
NOAA9 data with the Ca K index, together with the regression analysis of the Nimbus-7

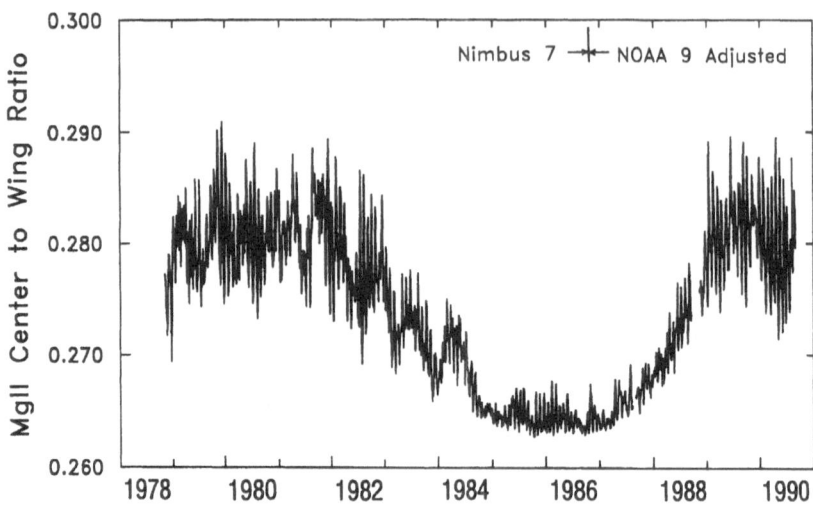

Fig. 6. Combined time series of the Nimbus-7 core-to-wing ratio of the Mg II h and k lines and the
equivalent Nimbus-7 values derived from the NOAA9 modified ratio (Donnelly, 1991).

R_{Mg} as a function of the Ca K index, make it possible to derive equivalent Nimbus-7 values
of R_{Mg} from the NOAA9 data and thereby extend the Nimbus-7 results to later times.
These R_{Mg} values, together with the wavelength scaling function, provide estimates of
the solar UV flux in the 170–290 nm range as a function of time since November 7, 1978,
the start of Nimbus-7 measurements.

In effect, the Ca K data set the magnitude of solar cycle 22 R_{Mg} data with respect to
cycle 21 in Figure 6; the satellite measurements provide the cycle 21 results and the day-to-
day variations in cycle 22. The high correlation of Nimbus-7 R_{Mg} values with the Kitt Peak
Ca K index, the regression relation in equation (3), and the wavelength scaling function also
allow ground-based measurements of the Ca K index to be used to estimate the solar UV
flux. Other observatories attempting to measure the Ca K 1 Å index should note the importance of keeping scattered light extremely low; the results must first be compared with the
Kitt Peak measurements before they are applied to equation (3).

References

Allen, M. S., McAllister, H. C., and Jefferies, J. T.: 1978, High Resolution Atlas of the Solar Spectrum
2678–2931 A, Inst. Astron., U. Hawaii, Honolulu, Hawaii, 58 pp.
Avrett, E. H.: 1992, in R. F. Donnelly (ed.), Proc. of the Workshop on the Solar Electromagnetic Radi
ation Study for Solar Cycle 22, Space Environment Lab., NOAA ERL, Boulder, Colo., 20.
Donnelly, R. F.: 1991, J. Geomagnet. Geoelectr. 43, Suppl., 835.
Fredga, K.: 1971, Solar Phys. 21, 60.
Hall, L. A., and Anderson, G. P.: 1988, Ann. Geophys. 6, 531.
Heath, D. F. and Schlesinger, B. M.: 1986, J. Geophys. Res.91, 8672.
Lemaire, P.: 1984, Adv. Space Res. 4, (8)29.
White, O. R. and Livingston, W. C.: 1981, Astrophys. J. 249, 798.

SOLAR EUV FLUX VARIATIONS NEAR
ACTIVITY MAXIMA

T.V. KAZACHEVSKAYA, A.I. LOMOVSKY and A.A. NUSINOV

*Fedorov Institute of Applied Geophysics, Rostokinskaya, 9,
Moscow 129226, RUSSIA*

Abstract. Data on solar emission variations in the extreme ultraviolet range $\lambda < 1300$ Å (EUV-range) performed on board the "Prognoz" satellites and the "Phobos" spacecraft by the thermoluminescent method are presented. Flux variations from the 11-years cycle are factors of 2-2.5, and that by the 27-days cycle do not exceed 30%.

A Solar Ultraviolet Radiometer (SUFR) designed for measuring the integral flux in the region $\lambda < 1300$ Å has been developed and manufactured at the Institute of Applied Geophysics (Kazachevskaya Ivanov-Kholodny, and Gonyukh, 1985). SUFR equipment is based on the thermoluminescent technique, which provides absolute measurements within the above spectral range using $CaSO_4(Mn)$ thermophosphorus. The thermophosphorus sensitivity defined (Arkhangelskaya and Razumova 1963; Kazachevskaya et.al,1976) in the region 1-1300 Å has a few maxima. The main one includes a 1-350 Å bandpass. In the operation of the device both detectors without filters and a MgF_2 filter, thin aluminium foil, thin Mylar film was applied. As for the MgF_2 filter, it allows one to distinguish the L_α- line. The contribution of different spectral portions to the total measured flux was determined by using the real Sun spectrum and the sensitivity of the thermophosphorus along λ. The portion of the flux from the 1-350Å region to the sum of the flux of the "quiet" Sun is 35% and rises to 40% when activity increases, but the portion of L_α -line changes from 63% to nearly 58% at the same time. The total error including the absolute calibration error, as well as the calculation errors does not exceed 15% of the measured value.

The SUFR equipment has been used more than once for recording the solar ultraviolet flux variations at $\lambda < 1300$ Å on board the "Prognoz" Earth's satellites as well as the "Phobos" interplanetary spacecraft (see Table 1).

TABLE 1: Time period of SUFR measurements

Spacecraft	Starting	Ending
Prognoz 7	October 30, 1978	February, 1979
Prognoz 10	April, 1985	May, 1985
Phobos 1,2	July, 1988	March, 1989

Solar Physics **152**: 77–80, 1994.
© 1994 *Kluwer Academic Publishers.*

Solar activity measurements by the "Prognoz 7" were described in detail by Ivanov-Kholodny and Kazachevskaya (1981). An arithmetic mean value of all measurements carried out on a given day (the mean of 180 measurements) was as a value characterizing the flux of the Lα-line radiation on that given day ($F_{L\alpha}$) normalized to 1 A.U.

Variations in solar activity level for the whole time period under consideration can be seen in Figure 1. Curves represent the values of $F_{L\alpha}$ (a), solar radio flux at 10.7 cm ($F_{10.7}$) (b) from the Ottawa station and the E-region ionization index J_E (c), smoothly averaged over 7 days. Index $J_E = f_0E^4/\cos\chi$, where f_0E is the noon E-region critical frequency from Krasnaya Pakhra ionospheric station data, and χ is the zenith angle of the Sun. Straight lines represent mean monthly values of these parameters.

As seen in Figure 1, variations in $F_{L\alpha}$ and $F_{10.7}$ are generally similar, however,the scales of the variations are different: the radio flux varied by approximately 40%, while Lα line flux variations did not exceed 10%. The comparison of $F_{L\alpha}$ and $F_{10.7}$ variations over the above time period shows that, along with the temporal increase in both parameters after the profound minimum in mid-November, one can also note differences both in the scale of variations and in certain details of temporal variations.

The nature of J_E variations is seen to be qualitatively similar to that of the temporal variations in $F_{L\alpha}$ and $F_{10.7}$. In particular, a gradual increase of these values is noted from November until February. The solar activity varies appreciably during the time period under consideration and causes significant variation in $F_{10.7}$, $F_{L\alpha}$ and J_E parameters.

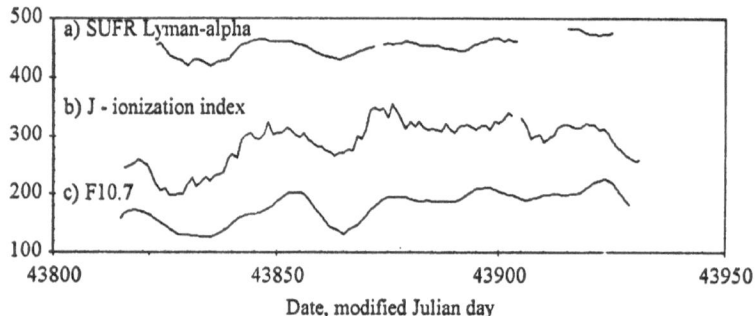

Figure 1. Variations in the Solar fluxes and ionospheric index from Nov. 1978 through
        Feb.1979.

a) $F_{L\alpha}$ - flux in the Lα line, 10^9 photon cm^{-2} s^{-1}(SUFR,"Prognoz 7").

b) J_E -ionization index of the ionospheric E-region, MHz^4 (from the Krasnaya Pakhra
   station).

c) $F_{10.7}$-radio flux at λ=10.7 cm, 10^{-22} Wt m^{-2} s^{-1} (from the Ottawa station).

Our next observations near the solar cycle minimum were undertaken by the SUFR instrument on board the "Prognoz 10" satellite, when $F_{10.7} \sim 88$. Those solar cycle minimum fluxes were F_{EUV} = 6 to 6.5 ergs cm^{-2} s^{-1} and $F_{L\alpha}$ =3.5 to 3.7 ergs cm^{-2} s^{-1}. The measurements of solar EUV emission flux by the SUFR equipment installed aboard "Phobos 1 and 2" were taken at the growth phase of the 11- year solar cycle. The information from the Phobos instrument was received over the whole flight route from the Earth to Mars and near Mars and Phobos. According to the flight program, measurements were taken every day from 18 hours 25 minutes to 19 hours 25 minutes UT.

Daily solar EUV flux was measured over 8 months, at a distance of up to 239.1 million km from the Sun and 265 million km from the Earth. Emission flux, corrected to the varied distance to the spacecraft and reduced to 1 AU, amounted to F_{EUV} (9-13) ergs cm^{-2} s^{-1} at λ<1300 Å, and $F_{L\alpha}$ =(5-7) ergs cm^{-2} s^{-1} in the L_α line.

The totality of our measurements allowed us to affirm that solar EUV intensity was rather high, 2.5 times greater than during solar minimum.

Figure 2 represents temporal variations in the H_α - line (λ=1216 Å) (SME satellite data were kindly provided by G. J. Rottman), X-ray intensity within 1-8 Å (GOES satellite data taken from SGD,1989) and EUV flux at λ<1300 Å obtained by the SUFR equipment, smoothed over 3 days.

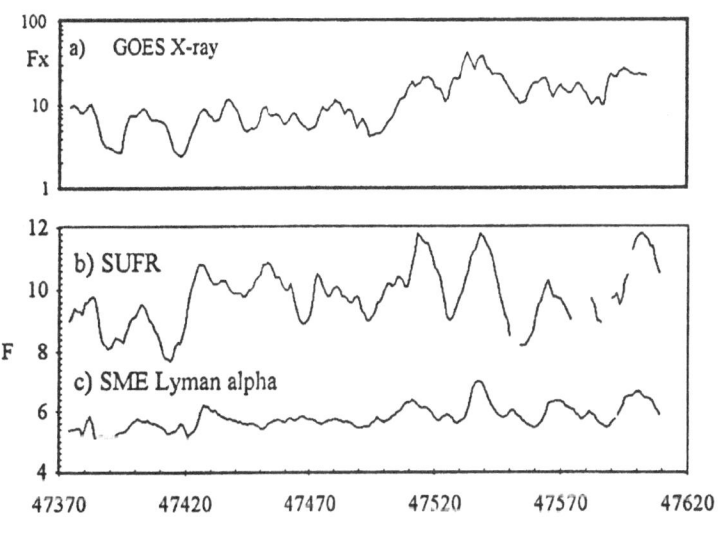

Date, modified Julian day

Figure 2. Variations of solar emission normalized to 1 A.U. from August 1988 through March 1989.
a) X-ray flux in the region 1-8 Å from GOES data, 10^{-4} ergs cm^{-2} s^{-1}
b) SUFR measurements of EUV flux at λ<1300 Å from "Phobos 2", ergs cm^{-2} s^{-1}.
c) L_α -line flux from SME data, ergs cm^{-2} s^{-1}

Due to the constantly varying angle between the lines of sight from the Earth and the spacecraft, one should take into account the temporal shift in the observations of the same active regions. The maximum spacecraft-Sun-Earth angle in February-March 1989 was over 60°. Knowledge of the angle at a given time allows one to compare observations from the spacecraft to the ones performed at the same time from the Earth. Figure 2 shows corrected data obtained by the SUFR, taking the temporal shift into consideration.

A comparison of the smoothed curves is indicate of rather good agreement between the experimental data obtained in various ranges of electromagnetic emission and the variability period of 27-28 days, and may reveal more delicate effects related to the stratification of emission regions at various wavelengths. An intermediate position of the SUFR data between the data on the L_α -intensity and the data on the X-ray is explained by the strong contribution of radiation of the shortwave end of the registered region to the entire flux at $\lambda < 1300$ Å.

Several words about solar flares: large ones were recorded in March 1989 by the SUFR equipment aboard the "Phobos 2", which was then orbiting Mars. At that time the solar EUV emission flux was 11.5-12.5 ergs cm^{-2} s^{-1} and more then 7 ergs cm^{-2} s^{-1} in the L_α -line (Kazachevskaya, Lomovsky,1992). The increase of radiation at $\lambda < 1300$ Å was 11%- 17% for series of flares, and during the flare on March 10, 1989 - more then 2 times.

The SUFR detector signal did not exceed background values when the non-flaring Sun was observed with using aluminum and Mylar filters. Filter capacity, taking into account the photocathode spectral sensitivity, allows observations of > M4-M5 class flares in the region $\lambda < 100$ Å. Thus, for the above-mentioned flare a large increase in the signal from the filter-free detector at $\lambda < 1300$ Å resulted from the considerably increased emission from the X-ray spectral region. Measurements show that the emission in the latter flare increased within the range 1-10 Å and 1-22 Å by 2.5 and 7-8 times, respectively, and by 19% in the L_α -line, as compared to the "quiet" Sun. These data refer to the initial phase of flare development and, unfortunately, do not include the maximum.

References

Arkhangelskaya, V.A. and Razumova, T.K.: 1963, *Optics and Spectroscopy (in Russian). Separate issue 'Luminiscence'*, 299.

Ivanov-Kholodny, G.S. and Kazachevskaya, T.V.: 1981, *God Solnechnogo Maximuma, Proc.Intern.Conf.,Simferopol.* 11, 390.

Kazachevskaya, T.V., Ivanov-Kholodny, G.S. and Gonyukh, D.A.: 985, *Geomagnetizm and Aeronomia* 25, 995.

Kazachevskaya, T.V. and Lomovsky, A.I.: 1992, in R. Donelly (ed.), *Proc. Workshop. SOLERS 22*, Boulder, 319.

THE XRAY-EUV SPECTRUM OF OPTICALLY THIN PLASMAS

B.C. MONSIGNORI FOSSI[1] and M. LANDINI[2]

[1] Astrophysical Observatory of Arcetri, Florence, Italy
[2] Department of Astronomy and Space Science, University of Florence, Italy

Abstract: The new observations of the Extreme Ultraviolet Explorer Spectrometers on EUVE and the future high resolution observation by Coronal Diagnostic Spectrometer (CDS) and SUMER on SOHO have suggested the revision of the Xray-EUV spectral code of the authors (1990) for optically thin plasmas. More accurate atomic data computations are now available and several lines have been added. Work is in progress to update the Xray-EUV code following the suggestions of the reviewers of the workshop on "Atomic data assessment for SOHO" held in Abingdon (March 1992). Special care is given to the highly ionized iron lines in the EUV spectral region.

1. Introduction

The Xray-EUV spectral code by Landini and Monsignori Fossi (1990) includes, between 0.1 and 2000 Å, about 2000 lines of the most abundant elements, together with free-free, free-bound and two photons continuum. It evaluates the emitted power from thin plasmas for temperatures between 10^4 and 10^8 K. The code has recently been revised using more accurate atomic data computations and splitting some multiplets that in the previous code were lumped.

Special care has been given to the iron line evaluation. Highly ionized iron lines are very common features in high-temperature, thin-plasma radiation. Observations of these lines provides inputs for plasma diagnostics in very broad temperature ($5.9 < \log T < 7.5$) and density ($10^9 < \log N_e < 10^{15}$) regimes and allows investigation of astrophysical plasmas in very different conditions.

A general review has been performed concerning the evaluation of atomic models and main atomic processes (collisional excitation rates, radiative decays, ...) for ions from Fe IX to Fe XXIII.

Stationary balance has been assumed to compute the level population for each ion and the intensity of the emission lines has been evaluated as a function of temperature and density. The ionization balance for iron, provided by Arnaud and Raymond (1992), has been adopted.

Systematic analysis has been performed in the spectral bands covered by CDS on SOHO to identify the best candidates for temperature and density diagnostics.

Solar Physics **152**: 81–86, 1994.

2. The atomic data for iron

Following the reviewers suggestions of the workshop on "*Atomic data assessment for SOHO*" held in Abingdon, the most updated atomic models available in the literature have been selected for each ion; the observed values for the energy of the levels have been used in order to make an easier comparison with the observations; radiative decay probabilities and electronic collision strengths have been collected for any pair of levels. The selected references are given in Table I.

The evaluation of the effective collision strengths over a Maxwellian distribution has been performed following a slight modification of the method suggested by Burgess and Tully (1992); a cubic spline approximation of the effective collision strength versus reduced temperature, kT/E_{ij}, has been computed for each pair of levels i and j (Monsignori Fossi and Landini , 1994).

Level population has been evaluated for temperatures between 10^4 and 10^8 K and densities between 10^8 and 10^{15} cm^{-3} by balancing for each level radiative decay and collisional processes; the power emitted per unit emission measure has been computed for any available radiative transition.

Using the numerical code developed by the authors (1990), free-bound and free-free continuum has also been computed in order to have a better selection of the bright line contribution.

Table I

Fe IX
Fawcett B.C: and Mason H.E.: 1991,Atomic Data Nuclear Data Tables 47, 17
Flower D.R.: 1977,Astron. Astrophys., 56, 451
Fe X
Corliss C. and Sugar J.: 1985,in Spectroscopic Data for Iron, ed Wiese W.L.
Mason H.E. : 1975,Monthly Notices Roy. Astron. Soc. , 170, 651
Fe XI
Mason H.E. : 1975,Monthly Notices Roy. Astron. Soc. , 170, 651
Fe XII
Bromage G.E., Cowan R.D., Fawcett B.C.: 1978,Monthly Notices Roy. Astron. Soc. , 183, 19
Flower D.R.: 1977,Astron. Astrophys., 54,163
Tayal S.S., HenryR.J.W.: 1986, Astrophys. J., 302, 200
Tayal S.S., Henry R.J.W., Pradhan A.: 1987,Astrophys. J., 319, 951
Tayal S.S., HenryR.J.W.: 1988,Astrophys. J., 329, 1023
Tayal S.S., Henry R.J.W., Keenan F.P., Mc Cann S.M., Widing K.G.: 1991,Astrophys. J., 369, 567
Fe XIII
Fawcett D.R. and Mason H.E.: 1989,Atomic Data Nuclear Data Tables, 43, 245
Mc Kim Melville J., Berger R.: 1965,Planet. Space Sc., 13, 1131
Fe XIV
Dufton P.L., Kingston A.E.: 1991,Physica Scripta, 43, 386
Froese Fisher C., Liu B.: 1986,Atomic Data Nuclear Data Tables, 34, 261
Mason H.E. : 1976,Monthly Notices Roy. Astron. Soc. , 170, 653

Fe XV
Christensen R.B., Norcross D.W., Pradhan A.K.: 1985,Physical Review A, 32, 93
Pradhan. A.K.: 1988,Atomic Data Nuclear Data Tables, 40, 335
Fe XVI
Flower D.R. et Nussbaumer H.: 1975,Astron. Astrophys., 42,265
Sampson D.H., Zhang H.L., Fontes C.J.: 1990,Atomic Data Nuclear Data Tables, 44, 210
Fe XVII
Bhatia A.K. and Doschek G.A.: 1992,Atomic Data Nuclear Data Tables, 52,1
Fe XVIII
Cornille M., Debau J., Loulergue M., Bely-Debau F., Faucher P.: 1992,Astron. Astrophys., 259, 669
Fe XIX
Loulergue M., Mason H.E., Nussbaumer H., Storey P.J.: 1985,Astron. Astrophys., 150, 246
Fe XX
Bhatia A. K.and Mason H.E.: 1980,Astron. Astrophys., 83, 380
Fe XXI
Mason H.E., Doschek G.A., Feldman U., Bhatia A.K.: 1979,Astron. Astrophys., 73, 74
Fe XXII
Bhatia A. K., Feldman U., Seely J.F.: 1986, Atomic Data Nuclear Data Tables, 35, 319
Edlen B.: 1983, Physica Scripta, 28, 483
Fe XXIII
Bhatia A.K. and Mason H.E.: 1981, Astron. Astrophys., 103, 324

3. Results

The comparison of results for two selected temperatures and densities are shown in Figures 1- 4 for the spectral regions 70-200 Å and 140-380 Å using a spectral resolution of 1 Å .

The ratio of lines which do not change with density and are produced by ions of subsequent degrees of ionization is a powerful tool in order to perform accurate temperature diagnostic in this temperature regime.

In the figures thin and thick lines allow one to easily identify the density sensitive features.

In fig. 1a and 1b the results for the "*quiet corona*" temperature condition are shown (log T = 6.2) . A "crowd" of bright lines is present; most of them originate by transitions between levels of configurations $3s^23p^4$ - $3s^23p^33d$ and $3s^23p^4$ - $3s3p^5$ of Fe XI and $3s^23p^3$- $3s^23p^23d$ and $3s^23p^3$ - $3s3p^4$ of Fe XII.

The main ions which are useful for density diagnostics are Fe XII and Fe XIII. For instance, Fe XII 195.14, Fe XII 193.53, Fe XII 202.05 which decrease with increasing density and Fe XIII 186.87 and Fe XI 316.96 which increase with density.

In fig. 2a and 2b the results for "*very active*" or "*flare*" conditions are shown (log T = 7.2). Rather few lines are present; the main contribution comes from Fe XXI, Fe XXII and Fe XXIII. Few Fe XXI and Fe XXII lines could be used for the density diagnostic. For instance Fe XXI 128.73, which decreases with increasing density and Fe XXI 142.14, 145.66 and Fe XXII 155.94, which increase as density increases.

In Figure 3 the ratio of the pairs of lines Fe XXI 142.14/128.7 and Fe XXII 114.4/135.6 at the peak temperature is shown in the density interval 10^{10} - 10^{16} cm^{-3}.

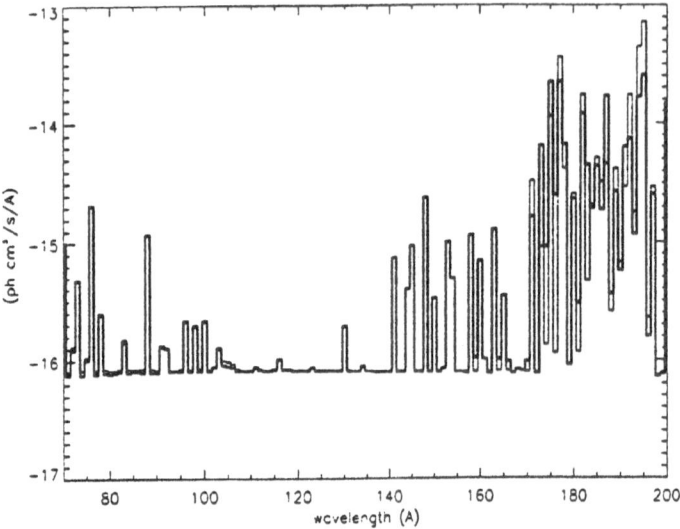

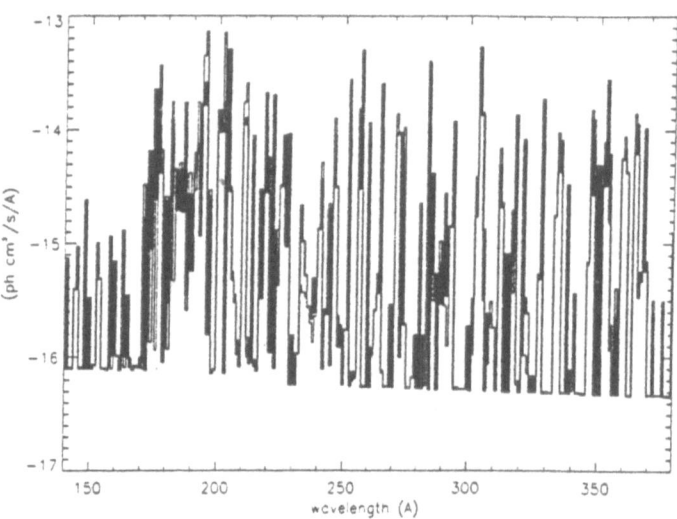

Fig 1 . a) Number of photons (log units) emitted per unit emission measure per second and per Å by a thin plasma with log temperature = 6.2 and electron density =10^9 (thin line) and 10^{13} (thick line) cm^{-3} in the spectral interval 70 Å - 200 Å . b) The same for the spectral region 140 Å - 380 Å.

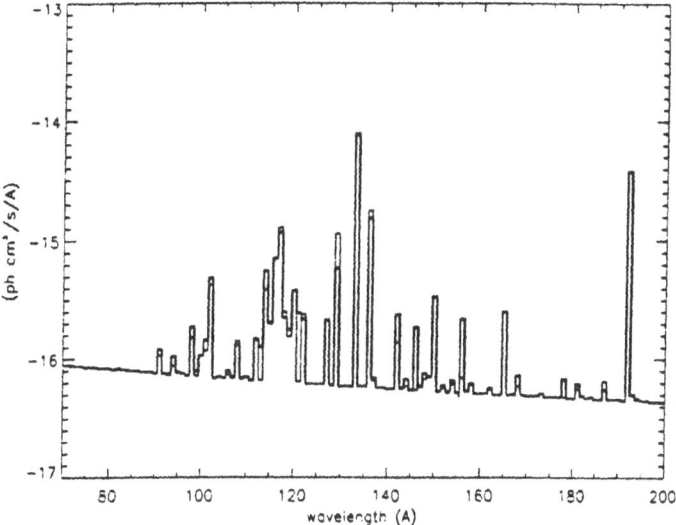

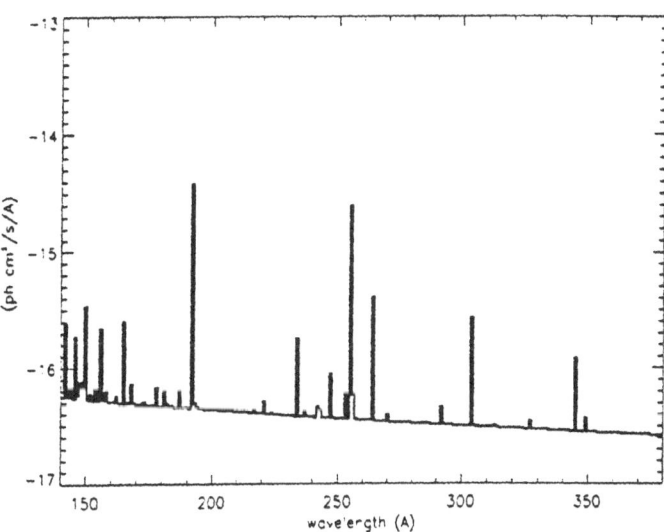

Fig 2 . a) Number of photons (log units) emitted per unit emission measure per second and per Å by a thin plasma with log temperature = 7.2 and electron density = 10^9 (thin line) and 10^{13} (thick line) cm^{-3} in the spectral interval 70 Å - 200 Å . b) The same for the spectral region 140 Å - 380 Å.

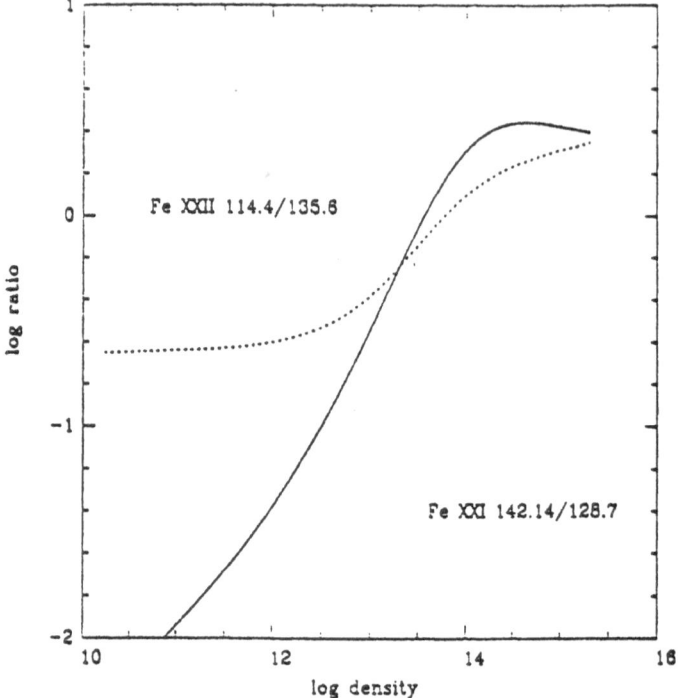

Fig. 3 . The ratio of the pairs of lines Fe XXI 142.14/128.7 and Fe XXII 114.4/135.6 at the peak temperature is shown in the density interval 10^{10} - 10^{16}.

4. Conclusions

The spectral region covered by CDS on SOHO (100 - 400 Å) includes highly ionized iron lines; they are very useful to make temperature and density diagnostics for a broad selection of astrophysical thin plasmas.

The best available atomic data for iron ions have been inserted in the updated version of the spectral code developed by the authors to produce synthethic spectra of plasma emission for temperatures ($5.9 < \log T < 7.5$) and densities ($10^9 < \log N_e < 10^{15}$). Numerical results will be available upon request by e-mail (*brunella@arcetri.astro.it*)

References

Arnaud M. and Raymond J.: 1992, *Astrophys.J.*, **398**, 394.
Burgess A. and Tully J.A.: 1992, *Astron. Astrophys.*, **254**, 436.
Landini M. and Monsignori Fossi B.C.: 1990, *Astron. Astrophys. Suppl.Ser.*, **82**, 229.
Monsignori Fossi B.C. and Landini M. : 1994, *Atomic Data Nuclear Data Tables*, (in press).

LIST OF STARS RECOMMENDED AS
SPECTROPHOTOMETRIC STANDARDS

E. I. TEREZ

*Simferopol State University, Department of Astrophysics
and Atmospheric Physics, 333036, Simferopol, Ukraine*

Abstract. Based on long-term observations of the absolute energy distribution in stellar spectra (starting in 1975), and from a survey of the literature, 14 stars of early spectral type (B7 III–A7 V) were selected. The constancy of their brightness is beyond any doubt now. These stars, with visual magnitudes 0.0 to 4.8, can be recommended as spectrophotometric standards for the spectral region 310–750 nm.

1. Results

The problem of absolute spectral energy distribution determination of the objects studied in observational astronomy consists in comparing their spectra with a spectrophotometric standard star spectrum. The spectrophotometric standards are divided into two groups: primary and secondary standards. The primary standards are basic stars for which the absolute energy distributions are determined by direct comparison of their spectra with the spectrum of a laboratory standard light source. Usually α Lyr is used as a primary spectrophotometric standard. The conversion to the absolute energy distribution, however, can only be made using the so-called secondary standards. The secondary standards are determined by relative observations, comparing their spectral energy distribution with the spectrum of Vega. But the comparison deteriorates the accuracy of the result to a great extent. The matter is made worse by the fact that at present there is not a unified list of secondary standards. Every group of astronomers concerned with spectrophotometric standards uses its own group of stars to compare with Vega. Moreover, the absolute Vega spectrum is sometimes used from data of different calibrations. For example, the catalog of stars obtained by Jacoby, Hunter, and Christian (1984) is referred to Vega by Hayes and Latham (1975); the catalog of stars, obtained by Alekseev, et al. (1978) is referred to Vega by Oke and Schild (1970), etc. In 1984 Hayes (1985), on the basis of the best Vega calibrations, proposed that a certain average curve of the absolute energy distribution be used as a standard. The error in this average curve evidently does not exceed 1–2% in the spectral region from 320 to 1050 nm. However, the question about the constancy of Vega is not sufficiently clear. For example, in the paper of Vasilyev et al. (1989) a slight variability of Vega with a period of 23–35 years is suggested.

It is evident that to solve the problems of stellar spectrophotometry it is necessary to construct a network of primary standard stars, the spectral

Solar Physics **152**: 87–90, 1994.
© 1994 *Kluwer Academic Publishers.*

energy distributions of which are determined with high accuracy. The requirements for these standards are: 1) They must not be variable; 2) the total number of stars must not be large, but must be large enough to allow the use of the same standards by observers in both the northern and southern hemispheres; 3) the stars must be rather bright (0^m–4^m), because only from bright stars can absolute fluxes be accurately measured with the help of a laboratory standard source; 4) it is desirable to use stars of early spectral types because their continua are not strongly distorted by the absorption lines; 5) data on absolute energy distributions in the spectra of the standards are reduced to the outer boundary of the earth's atmosphere, not to the stars themselves, i.e. the reddening effect is absent.

Investigations on the development of a group of primary standard stars began at Simferopol State University in 1975. For the observations a special dome, with a 22-cm telescope, a scanner and calibration instrumentation were used. The calibration equipment consisted of a diffuser screen, illuminated by a standard source (Terez, 1979). Observations from 1975 to 1984 were made in the Ararat expedition of the Main Astronomical Observatory of the Academy of Sciences of the USSR (at an altitude of 2100 m).

On the basis of the published literature, a list including 24 stars was compiled. The results of the absolute energy distribution measurements in the spectra of these stars, made in 1975–1980, showed that absolute fluxes from some stars (for example, α And, β Ari), obtained during different observational periods, differ by 5–10%. At the same time, fluxes from some other stars were measured with an accuracy of 2–3%. This effect may be explained by the non-constancy of some stars. Therefore, in 1980 the primary star list was revised. Finally, there remained 14 stars, which are listed in Table I.

Absolute fluxes measured from these stars by the method of data averaging for the observational period from 1975 to 1984 in the spectral range 320–750 nm are given in papers by Terez (1985; 1987).

Since 1990, investigations of the spectral absolute calibrations of stars have been continued in the Crimea at the Karadag Station of Simferopol State University. Also, it was decided to reduce the list of primary standard stars, having excluded from it a weak star, 29 Psc, and also γ Ori, in which according to Knyazeva (1991) a certain variability is possible.

Observations for the primary standard star program are being made at Simferopol State University at present. It would be reasonable to revise the list of standard stars again, coordinating it with the lists of other astronomical institutions. It seems important to include the Sun into the standard star list. The fact is that at present the solar spectral energy distribution is measured with a larger error than the absolute energy distribution in the standard star spectra. In the recently completed investigations of Burlova, et al. (1994) it is shown that the absolute flux from the Sun, measured by Neckel and Labs (1984) is possibly low by 5–8% in the region from 310 to

400 nm, and may be overstated by 2–2.5% in the spectral region from 600 to 680 nm. A unified solar-stellar absolute spectrophotometric scale will allow us to give an answer to the question: Why do color indices of the Sun, obtained from the energy distribution in its spectrum, differ from color indices of a G2 V star?

The inclusion of the Sun into the program of spectral absolute calibration of standard stars requires an equipment modernization. But it is absolutely feasible as is shown in the paper of Lockwood et al. (1992), where a comparison of the absolute solar spectrum with the Vega spectrum was made using one telescope.

TABLE I

Stars selected as spectrophotometric standards

N	BS	Star	M_v	Sp
1	718	ξ Cet	4.28	B9 III
2	1122	δ Per	3.01	B5 III
3	1791	β Tau	1.65	B7 III
4	2421	γ Gem	1.92	A0 IV
5	3454	η Hya	4.30	B3 V
6	3982	α Leo	1.35	B7 V
7	4357	δ Leo	2.56	A4 V
8	5191	η Uma	1.86	B3 V
9	5511	109 Vir	3.73	A0 V
10	6629	γ Oph	3.75	A0 V
11	7001	α Lyr	0.03	A0 Va
12	7557	α Aql	0.76	A7 IV–V
13	7950	ϵ Aqr	3.77	A1 V
14	8781	α Peg	2.48	B9.5 III

References

Alekseev, N. L., Alekseeva, G.,Arkarov, A., *et al.*: 1978, *Trudy Glavnoy Astronomicheskoy Observatorii v Pulkovo* **63**, 4.

Burlov-Vasilev, K. A., Gourtovenko, E. A., Mateev, Yu. B.: 1994, *Fizika i Kinematika Nebesnyh Tel* **10**, in press.

Hayes, D. S.: 1985, in D. Hayes, L. Pasinetti, and A. David-Philip (eds.) *Proceedings of IAU Symposium No. 111*, 225.

Hayes, D. S., and Latham, D. W.: 1975, *Astrophys. J.* **197**, 593.

Jacoby, G. H., Hunter, D. A., and Christian, C. A.: 1984, *Astrophys. J. (Suppl.)* **56**, 257.

Knyazeva, L. N., and Kharitonov, A. V.: 1991, *Soviet Astron. J.* **68**, 501.

Lockwood, W., Tug, H., and White, N. M.: 1992, *Astrophys. J.* **390**, 668.

Neckel, H., and Labs, D.: 1984, *Solar Phys.* **90**, 205.

Oke, J. B., and Schild, R.: 1970, *Astrophys. J.* **161**, 1015.

Terez, E. I.: 1979, *Soviet Astron. J.* **56**, 205.

Terez, E. I.: 1985, in *Fotometricheskie i Polyarimetricheskie Issledivaniya Nebesnyh Tel*, Naukova Dumka Press, Kiev, Ukraine, p. 55.

Terez, E. I.: 1987, *Kinematika i Fizika Nebesnyh Tel* **3**, 17.

Vasilyev, I. A., Meregin, V. P., Nalimov, V. N., and Novoselov, V. A.: 1989, *Trudy Kazanskoy Gorodskoy Astronomicheskoy Observatorii*, **No. 52**, 25.

VARIABLE STARS WITH THE HIPPARCOS SATELLITE

L. EYER and M. GRENON

Observatoire de Genève, 51 ch. des Maillettes, CH-1290 Sauverny, Switzerland

and

J.-L. FALIN, M. FROESCHLÉ and F. MIGNARD

OCA/CERGA Av. Copernic, F-06130 Grasse, France

Abstract. We present the photometric results of Hipparcos satellite for the first 1.5 year of the mission. For this period, the satellite gave numerous (20 to 120) photometric measurements of 120000 stars. More precisely, we describe: the precision, the capability to find periods with respect to the specific sampling of the satellite (based on simulated data), some examples of variable stars, and finally a general description of stellar variability across the HR diagram.

1. Introduction

1.1. HIPPARCOS PHOTOMETRY

The Hipparcos satellite was designed for astrometric measurements, but during its development it was realized that with some minor modifications it could perform accurate photometric measurements. The Hp pass-band has a wide extent from 3500 to 8500 Å. The results obtained are indeed exceptional mainly for two reasons:

(1) Hipparcos achieves a complete and long-term photometric survey. Between 45 and 495 measurements were collected over 3.4 years for each star depending on ecliptic latitude, with a mean number of 125 measurements.

(2) A very high precision is attained, because the instrumental sensitivity drifts are continuously monitored through a detailed photometric calibration (Mignard *et al.*, 1992; Evans *et al.*, 1992). The main limitation is the photon noise.

1.2. DATA ACQUISITION

The satellite superposes the images of two distinct fields of the sky (separated by an angle of 58°) on a grid. The satellite is rotating, so that the stars pass by on the grid. A star is measured 9 times per transit. The mean instrumental magnitude Hp_i over a transit, its associated accuracy, and the epoch of the observation t_i are stored.

1.3. EXPECTED PERFORMANCE

The allocation of observing time per star is magnitude dependent and ensures an optimisation of the photometric and positional accuracies. The photometric precision for one field transit and time allocation t is given in Table I (Mignard *et al.*, 1992).

Solar Physics **152**: 91–96, 1994.
© 1994 *Kluwer Academic Publishers.*

TABLE I

Photometric precision and time allocation as a function of the magnitude

Hp mag:	4	5	6	7	8	9	10	11	12
t[sec]:	1	1	2	2	3	4	6	6	10
precision[mmag]:	2	4	4	7	10	14	17	30	40

This achieved precision is higher than that obtained by most ground-based observations. The achievable precision, the number of measurements and the expected duration of the mission allows a detection of variable and microvariable stars with an unprecedented efficiency and accuracy.

2. Sampling by Hipparcos

The sampling by Hipparcos is quite peculiar and introduces aliasing peaks, which must be taken into consideration when trying to find periods. The usual pattern is as follows: a sequence of 4 to 6 transits separated by 20 minutes and 108 minutes which correspond to the time elapsed from one field to the other through the separation of 58° and 302° respectively. The rotation period of the satellite is 2h08m. The next group of transits appears 3 to 5 weeks later, depending mainly on the ecliptic latitude. When a star is temporarily at the node of the great circle of scanning, it is observed continuously for several days.

The period of rotation of the satellite around the earth is 10h42m with an interruption for data collection at perigee lasting about 4 hours.

3. Capability for the Recovery of Periods and Amplitudes

When stellar variation is expected to be periodic with one or several modes, it is important to define the sensitivity of the search of periods and amplitudes. To reach this goal, we produce simulated data with the sampling of one representative star (having 145 transits), analyse them and compute the chance to find again the right period. In detail, we take as a signal for single mode variables $y(t) = a \sin 2\pi\nu t + nX$, with the three following characteristics: a period $P = 1/\nu$, a ratio $Q = amplitude/noise = a/n$ and a gaussian random noise $X \sim N(0, 1)$.

These simulated data were analysed with the method proposed by Stellingwerf (1978) with $(N_b = 6, N_c = 5)$, in an interval of frequency ranging from 10^{-3} to 37 [1/day]. A period was extracted from all the samples, spurious or not. The rate of correct detection of the period R is a function of the period P and of the ratio $Q = amplitude/noise$.

TABLE II
Rate R in % of correct detection as a function of P and Q

$Q\backslash P$	34.9	30.0	26.2	17.5	10.0	2.0	0.43	0.06
0.75	50	42	22	30	44	52	56	50
1.00	88	82	66	52	80	82	92	92
1.25	92	92	78	84	92	100	100	96

Table II shows the rate of correct detection for the first 18 months of the mission. As soon as Q exceeds 1.25 (or the amplitude $> 1.25\sigma$), the rate of correct detection is above 90% except for periods close to the mean intervals between two groups of transits (20 − 40 days). For stars brighter than 7th magnitude, the minimum detectable amplitude, a, is below 0.01 mag. The capability to recover periods and amplitudes will be greatly improved when the whole set of data will be available at the end of the mission.

4. Variable Stars

At the time of the Input Catalogue compilation 5830 stars were known as variables or suspected variables among 118209 programme stars. Only 2031 variables had a determined period. On the basis of 18 months of Hipparcos mission a χ^2 test provides 38837 variables, periodic or not. The gain on variable detection is still enormous. With the whole set of photometric data a large fraction of former and new variables will have their amplitudes and, in most cases, their periods, determined unambiguously. For short period bright stars, the first two or three pulsation modes may be determined. A major gain in variability detection is expected for luminous stars, irregulars and main-sequence spotted stars. Two examples are presented in Figure 1 and 2.

5. Stellar Microvariability and Variability across the HR-Diagram

Hipparcos offers a unique opportunity up to now to describe the variability without the classical daily and yearly biases. A systematic investigation of the amount of intrinsic stellar variability was performed by Grenon (1993) on the basis of high precision Geneva photometry. Here we will investigate the stellar variability, as a function of MK spectral type and luminosity class, with magnitudes produced by Hipparcos during the first 1.5 year of the mission. Cases where the variability is due to the effect of a companion or a circumstellar envelope are rejected. In order to remove accidental instrumental errors due to satellite mispointing, image overlapping or erratic

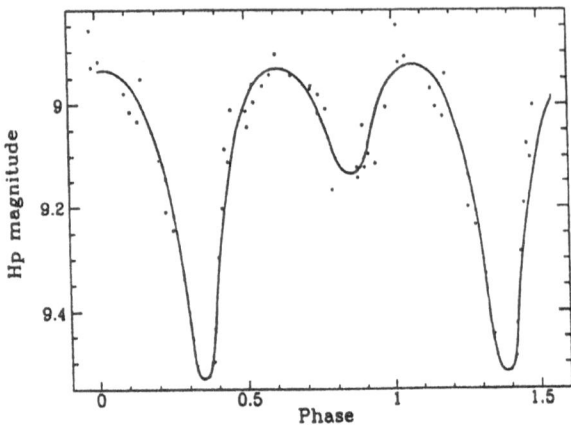

Fig. 1.   Phase diagram of a previously known eclipsing binary (period = 0.610 day).

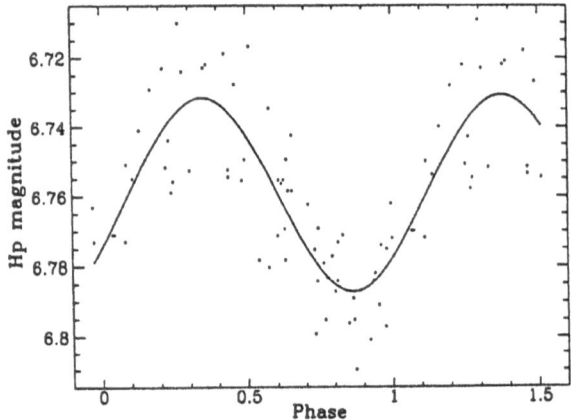

Fig. 2.   Phase diagram of a new variable, detected by Hipparcos. The amplitude of the sinusoidal light curve of this F5II supergiant is 0.028 mag (period = 0.109 day).

stellar behaviour as flares, interdeciles are used instead of standard deviation. The results are given in terms of σ = interdecile interval divided by 2.6.

5.1. SELECTION CRITERIA

Since we consider here the variability of single stars with well-defined temperature and gravity, we rejected visual doubles, eclipsing binaries, composite and weak-lined spectra. And we selected only stars with accurate colour indices (an erroneous colour index provisionally produces a magnitude trend in Hp magnitude, thus a spurious spread), with MK classification and brighter than $Hp = 8$. So we are left with 11663 stars.

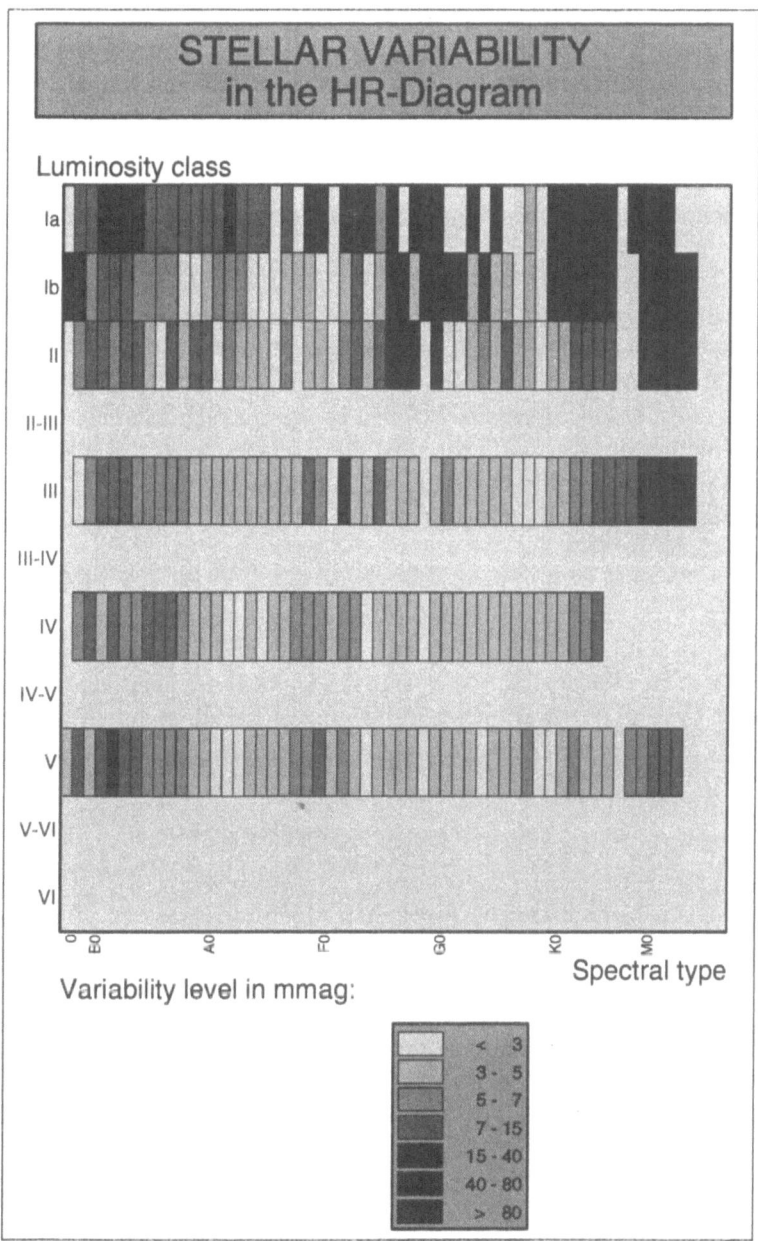

Fig. 3. Mean stellar variability in the HR diagram, expressed as mean standard deviation. Insufficiently documented boxes are blank.

5.2. NOISE SUBTRACTION

Photon noise. The precision of the magnitude strongly depends on the magnitude itself. So we have first subtracted the contribution to the noise due to the magnitude in order to make the stars of different magnitudes comparable. We found for the spread due to photon noise s_{ph} the following law:
$s_{ph} = 0.223 \times 10^{0.2 \times Hp} mmag$.

Reduction noise. We adopt the residual noise of the most stable stars as the reduction noise. This noise $s_{red} = 3.44$ mmag was subtracted for all stars.

5.3. VARIABILITY IN THE HR-DIAGRAM

Figure 3 displays the provisional status of the amount of variability in the HR-diagram. On both sides of the instability strip, two areas of high stability extend from main sequence to supergiants, narrowing towards high luminosity. The most stable areas are occupied by A2-A5 subgiants and dwarfs and F5-K0 main sequence to giants stars. G8III giants appear to be among the most stable stars. In late-type stars the amount of variability increases with decreasing temperature and gravity. The gravity effect is very conspicuous at K1-K5 class, III to Ia.

When Be stars with large amplitude are removed, the κ-mechanism effect for early B-type stars is restricted to main sequence B3V stars.

The instability strip is still poorly defined and broader than expected mainly because the available spectral type is not the type at mean temperature. At the end of the Hipparcos mission, the HR-diagram will be labelled in terms of iso-amplitudes and iso-periods as a result of the ongoing investigation.

6. Conclusion

The provisional photometric performance of the Hipparcos main mission already demonstrates that this mission will have a major impact in the field of variable stars, for the completeness of their detection, for the low detectable amplitude and appreciably for the very large number of precisely determined light curves. Major improvements are expected, namely in the fields of stellar interior structure for massive and evolved stars, on the limits of instability zones, and on the effects of rotation for the low main sequence.

References

Evans, D.W., van Leeuwen, F., Penston, M.J., Ramamani, N., Høg, E.: 1992, *Astron. Astrophys.* **258**, 149.

Grenon, M.: 1993, *Astron. Soc. Pacific. Conf. Series* **40**, 693.

Mignard, F., Froeschlé, M., Falin, J.-L.: 1992, *Astron. Astrophys.* **258**, 142.

Stellingwerf, R.F.: 1978, *Astrophys. J.* **224**, 953.

VARIATION OF THE SOLAR DIAMETER FROM SOLAR ECLIPSE OBSERVATIONS, 1715–1991

ALAN D. FIALA
U.S. Naval Observatory, Washington, DC 20392-5420

DAVID W. DUNHAM
*International Occultation Timing Association and
Johns Hopkins University
Applied Physics Laboratory, Laurel, MD 20723-6099*

and

SABATINO SOFIA
Yale University, New Haven, CT 06520

Abstract. The diameter of the Sun may be measured at the time of a solar eclipse. We have performed an exhaustive search of the astronomical literature to find all existing observations of solar eclipses suitable for this purpose. We have also taken new observations by new techniques. We have undertaken a project to reduce them systematically, and in an automated, self-consistent way. This will produce determinations of the solar radius at the times of solar eclipses from 1715 to the present. Re-reduction, using newer ephemerides, of observations made in 1984 shows that the component of the residuals caused by the ephemeris is substantially reduced. This paper summarizes the research plan, outlines the detailed astronomical features included in the calculations, and presents the results available.

1. Introduction

The question of whether or not the solar diameter is constant had been neglected for many years until early 1979, when Sofia, O'Keefe *et al.* noted that the solar constant could undergo variations if the solar diameter changed. Later that year, Eddy and Boornazian (1979) reported an apparent significant change over a few centuries, based upon analysis of a long series of meridian observations taken at Greenwich. Following the work of Smith and Messina (1981), this conclusion was shown to be unwarranted because systematic effects in the observations had not been treated correctly. By that time, however, the suggestion had inspired other lines of investigation.

Observations of a solar eclipse from the edges of its predicted path, taken to improve the parameters used in the calculation, were first attempted in 1715. Newcomb noted in 1870 that while the idea was good in concept, the execution was practically impossible at that time because of the effects of the rough limb of the Moon on contact times. It was not until 1963 that the terrain of the apparent limb of the Moon was known well enough to allow the effects to be calculated with useable accuracy.

Solar Physics **152**: 97–104, 1994.
© 1994 *Kluwer Academic Publishers.*

Late in 1979, Sofia, Dunham, and Fiala reported that, in effect, a solar eclipse could be observed in the same way as a grazing occultation of a point-source star. Observers station themselves along the predicted edges of the track of the umbra at a solar eclipse. The observations consist of timings of the formation and disappearance of individual Baily's Beads. Each timing is actually a measurement of the direction in space of a point on the apparent limb of the Sun. It is compared to a predicted sequence of Bead phenomena based on the calculated limb of the Moon for that instant as seen from each site, and identification made with a known limb feature (mountain peak or valley). As long as the position of each observing site is known to an accuracy within 50 feet, the only adjustable parameters in the calculation that affect the predicted times are the apparent diameter of the Sun, and the Sun's position relative to the Moon. Observations of the times of bead phenomena can be analyzed for these parameters. Timings from both edges of the predicted path are required in order to separate the effects of a change in the width of the path from a shift in its longitude.

Several reviews have evaluated and compared the various methods used for monitoring the solar diameter (Wittmann and Debarbat, 1990; Ribes, 1990; Ribes *et al.*, 1991). In general, the method we employ has been noted as being very accurate, but then set aside because the frequency of measurement is so low. Each eclipse observed provides a single measurement of the solar diameter, hence under the very best of conditions no more than three per year can occur.

While it is true that measurement by means of eclipses will not detect fluctuations with periods less than a few years, it can provide very precise calibration points for comparison to other data bases. Furthermore, it is the only means of accurate calibration, comparable to the accuracy of modern methods, which can use observations extending back in time for several decades.

2. Objectives

In long-term studies of global climate variation, it would be extremely valuable to identify a proxy indicator of variation of solar energy production, tied to an identifiable physical condition. Measurements of such a quantity could then be correlated to measurements of the solar constant over a period of decades to centuries. Variation of the diameter may be such a proxy (Sofia *et al.* 1991).

The solar ephemeris being used for predictions at the time we started was still based on Newcomb's theory of the Sun, and was suspected to be less accurate than that represented by DE200. Thus, accurate measurements of the relative position of the Sun and Moon would help check the accuracy of the ephemerides, the reference system, and the eclipse prediction procedure.

The precision and accuracy of the observations depends upon Watts' limb profile data (Watts, 1963). These observations also serve to evaluate the consistency of those data.

The precision of the observations depends on the detection system used. While it is necessary to have a system which functions in the same wavelengths as the human eye, sensitivity and recordability are important, and we have experimented with a variety of detectors for comparative evaluation. At the same time, we intercompare observing techniques in order to better evaluate the archival observations.

3. Implementation

When we began our current investigation in 1979, one of the guiding principles was to encourage as many people to observe as possible. It was also a requirement that the method used be comparable to the visual observations used in the past, and if possible, provide a record for remeasurement.

To this end, we used equipment that, while providing the necessary precision of observation, would be relatively inexpensive, very portable, and self-contained. For our first expeditions, use was made of 8-mm movie film and audio tape recorders. Subsequently, video cameras of increasing sensitivity were employed, progressing from vidicon to CCD.

The task of obtaining precise positions has progressed from using small scale topographic maps to using inexpensive, portable GPS receivers.

Time signals were obtained from standard radio broadcasts, and corrected for ΔT (variations in the Earth's rotation) with data from the U.S. Naval Observatory's Earth Orientation Division of the Time Service Department.

4. Current Status

Up to the current time, reduction of such observations has remained tedious, and still requires many hours of playing tapes over and over, identifying lunar features, distinguishing beads, and deciding the instants at which various phenomena occurred. Many videotapes of several eclipses remain unanalyzed for this reason. Although the capability of digitizing images has existed for several years, an affordable means of storing a sufficient quantity simultaneously, rectifying the orientation, and synchronizing the time signals from the sound track has remained unsolved. Table I lists the available data still to be analyzed.

Several advances in data and techniques promise to speed up the analysis as well as improve the accuracy. We have now integrated the DE200/LE200 ephemerides into our software, and JPL has extended the range of those ephemerides back to 1600. This provides a consistency which was previously uncertain.

5. Future Plans

Graphics capabilities now exist for direct comparison of the video records
to predicted bead phenomena. We have already implemented a system that
allows generation of the predicted patterns onto transparencies at the same
scale as the video image. The predicted pattern can be laid directly over the
video screen, and the tape stepped frame by frame until a match is reached.
The problems remaining with that are that the video image still suffers from
"blooming," because the filters were not quite correct; and the time signal
is an estimate from a superimposed digital timer coordinated to the sound
track by eye-and-ear estimation. This technique was used to re-analyze the
data from 1984 (Section 6). A new workstation is available that has the
capability of reading in several frames at once and displaying on the same
screen not only several images but also the associated sound track, so that
the time ticks are visual traces. The images can be enhanced to reduce the
bloom. Predicted profiles for each frame can be displayed alongside as well.

We have collected 208 contact timings from central line stations for 59
eclipses over the years 1806–1979. We suspect that these are less accurate
than timings from the edge, because the limb features in the central limb
regions change dramatically from eclipse to eclipse. However, two developments may make these observations worth examination if the positions of the
observers can be determined with sufficient accuracy and the Watts' data
verified through independent means. First, improved lunar limb data may
be generated from lunar orbiters with laser ranging equipment, such as the
Clementine mission (Fisk, 1993; Dunham and Farquhar, 1993). Second, the
precision becoming available with the new graphics methods may overcome
the uncertainties of identification.

We feel that it is time for a comprehensive re-examination and re-reduction
into one solution of all the eclipses listed in Table II. This will form the basis
for adding in the reductions for eclipses listed in Table I.

6. Results

This method of using observations of eclipses has been applied for 9 eclipses
(Table II). Each eclipse was treated independently. The result for 1715 (Dunham et al., 1980) depends critically on the location of one of the three observers, and new information has been found (Morrison, Stephenson, and
Parkinson, 1988). The result for 1987 (Wan et al., 1990) was calculated with
our old software system.

The result for 1984 (Fiala et al., 1985, shown as 1984[a] in Table II) has
just been done anew (Table II, 1984[b]). The original 51 observations were re-
reduced with the new ephemerides. In the original reduction, plots to identify
limb features were linear, as shown in Figure 1. A new configuration plot

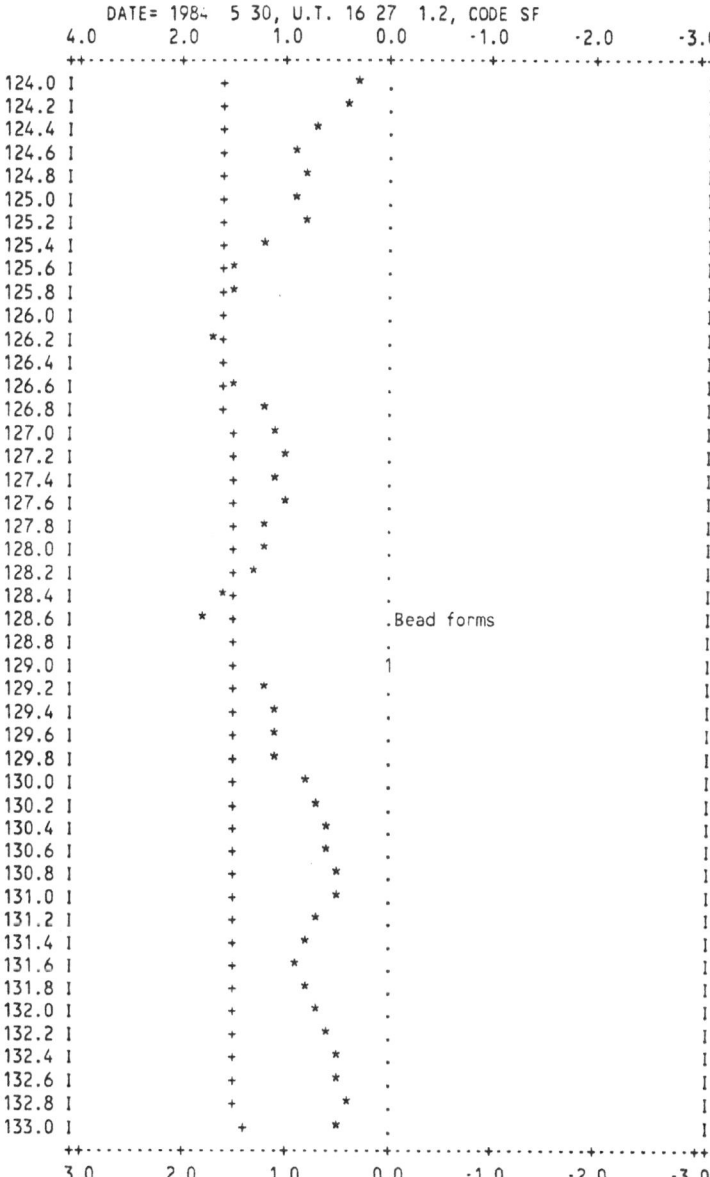

Fig. 1.   Line profile of portion of limb. Watts angle vs. residual in seconds of arc. Angle measured from north to east. +++++ = solar limb. **** = lunar limb, ... = Watts mean datum.

TABLE I
UNREDUCED OBSERVATIONAL MATERIAL
Archival data, observations from edge of path:

1869 August 8	Total	Eastern U.S.
1878 July 2	Total	Western U.S.
1922 September 21	Total	Australia.
1988 March 18	Total	Indonesia.

Videotapes, from edge of path:

1984 May 30	Annular	Eastern U.S., several
1984 Nov. 22	Total	Papua New Guinea.
1987 Sept. 23	Annular	China, two locations

TABLE II
RESULTS FROM SOLAR ECLIPSE OBSERVATIONS

Date	Eclipse Type	No. Obs	Solar Radius[1]	Correction to Lunar Ecliptic Long	Lat
			(All in seconds of arc)		
1715 May 3	Total	3	$+.48 \pm .2$		
1925 Jan. 24	Total	8	$+.51 \pm .08$		
1976 Oct. 23	Total	43	$+.04 \pm .07$	$+.65 \pm .10$	$-.45 \pm .09$
1979 Feb. 26	Total	47	$-.11 \pm .05$	$+.52 \pm .09$	$+.25 \pm .05$
1980 Feb. 16	Total	232	$-.03 \pm .03$	$+.32 \pm .03$	$+.07 \pm .04$
1981 Feb. 4	Annular	153	$-.02 \pm .03$	$+.02 \pm .05$	$-.02 \pm .04$
1983 June 11	Total	201	$+.09 \pm .02$	$+.65 \pm .03$	$-.19 \pm .02$
1984 May 30[2]	Annular	51	$+.23 \pm .04$	$+1.01 \pm .05$	$-.47 \pm .06$
1984 May 30[3]	Annular	51	$+.09 \pm .04$	$-.12 \pm .06$	$-.03 \pm .06$
1987 Sep. 23[4]	Annular	123	$-.11 \pm .03$	$+.57 \pm .09$	$+.02 \pm .02$

Notes.
1: Standard value 959.63 seconds of arc.
2: Old analysis, Dunham site (Fiala *et al.*, 1985).
3: New analysis, Dunham site, with LE200/DE200 ephemerides.
4: Shanghai Observatory expeditions to Funing and Tenchang, China (Wan *et al.*, 1989).

of the predicted appearance for each observation was prepared (Figure 2), and compared with the video tape. Figure 3, a digitized image of a video frame made several years ago, illustrates how imprecise the video image is by comparison to the prediction. About one-third of the identifications of limb features were changed. The result is quite different, but with nearly the same mean error as before. What is not shown is the great improvement in the residuals overall. The correction seems to have been shifted from the ephemeris to the solar diameter. This is consistent with the suspicion that the previously used solar ephemeris was indeed drifting in accuracy as

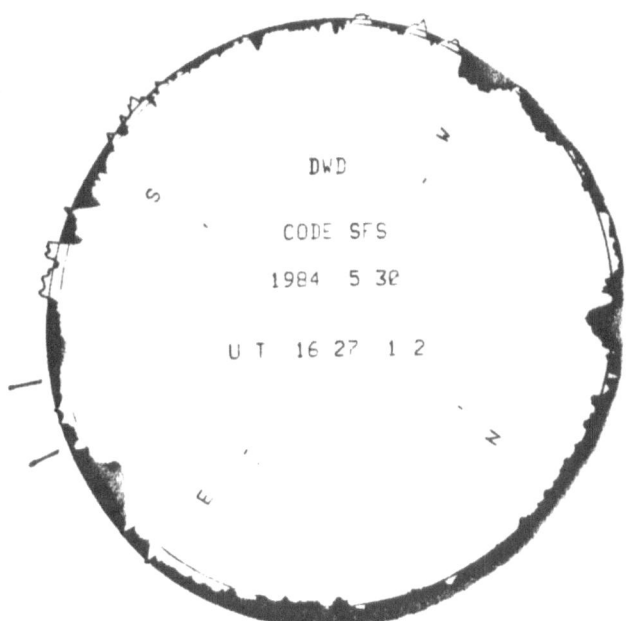

Fig. 2. Predicted profile for one observation. The region corresponding to Figure 1 is marked.

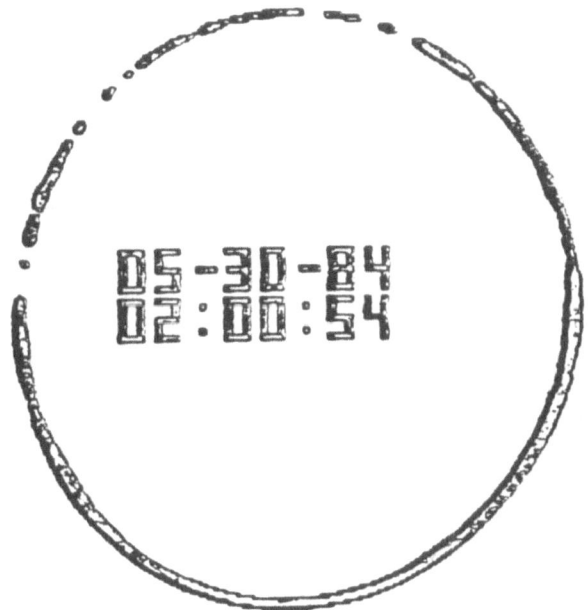

Fig. 3. A contour plot of a video image corresponding to Figure 2, in the same orientation.

time progressed from the epoch, and that made the correct identification of the limb features difficult. The effect of the errors in the ephemeris was probably magnified in this particular eclipse, it being unlike all the others in that many bead events were occurring nearly simultaneously around the complete limb of the Moon. Revisions in the values in Table II will probably not be as large for the other eclipses, but this result for the eclipse of 1984 really emphasizes the need to re-reduce those other data.

7. Acknowledgments

We thank Mitsuru Sôma at the National Observatory in Japan for performing some of the calculations used in the new analysis. We thank Marion Schmitz for collecting and analyzing the central eclipse data. We thank Richard Schmidt for Figure 3.

References

Dunham, D.W. and Farquhar, R.W.: 1993, Amer. Astronaut. Soc. Paper 93-145, Pasadena, CA.
Dunham, D.W., Sofia, S., Fiala, A.D., Herald, D., and Muller, P.M.: 1980, *Science* **210**, 1243.
Eddy, J.S. and Boornazian, A.A.: 1979, *Bull. Amer. Astron. Soc.* **11**, 437.
Fiala, A.D., Dunham, D.W., Dunham, J.B., and Sofia, S.: 1985, *Bull. Amer. Astron. Soc.* **17**, 624.
Fisk, L.A.: 1993, NASA Research Announcement NRA-93-OSSA-2.
Morrison, L.V., Stephenson, F.R., and Parkinson, J.: 1988, *Nature* **331**, 421.
Newcomb, S.: 1871, *Washington Observations for 1869*, GPO: Washington, Appendix I.
Ribes, E.: 1990, *Phil. Trans. R. Soc. Lond.* **A 330**, 487.
Ribes, E., Beardsley, B., Brown, T.M., DeLache, P., Laclare, F., Kuhn, J.R., and Leister, N.V.: 1991, in C. P. Sonett, M. S. Giampapa, M. S. Matthews, eds., *The Sun in Time*, University of Arizona Press: Tucson, 59.
Smith, C., and Messina, D.: 1981, in S. Sofia, ed., *Proceedings of a Workshop on Variations of the Solar Constant*, Nov. 5-7, 1980, NASA:Washington CP-2191, pp. 39 and 111.
Sofia, S., Dunham, D.W., and Fiala, A.D.: 1980, in R.O. Pepin, J.A. Eddy, and R.B. Merrill, eds., *Proc. Conf. Ancient Sun*, Lunar and Planetary Science Institute: Houston, Texas, 147.
Sofia, S., Fox, P., with contributions from Maier, E., Schatten, K., Chiu, H-Y., Heaps, W., Chan, K., Twigg, L., Theobald, M., and Lydon, T.: 1991, "Solar Variability with Consequences for Terrestrial Climate", CSSR Technical Report 91-01, Yale U.
Sofia, S., O'Keefe, J., Lesh, J.R., and Endal, A.S.: 1979, *Science* **204**, 1306.
Wan, L., Zhao, J-L., Chu, Z-Y., Wang, R-Y., Tan, Z-X., Zhao, G., and Cheng, Z-Y.: 1990, *Chin. Astron. Astrophys.* **14/2**, 213.
Watts, C.B.: 1963, *Astronomical Papers of the American Ephemeris*, *XVII*, GPO:Washington.
Wittmann, A.D. and Debarbat, S.: 1990, *Sterne und Weltraum* **29**, 420.

LATITUDINAL VARIATION OF THE SOLAR
LIMB-DARKENING FUNCTION

RONALD J. KROLL

University of Arizona, Dept. of Physics, Tucson, AZ 85721

Abstract. In an effort to monitor solar limb-darkening variability, the continuum radiation intensity at 550 nm over the outermost 32 arcseconds of the limb is measured at various solar latitudes. Using the Finite Fourier Transform Definition, the edge location of the Sun is determined for a series of scan amplitudes at each of the observed positions. The differential radius is the difference between edge locations for a fixed pair of scan amplitudes, and is a quantity which characterizes the slope of the solar limb-darkening function. Utilizing the differential radius, such observations offer the possibility of revealing a latitudinal variation of the photospheric temperature gradient and could provide clues to the mechanisms and variability of energy transport out of the Sun. These observations began in 1988 with measurements at 24 separate limb positions and include observations since 1990 when 36 positions were observed. The daily differential radius measurements for each position that is free of contamination from solar active regions are weighted according to the corresponding daily variance and averaged to obtain an overall value at each position for the observing season. The results indicate that during the 1991 observing season, there were regions near 20°N latitude and 30°S latitude on the Sun where the differential radius values were significantly greater than surrounding regions. This suggests that perturbations to the temperature gradient occur in latitudinally localized regions and persist for at least several months. It is shown that this phenomenon could have the same origin as the observed latitudinal variations of surface temperature and could also speak to the question of a lag time between the cycles of irradiance and magnetic variation.

1. Introduction

Solar limb-darkening observations provide information about the temperature gradient in the solar atmosphere which is determined by the radiative energy flux penetrating the quiet photosphere. Variations in the magnitude or distribution of this energy flux would contribute to total irradiance variability in a manner very different from sunspot and facular contributions. Ground-based limb-darkening observations are thus a valuable complement to satellite observations of total irradiance in the effort to understand solar irradiance variability. Efforts to monitor the solar limb darkening, however, have given a variety of results (see, e.g., Petro et al., 1984; Brown, 1988; Beardsley et al., 1988; Kroll, Hill, and Beardsley, 1990).

Limb-darkening observations are also useful for ascertaining the spatial uniformity of the solar temperature structure which could provide clues to mechanisms affecting the transport of energy through the convection zone and atmosphere. A temperature difference between the pole and equator was measured by Altrock and Canfield (1972); and more recently, Kuhn, Libbrecht, and Dicke (1985) have measured a time-varying latitudinal temperature distribution. Hill et al. (1974) observed a difference between the

polar and equatorial limb-darkening functions, which was seen to change
over time. Using a slightly modified version of their apparatus, observations
of the solar limb darkening have been made at numerous solar latitudes. Beginning in 1988, limb-darkening observations were obtained at 24 locations
around the solar limb, and in 1990 the number was augmented to 36. The
following sections describe the manner of observation and data analysis. A
discussion of the results is presented in the context of the suggestion by Kuhn
(1991) that magnetic fields at the base of the convection zone modulate the
radiative flux to the convective zone which is manifested as a latitudinal
temperature variation in the photosphere.

2. Observations and Data Analysis

The design of the SCLERA astrometric telescope facilitates limb-darkening
measurements in a manner very different than the standard drift scan method.
The telescope continuously tracks the Sun and projects the solar image onto
a horizontal focal plane. In this plane is the detector assembly, which consists of two slit blocks positioned near the edge of the solar image at opposite
ends of a solar diameter. Both slit blocks contain three apertures, or slits,
each covering approximately 5.7°of the solar circumference and 1 arcsecond
of radius. The centers of the slits are separated by one-eighth radian, or
about 7.2°, along the solar circumference. The entire detector assembly can
rotate to coincide with any chosen diameter of the image. In order to sample
a radial portion of the limb, one of the mirrors in the optical path oscillates
sinusoidally so that about 32 arcseconds of the limb passes over the slits.
The scan frequency is 1.6 Hz, and thus as many as 10^5 scans can be obtained
in a 10-hour observing day.

The light falling on the slits encounters an interference filter which passes
light in a 8-nm band of the continuum centered at 550 nm. The intensity
of this radiation is digitized every 2.5 ms. Each non-overlapping series of 11
scans is averaged to give a recorded limb profile. Throughout the observing
day, the detector is rotated through a predetermined sequence of angles
relative to the equator obtaining data at each position for approximately 64
seconds.

The observations obtained are the intensity profiles of approximately 32
arcseconds of the extreme solar limb at particular solar latitudes. In order
to characterize the slope of the limb-darkening function, the location of the
solar edge is determined by applying the Finite Fourier Transform Definition (FFTD) to each of the limb profiles (Hill, Stebbins, and Oleson, 1975).
Specifically, the edge location ρ is found when

$$\int_{-1/2}^{1/2} I(\rho + a\sin\pi s)\cos 2\pi s \, ds = 0, \tag{1}$$

where I is the limb intensity as a function of radius and a is the scan amplitude used. This definition has several particularly favorable properties. First, the edge location is only weakly dependent on atmospheric seeing when the scan amplitude is greater than the width of the seeing function. Second, atmospheric transparency has no effect on the edge location, since scaling I by any factor will not alter the integral. Third, the edge location is sensitive to changes in the shape of the limb-darkening function.

For a given limb profile, the edge location is also dependent on the scan amplitude which can in effect be changed by using a truncated limb-darkening profile in Equation (1), i.e., $\rho = \rho(a)$. Exploiting this fact provides a measure of the slope of the limb-darkening function. This measure, the differential radius, is given by

$$\Delta r(a_i, a_j) = \rho(a_i) - \rho(a_j), \tag{2}$$

where a_i and a_j are two different scan amplitudes with $a_i < a_j$. Qualitatively, the value of the differential radius for a limb-darkening function described by a step function is $\Delta r = 0$. The more a limb-darkening function deviates from a step function, the more positive Δr becomes.

The data analysis was carried out for ten seasons of observations, however, the presence of certain instrumental effects limits the utility of portions of the data set. At this time the data from slits three and four, as well as the fall observing runs, are not considered. The data from locations where solar active regions are present are eliminated from the analysis, so the results should not reflect the effects of transient solar activity.

3. Results and Discussion

The results of the spring 1991 observations are of particular interest and will constitute the focus of the remaining discussion. The data for this season are from 26 days of observations made over a period of two and one-half months. It is seen from Figure 1 that in the spring of 1991 differential radius values are greater in the regions of about 20°N and 30°S latitudes on the Sun. The differences between the differential radius values at these latitudes and the nearby differential radius values at lower latitudes are statistically significant at the 99% level. Thus, a limb-darkening function which is distinctly variable with latitude was detected in 1991.

This result indicates that for at least several months near the maximum of solar cycle 22, the temperature gradient in the upper photosphere of the Sun was in some way perturbed in localized regions of latitude in both northern and southern hemispheres. The radiative flux from these regions would also have been modified, possibly to the extent that the solar total irradiance could have been affected. This is an interesting conclusion, since proxy models of solar total irradiance whose parameters depend largely on

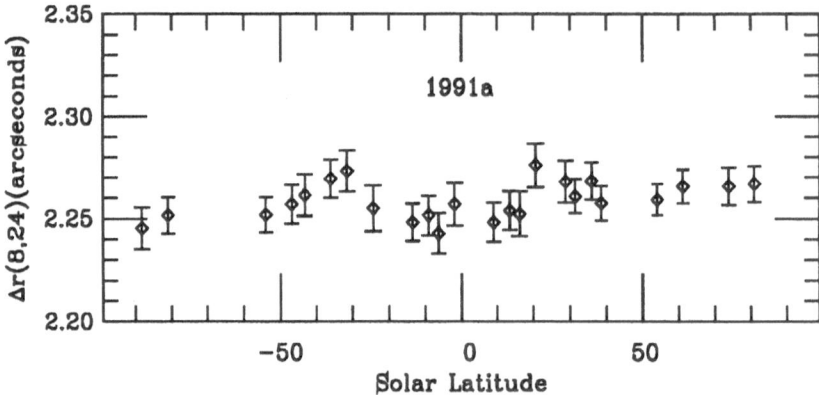

Fig. 1.   Differential radius $\Delta r(8, 24)$ measured during the spring 1991 observing season. Data have been corrected for offsets among the detectors, and data from slits 3 and 4 are not included.

the presence of solar active regions seem to inadequately match the observed total irradiance during solar maximum periods.

It is instructive at this point to consider the observations of Kuhn and Libbrecht (1991), which consist of color difference measurements spatially resolved around the solar limb. From their observations, they are able to infer latitudinal variations in the effective temperature of the photosphere which arise from both facular contributions and what they term "smooth" flux contributions. They construct a model in which these two contributions to the solar flux can be used to calculate the solar cycle irradiance variability, leading them to interpret this "smooth" flux contribution as the "third component" of the total irradiance variation. Their model predicts an irradiance increase from 1989 to 1990 when the magnetic activity measured by sunspot number decreases, which implies that the irradiance maximum lags the magnetic activity maximum and that the irradiance variation is not explained by magnetic phenomena alone.

There are some suggestive similarities between the effective temperature observations of Kuhn and Libbrecht (1991) and the limb-darkening observations under discussion. The latitudinal distribution of differential radius in 1991 is reminiscent of the "smooth" temperature distribution observed in 1990. The latitudes where the differential radius is greater might correspond to latitudes at which there is a "smooth" temperature excess. If this is the case, then a large "smooth" temperature excess would have been present in 1991 with the associated consequences pertaining to the total irradiance.

Is this connection physically plausible? To begin an investigation of this question, consider the suggestion of Kuhn (1990;1991) that magnetic fields at the boundary between the radiative and convection zones are the source of the deep latitudinal variations in the radiation incident on the bottom of the convection zone. Numerical modeling results show that small temperature differences at the base of the convection zone are magnified into much larger temperature differences in a shallow layer several hundred kilometers below the base of the photosphere. This shallow-layer temperature perturbation also serves to explain the shifts in p-mode frequencies measured over the solar cycle.

Given that shallow-layer temperature perturbations appear at certain solar latitudes during the solar cycle, there should be an associated change in the limb-darkening function at those latitudes. The responses of the limb-darkening function to impulsive variations in the source function are modeled by Petro, Foukal, and Kurucz (1985). Such impulsive variations could approximate the effects of temperature perturbations confined to a shallow photospheric layer. Using the limb-darkening function at 4451Å determined by Pierce and Slaughter (1977), Petro, Foukal, and Kurucz (1985) find that an impulsive variation of the source function S below $\tau = 1$ such that $\Delta S < 0$ will increase the limb-darkening function. Conversely, therefore, a region of enhanced temperature with $\Delta S > 0$ in a shallow sub-photospheric layer produces a negative change in the limb-darkening function. The shapes of the changes in limb-darkening function for various depths of impulsive variations are illustrated in their paper.

The change in differential radius can be found, given the change in shape of the limb-darkening function. This has been done by perturbing the Pierce and Slaughter (1977) limb-darkening function at 4451Å with a function of the form shown by Petro, Foukal, and Kurucz (1985) for an impulsive variation at depth $\tau = 7.4$ and scaled by a factor of -5.0. The resulting change in differential radius is 0.010 arcseconds. The assumptions made here are numerous and broad, *viz.* that the shallow-layer temperature perturbation causes an impulsive source-function variation, that limb-darkening change simply scales with the magnitude of the impulse, and that the results modeled at 4451Å are qualitatively similar to the response at 5500Å. If these assumptions are approximately correct, however, an increase of the temperature and source function within a shallow layer below the photosphere decreases the limb-darkening function which in turn increases the magnitude of the differential radius. Therefore, the postulated temperature perturbation that can give rise to the "smooth" flux contribution could modify the local temperature gradient in a manner which will enhance the differential radius values similar to what was observed in 1991.

4. Conclusion

The differential radius measurements obtained in 1991 indicate that there are regions in the low latitudes of both the northern and southern solar hemispheres where the photospheric temperature gradient is perturbed relative to the surrounding regions. The latitudinal variation of differential radius values is similar to the distribution of photospheric temperature measured by Kuhn and Libbrecht (1991) in 1990. Despite the difference of about a year in the time that the observations were made, the two phenomena are shown to be related. In fact, the temperature perturbations suggested by Kuhn (1991) can give rise to both an effective temperature enhancement and an increase in the differential radius. The regions of enhanced differential radius in 1991 could be evidence that an associated "smooth" flux component was also present at that time. If the model of Kuhn and Libbrecht (1991) correctly establishes the relationship between the facular and "smooth" flux contribution and the solar irradiance, then a large "smooth" flux contribution in 1991 would contribute significantly to the total irradiance for that year. Evidence such as that presented at this meeting by Pap et al. (1993) indicates that the observed irradiance in 1991 was indeed greater than predicted by models accounting for magnetic effects alone and that the peaks of solar magnetic activity and total irradiance are not coincident.

Acknowledgements

This work was supported in part by the U.S. Department of Energy

References

Altrock, R. C., and Canfield, R. C.: 1972, *Astrophys. J.* **171**, L71.
Beardsley, B. J., Czarnowski, W. M., Kroll, R. K., Cornuelle, C. S., Oglesby, P. H., Yi, L., and Hill, H. A.: 1988, *B.A.A.S.* **20**, 1011.
Brown, T. M.: 1988, in Solar Radiative Output Variation, P. Foukal, ed.
Hill, H. A., Clayton, P. D., Patz, D. L., Healy, A. W., Stebbins, R. T., Oleson, J. R., and Zanoni, C. A.: 1974, *Phys. Rev. Lett.* **33**, 1497.
Hill, H. A., Stebbins, R. T., and Oleson, J. R.: 1975, *Astrophys. J.* **200**, 484.
Kroll, R. J., Hill, H. A., and Beardsley, B. J.: 1990, in *Climate Impact of Solar Variability,* NASA C.P. 3086, K. H. Schatten and A. Arking, eds.
Kuhn, J. R.: 1990, in *Seismology of the Sun and Stars,* Y. Osaki and H. Shibahashi ed., pp. 157-162.
Kuhn, J. R.: 1991, *Adv. Space Res.* Vol. **11**, No. 4, p(4)171.
Kuhn, J. R., and Libbrecht, K. G.: 1991, *Astrophys. J.* **381**, L35.
Kuhn, J. R., Libbrecht, K. G., and Dicke, R. H.: 1985, *Astrophys. J.* **290**, 758.
Pap, F., Willson, R. C., Fröhlich, Donnelly, R. F., and Puga, L., this conference.
Petro, L. D., Foukal, P. V., Rosen, W. A., Kurucz, R. L., and Pierce, A. K.: 1984, *Astrophys. J.* **283**, 426.
Petro, L. P., Foukal, P. V., and Kurucz, R. L.: 1985, *Solar Phys.* **98**, 23.
Pierce, A. K., and Slaughter, C. D.: 1977, *Solar Phys.* **51**, 25.

IMPROVEMENT OF THE PHOTOMETRIC SUNSPOT
INDEX AND CHANGES OF THE DISK-INTEGRATED
SUNSPOT CONTRAST WITH TIME

CLAUS FRÖHLICH

Physikalisch-Meteorologisches Observatorium Davos, World Radiation Center
Davos Dorf, Switzerland

JUDIT M. PAP

Jet Propulsion Laboratory, California Institute of Technology
Pasadena, California, U.S.A.

and

HUGH S. HUDSON

Institute for Astronomy, University of Hawaii
Honolulu, Hawaii, U.S.A.

Received 30 September, 1993; in revised form 4 March, 1994

Abstract. The photometric sunspot index (*PSI*) was developed to study the effects of sunspots on solar irradiance. It is calculated from the sunspot data published in the *Solar-Geophysical Data* catalogue. It has been shown that the former *PSI* models overestimate the effect of dark sunspots on solar irradiance; furthermore results of direct sunspot photometry indicate that the contrast of spots depends on their area. An improved *PSI* calculation is presented; it takes into account the area dependence of the contrast and calculates 'true' daily means for each observation using the differential rotation of the spots. Moreover, the observations are screened for outliers which improves the homogeneity of the data set substantially, at least for the period after December 1981 when NOAA started to report data from a few instead of one to two stations. A detailed description of the method is provided. The correlation between the newly calculated *PSI* and total solar irradiance is studied for different phases of the solar cycles 21 and 22 using bi-variate spectral analysis. The results can be used as a 'calibration' of *PSI* in terms of gain, the factor by which *PSI* has to be multiplied to yield the observed irradiance change. This factor changes with time from about 0.6 in 1980 to 1.1 in 1990. This unexpected result cannot be interpreted by a change of the contrast relative to the quiet Sun (as it is normally defined and determined by direct photometry) but rather as a change of the contrast between the spots and their surrounding as seen in total irradiance (integrated over the solar disk). This may partly be explained by a change in the ratio between the areas of the spots and the surrounding faculae.

1. Introduction

An important part of the modelling of solar total irradiance variations is the effect of the dark sunspots. The Photometric Sunspot Index (*PSI* of Hudson et al., 1982) has been developed for this purpose. It is similar to other sunspot deficit models such as the ones from Foukal & Vernazza, 1979, Hoyt & Eddy, 1982 and Hudson & Willson, 1982 and describes the change of solar irradiance during to the passage of sunspot groups over the visible disk. The values are based on observations of sunspots as published in the Solar Geophysical Data catalogue by NOAA. Former studies (e.g. Foukal & Lean, 1988, Livingston et al., 1988, Fröhlich & Pap, 1989, Pap & Fröhlich, 1992) show that the dark sunspots and bright magnetic elements, including faculae and the active network, explain a considerable amount of changes in total irradiance, but significant variations in total irradiance remain unexplained after removing the effect of sunspots and bright magnetic elements. The question whether this remaining variability in total irradiance is caused by superficial features other than sunspots and bright magnetic elements or simply by the uncertainties of the data used or by the method of calculating *PSI*. This latter concern was the motivation for the present work.

2. Photometric sunspot index

Following Hudson et al., 1982, the radiation deficit on the total solar irradiance is calculated from the following equations. The irradiance at 1 AU is described by

$$S_0 = \int_{\Omega_\odot} I d\Omega = 2\pi I_0 \cdot (\frac{R_\odot}{R_0})^2 \int_0^1 f(\mu)\mu d\mu \qquad (1)$$

with $R_\odot$ = radius of the Sun, R_0 = mean Sun-Earth-distance (1 AU), and $I = I_0 f(\mu)$ the angular dependence of the radiance with $\mu = \cos\vartheta$ and ϑ being the angle between the line of sight and the surface normal. A spot with the area A_S on the surface affects the irradiance by S_S

$$I_S = I_{S0} \cdot g(\mu)$$
$$S_S = \mu \cdot \frac{A_S}{R_0^2} \cdot I_S \qquad (2)$$

with I_S and $g(\mu)$ being the radiance in the spot and its angular dependence respectively. The relative change in irradiance ΔS is then described by

$$\frac{\Delta S}{S_0} = -\frac{\mu A_S[f(\mu)I_0 - g(\mu)I_{S0}]}{2\pi I_0 R_\odot^2 \int_0^1 f(\mu)\mu d\mu} \cdot \qquad (3)$$

In order to evaluate Eq. (3) the following assumptions have been made traditionally:
- All umbrae and penumbrae have the same temperature T_u and T_p respectively;
- The ratios of the area of umbrae and penumbrae to the total area $(A_u/A_S, A_p/A_S)$ are constant;
- Umbra, penumbra and quiet Sun have the same limb darkening function, that is $f = g$.

Usually the area of the spot A_S is given as relative number a_S in millionths of the hemisphere according to

$$a_S = A_S/(2\pi R_\odot) \cdot 10^6 \text{ (in millionths of the hemisphere)} \qquad (4)$$

and the contrast of the spot – the darkness relative to the surrounding photophere – is defined by $\alpha = 1 - I_S/I_0$. With these assumptions the relative sunspot deficit can be described by

$$\frac{\Delta S}{S} = PSI = \mu \cdot \frac{f(\mu)}{\int_0^1 f(\mu)\mu d\mu} \cdot a_S \cdot \alpha \text{ (in ppm)} \qquad (5)$$

which is the definition of the Photometric Sunspot Index (PSI). From Eq. (5) it follows that PSI varies strongly with the central meridian distance, mainly due to the projection effect, but also due to the limb darkening function. For simplicity Hudson et al., 1982, used the Eddington limb darkening function given by

$$f(\mu) = \frac{3\mu + 2}{5}, \qquad (6)$$

and PSI was calculated according to

$$PSI = \mu \cdot \frac{3\mu + 2}{2} \cdot a_S \cdot \alpha \text{ (in ppm)} . \qquad (7)$$

The units in Eq. (7), however, are different from Hudson et al., 1982: here PSI is given in relative units (ppm) whereas it was formerly given in Wm^{-2}. The relative units are more appropriate as the sunspot areas – the basis for PSI – are also reported in relative units. In the original work the contrast α was derived from solar data of Allen, 1973 and calculated according to

$$T_u = 4240 \text{ K, } T_p = 5680 \text{ K, } T_0 = 6050 \text{ K,}$$
$$A_u/A_S = 0.18, \ A_p/A_S = 0.82$$
$$\alpha = \frac{A_u}{A_S}[1 - (\frac{T_u}{T_0})^4] + \frac{A_p}{A_S}[1 - (\frac{T_p}{T_0})^4] = 0.32 , \qquad (8)$$

yielding a constant value of 0.32.

The calculation of PSI can be improved by several means:

- improve the data quality by screening against outlier data (typos, wrong association with regions, etc.);
- average an observation at a given time during the day over 24 hours from UT 00h00 – 24h00 by integration over the rotational movement of the spot group during that day;
- transform the reported heliographic coordinates of the spot group to the position on the observed disk.

The screening of the data is done region by region. The median of a set of data (either one day or the whole region) is calculated and the values lying outside ± 2.5 times the average absolute deviation are rejected. In a first run the 3–5 observations of each day of the region are screened. This is repeated for the region as a whole for the days where not enough data are available for doing it at the daily level. The calculation of the daily average takes the longitudinal extent λ_{ext} and the rotation at the heliographic latitude of the region ($\Omega = 14.37 - 2.60 \sin^2 \phi$ in deg/day; Tang, 1981) into account. A weighing factor w by which $A_S \cdot \alpha$ has to be multiplied in order to get the daily average PSI of the sunspot group can be calculated according to

$$
w = \frac{\int_{\Lambda_{00}}^{\Lambda_{24}} \left(\frac{\int_{\Lambda-\lambda_{ext}/2}^{\Lambda+\lambda_{ext}/2} (A+B\cos\lambda\cos\phi+C\cos^2\lambda\cos^2\phi)\cos\lambda d\lambda}{\int_{\Lambda-\lambda_{ext}/2}^{\Lambda+\lambda_{ext}/2} \cos\lambda d\lambda} \right) \cos\Lambda d\Lambda}{\int_{\Lambda_{00}}^{\Lambda_{24}} \cos\Lambda d\Lambda} , \qquad (9)
$$

with the limb darkening function $f(\theta) = A + B\cos\theta + C\cos^2\theta$ ($\cos\theta = \cos\lambda\cos\phi$; λ, ϕ: longitude, latitude on visible disk), the longitudes of the center of the region at 00h00 and 24h00 UT $\Lambda_{00,24}$ (calculated from the observation time with the rotation at the corresponding physical latitude) and the longitudinal extent λ_{ext}. Eq.(9) can be evaluated analytically and yields an expression for w which is a function of $(A, B, C, \Lambda_{00}, \Lambda_{24}, \cos(\lambda_{ext}/2))$ and is used in the 'new' algorithm.

Direct photometric measurements have shown that the contrast α depends on the size of the spot (Steinegger et al., 1990) and (e.g. Fröhlich & Pap, 1989, Steinegger et al., 1990, Beck & Chapman, 1993) show that $\alpha = 0.32$ overestimates in most cases the effect of the spot deficit. Recent observations (Chapman et al., 1993) show also, that α is not only varying from spot to spot, but also with time for a given spot. In the present algorithm the α dependence on the projected area μA_S can be modelled by (Brandt, 1992):

$$
\alpha = 0.2231 + 0.0244 \cdot \log_{10}(\mu A_S) . \qquad (10)
$$

With these assumptions the following PSIs are calculated and discussed:
- screened data, new algorithms and Eddington limb darkening function (labeled 'PSI-Edd');
- screened data, new algorithms and Pierce limb darkening function (from Allen, 1973 labeled 'PSI-Pie'); this limb darkening function can also be used to calculate spectral PSI with corresponding α;
- screened data, old algorithms according to Eq. (7) (labeled 'PSI-Hud');
- original data from Hudson, 1989 (labeled 'PSI-old').

In order to clarify the question which average contrast best represents the irradiance variability, bi-variate spectral analysis between the different PSI and irradiance observations from ACRIM I (Feb.1980-June 1989), ACRIM

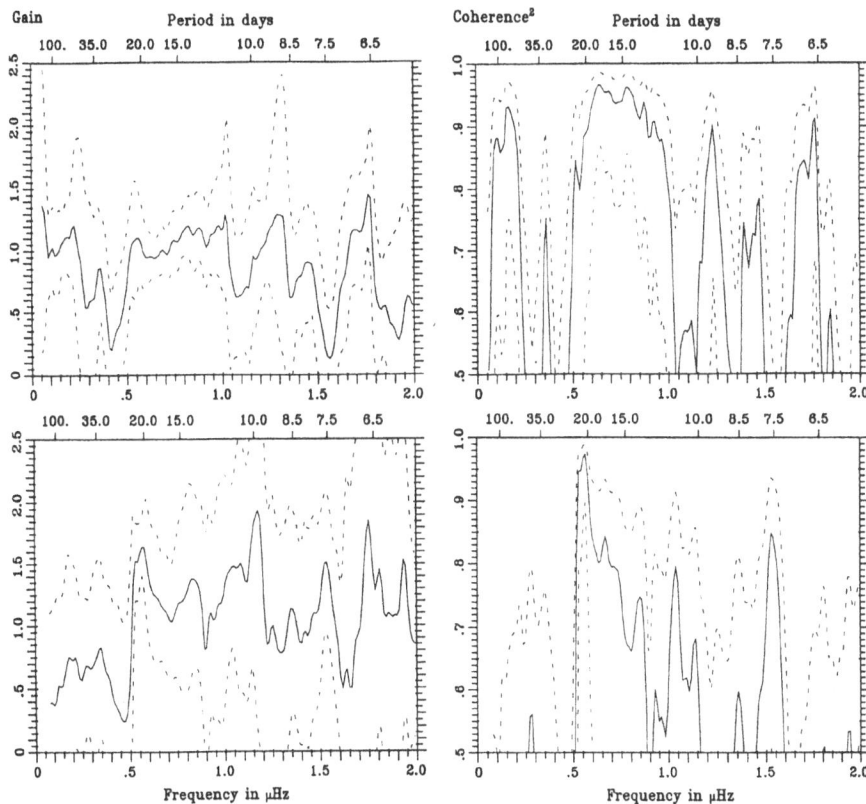

Fig. 1. Results from bi-variate analysis between ACRIM and PSI calculated with the new algorithm and the Eddington limb darkening function. The gain (left) and the coherence[2] (right) are shown as examples for 1985 (top) and 1992 (bottom). The broken lines indicate the upper and lower 90% confidence levels of the quantities shown.

II (since Sept.1991) and HF/NIMBUS7 (Nov.1978–Feb.1993) have been performed for periods of about one year from the end of 1978 to the beginning of 1993. Examples of the results for the gain – the factor by which PSI has to be multiplied to get the observed change in irradiance – and $C =$ coherence[2] – the relative amount of the irradiance explained by PSI – are shown in Figure 1 for the years 1985 and 1992. It is interesting to note that the highest coherence and thus the lowest uncertainty of the gain is in the range from 0.5 to 0.9 μHz ($\approx$ 13–23 days period). The period of the solar rotation is in a region of low coherence between irradiance and PSI indicating that the rotational time scale is less well determined than the evolutionary one. Lower coherence is found in the 1980 (not shown in Figure 1), but it is mainly due to the lower quality of the SGD data at that time. Compar-

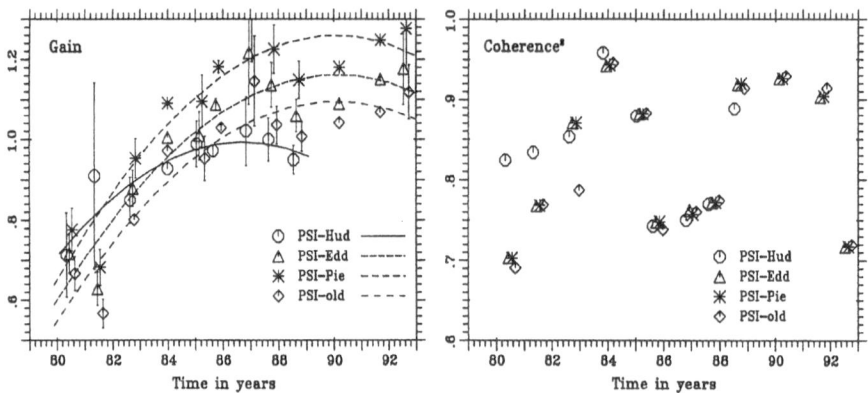

Fig. 2.   Gain between the solar irradiance and PSI (left panel) and coherence[2] (right panel). The gain is the factor to multiply the calculated PSI in order to yield 'true' sunspot influence; the error bars indicated are the extremes between the values calculated with the ACRIM and the HF values respectively. Note the low coherence in 1986–1988 and in 1992/93.

isons of the maximum coherence reached with the different methods show that screening the data has the largest effect (an increase of the peak C of 5-10%); the increase of C due to the more sophisticated methods is only 1-2% and the use of Pierce's limb darkening function instead of Eddington's $\leq 1\%$.

In order to enable use of these results to 'calibrate' PSI an average over the frequency is needed. Such an average gain is calculated by weighting the values with the inverse of the statistical uncertainty (depending mainly on the coherence) and only for the ranges from 4 to 1.5 and from 0.8 to 0.4 solar rotations ($0.1\cdots0.29$ and $0.54\cdots1.07\mu$Hz). This is to exclude from the beginning ranges with low coherence. The result is plotted for the different PSI (PSI-Hud, PSI-Edd, PSI-Pie, PSI-old) in Figure 2 together with the corresponding C values. All models show a strong variation of the gain with time; the period of change is certainly not that of the 11-year solar cycle, but could be of 22 years. It cannot, however, be interpreted by a change of the contrast relative to the quiet Sun (as α is normally defined and determined by direct photometry) but rather by a change of the contrast between the spots and their surrounding as seen in irradiance (integrated over the solar disk). The coherence plot shows that not in all cases the 'new' method improves the correlation: In 1980 and '81 the 'new' analysis had to be done with data which seem to be much less reliable than the ones used by Hudson, 1989, originally. From 1982 onward the data set looks quite homogeneous possibly because the daily data available are coming from at least 3 and usually 5 stations. This allows an efficient screening.

From the plot of the mean coherence the above statements about the improvement by using new algorithms are confirmed, although the differences in the averages are smaller than the ones stated above for distinct frequencies. The gain or 'calibration', however, depends significantly on the limb darkening function used in the calculation. This may indicate that the weighting on the solar disk depends on position in a way which is not completely described by the limb darkening function of the quiet Sun. This suggests that additional analysis may help to determine a PSI relevant limb darkening function.

3. Conclusions

The re-evaluation of PSI has shown three major findings:
- Screening for outliers improves the data set significantly;
- the integrated contrast of the spots changes with time;
- the integrated contrast depends on limb darkening function used.

Part of the change of the disk-integrated contrast with time can probably be interpreted as a change in the ratio of the areas of spots to those of the surrounding faculae. Due to the fact that no direct correlation with the 11–year cycle is found an interpretation in terms of the contrast changes observed and discussed by Maltby et al. (1986) is not valid. A 22–year period is suggestive, but obviously not conclusive – in any case the maxima of the two cycles are quite different (Figure 3). The observed change in contrast is important in view of the interpretation of the solar irradiance variations – mainly for those effects which are not related to sunspots because their influence is normally derived from time series for which the PSI effect has been removed. This is especially true for the influence of faculae and bright network, part of which may be already hidden in the observed change of the disk-integrated contrast.

In summarizing Figure 3 shows the 'calibrated' PSI in terms of an irradiance deficit for the period from 1978–1993. It is evident that the two cycles behaved quite different in amplitude and evolutonary patterns.

Acknowledgements

The research described in this paper was carried out by the Physikalisch-Meteorologisches Observatorium, partly funded by the Swiss National Science Foundation (C.F.) and by the Jet Propulsion Laboratory, California Institute of Technology (J.P.) under NASA grant 170-38-53-26, which are greatly acknowledged. The authors wish also to thank Helen Coffey, SEL NOAA, Boulder, for providing sunspot data before publication.

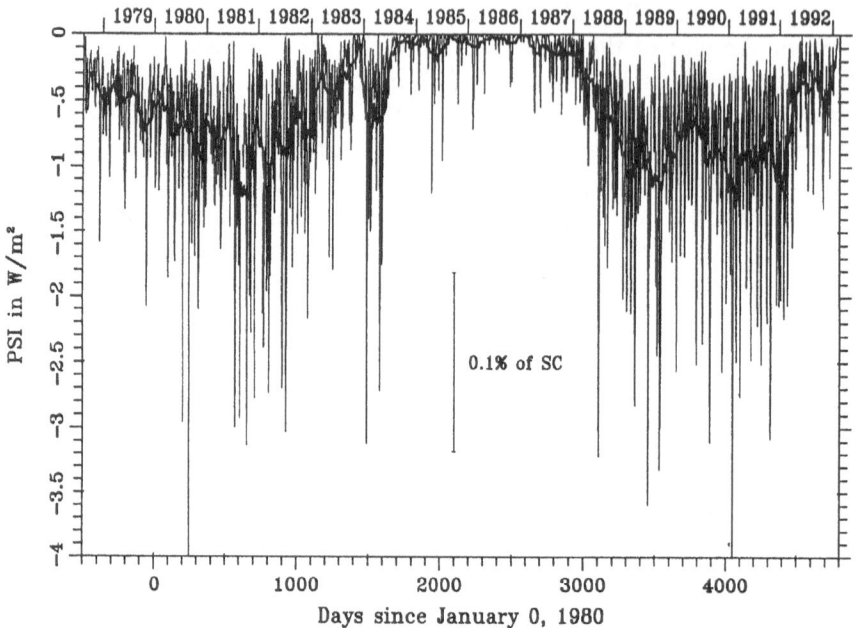

Fig. 3. *PSI* in Wm^{-2} plotted downward as irradiance deficit for the period 1978 – 1993. Note the difference between cycle 21 and 22 in amplitude and evolutionary patterns.

References

Allen, A.: 1973, *Astrophysical Quantities, 3rd Ed.*, Athlone Press: London.

Beck, J.G., and Chapman, G.A.: 1993, *Solar Phys.* , in press.

Brandt, P.N.: 1992, *private communication*.

Chapman, G.A., Cookson, A.C. and Dobias, J.J.: 1993, *Astrophys.J.* , submitted.

Foukal, P.V. and Vernazza, J.: 1979, *Astrophys.J.* **234**, 707.

Foukal, P.V. and Lean, J.: 1988, *Astrophys.J.* **328**, 347.

Fröhlich, C. and Pap, J.: 1989, *Astron. Astrophys.* **220**, 272.

Hoyt, D.V. and Eddy, J.A.: 1982, *An Atlas of Variations in the Solar Constant Caused by Sunspot Blocking and Facular Emissions from 1874 to 1981*, NCAR/TN194+STR, NCAR: Boulder, Co.

Hudson, H.S., Silva, S., Woodard, M. and Willson, R.C.: 1982, *Solar Phys.* **76**, 211.

Hudson, H.S. and Willson, R.C.: 1982, in Physics of Sunspots, ed(s)., *Cram,L. and Thomas,J.*, Sacramento Peak Obs: Sunspot, 434.

Hudson, H.S.: 1989, *private communication*.

Livingston, W., Wallace, L. and White, O.R.: 1988, *Science* **240**, 1765.

Maltby, P., Avrett, E.H., Carlsson, M., Kjeldseth-Moe, O., Kurucz, R.L., Loeser, R.: 1986, *Astrophys.J.* **306**, 284.

Pap, J.M. and Fröhlich, C.: 1992, in Proc.Solar Electromagnetic Radiation Study for Solar Cycle 22, ed(s)., *Donnelly, R.F.*, SEL NOAA ERL: Boulder, 62.

Steinegger, M., Brandt, P.N., Pap, J. and Schmidt, W.: 1990, *Astrophys.Space Sci.* **170**, 127.

Tang, F.: 1981, *Solar Phys.* **69**, 399.

MODELLING SOLAR IRRADIANCE VARIATIONS WITH
AN AREA DEPENDENT PHOTOMETRIC SUNSPOT INDEX

P.N. BRANDT, M. STIX, H. WEINHARDT

Kiepenheuer-Institut für Sonnenphysik, Freiburg, Germany

Abstract. The He 1083 nm line equivalent width and the 10.7 cm radio flux are employed to model the total solar irradiance corrected for sunspot deficit. A new "area dependent photometric sunspot index" (APSI) based on sunspot photometry by Steinegger et al. (1990) is used to correct the irradiance data for sunspot deficits. Two periods of time are investigated: firstly, the 1980–1989 period between the maxima of solar cycles 21 and 22; this period is covered by ACRIM I irradiance data. Secondly, the 1978–92 period which includes both maxima; here, the revised Nimbus-7 ERB data are used.

For both He 1083 nm and 10.7 cm radio flux irradiance models as well as ACRIM I and ERB irradiance data, the APSI yields an improved fit compared to the one obtained with the standard "Photometric Sunspot Index" (PSI) which uses a constant bolometric spot contrast α. With APSI, the standard deviation calculated from daily values is 0.461 Wm^{-2} for the period 1980–89 modelling ACRIM I vs. He 1083 nm, as compared to 0.478 when PSI is used, and to 0.531 for the uncorrected ACRIM series. A similar improvement is obtained for the same period modelling ERB vs. He 1083 nm, while there is almost no improvement for the long period.

As a general result the models provide a good fit with the spot-deficit-corrected irradiance only during the period between the maxima. If both maxima are included (period 1978–92) the He 1083 nm and 10.7 cm radio flux models show appreciably larger discrepancies to the irradiances corrected for PSI or APSI.

1. Introduction

Along with the space observations of total solar irradiance, efforts have been put forward to model the irradiance variability by estimating the effect of e.g. sunspots, faculae, and the network. Empirical model building (Hudson, 1988) aims at identifying the causes of particular variations, reducing the variance of the data in order to make subtler effects more visible, and providing a proxy for the estimation of variations in the absence of irradiance data. It is therefore desirable to model the various contributors to irradiance variability as precisely as possible.

Early models of the sunspot effect, such as the "Photometric Sunspot Index" (PSI) (Hudson et al., 1982), assume an area independent bolometric spot contrast α. Sunspot photometry (Steinegger et al., 1990), however, reveals that the effective temperature of umbrae and penumbrae, and the area ratio of umbrae and penumbrae, change from spot to spot. In order to improve the PSI function, we define an "area dependent photometric sunspot index" (APSI) by incorporating the spot contrast in the form

$$\alpha = 0.2231 + 0.0244 \cdot \log(A_s),$$

Solar Physics **152**: 119–124, 1994.
© *1994 Kluwer Academic Publishers.*

where A_s is the total sunspot area in millionths of a hemisphere (cf. Brandt et al., 1992).

The equivalent width of the He 1083 nm line and the 10.7 cm radio flux are representative of contributions from all bright magnetic elements on the disc. Both indices correlate well with the smoothed irradiance data if we first apply a correction for sunspot blocking to the irradiance (e.g. Willson and Hudson, 1991). We address the questions: Can this correlation be improved if we introduce the new APSI instead of the standard PSI? Is the correlation between index and spot-deficit-corrected irradiance equally good over a whole solar cycle or is it better for certain time periods and worse for others?

2. Observational Data

Sunspot areas and positions were used as published in the NOAA World Data Center (WDC) Solar Geophysical Data (SGD) catalogue. Total solar irradiance data are from the ACRIM I instrument on board the SMM satellite for the years 1980–89. For the period 1978–92 the revised total irradiance data from the ERB experiment on board the Nimbus–7 satellite were used (Hoyt et al., 1992). The full disk equivalent width of the He-line at 1083 nm is measured at the National Solar Observatory at Kitt Peak. Daily values of the 10.7 cm radio flux are measured at the Algonquin Radio Observatory and published by the NOAA/WDC.

3. Method

3.1. COMPARISON OF SUNSPOT DEFICITS YIELDED BY PSI, APSI

Daily PSI and APSI are calculated from sunspot data published in SGD for the years 1969–92. We use Mt. Wilson data until the end of 1981 and Learmonth data thereafter. The calculations were also carried out with data from five other reporting stations, but Learmonth yielded the most reliable results due to the highest number of reported days.

3.2. CORRELATION ANALYSIS

We investigate two time periods: a "short" period, 1980–89, covered by ACRIM I irradiance data; this is essentially the period between the maxima of solar cycles 21 and 22; a "long" period, 1978–92, covered by ERB irradiance data; this period includes both maxima.

The He 1083 nm equivalent width and 10.7 cm radio flux, respectively, are scaled by linear regression to the uncorrected ACRIM/ERB irradiance data, the PSI-corrected ACRIM and ERB irradiance data, and the APSI-corrected ACRIM and ERB irradiance data. We then calculate the following fit parameters: the linear regression coefficients, a, b, the correlation coeffi-

TABLE I

Fit parameters for the correlation ACRIM vs. He 1083 (1980-1989). The numbers in brackets give the standard errors of the regression coefficients (last digit).

He 1083 Irradiance model vs.	Regression coefficients (model = $a + b*$eqw)		Standard deviation		Mean absolute deviation	Corr. coeff.
			orig. data	smoothed data		
	a [Wm^{-2}]	b [Wm^{-2}(mÅ)$^{-1}$]	σ_1 [Wm^{-2}]	σ_{81} [Wm^{-2}]	d_{81} [Wm^{-2}]	c
ACRIM	1365.80(1)	0.027454(2)	0.531	0.199	0.139	0.56
ACR+PSI	1364.22(1)	0.062170(2)	0.478	0.171	0.129	0.86
ACR+APSI	1364.40(1)	0.058232(2)	0.461	0.165	0.124	0.86
Acrim+Psi*	1364.08	0.06361			0.178	

*Willson&Hudson, 1991

cient, c, and the standard deviation of the original unsmoothed data from the regression lines, σ_1. After applying a 81-day running mean smoothing, we calculate the standard deviation, σ_{81}, of the smoothed data from the regression lines, and the mean absolute deviation, d_{81}, of the smoothed data between model and the irradiance values.

4. Results

4.1. PSI, APSI SUNSPOT DEFICITS (1969–92)

Comparing the PSI and APSI sunspot deficit corrections over the time period 1969–92 we find that the PSI overestimates the spot deficit by up to 30% for the daily data and by 15–20% on average, depending on the phase of the activity cycle.

4.2. MODELLING HE 1083 NM VS. ACRIM (1980–89)

A correlation of the uncorrected ACRIM irradiance versus He 1083 equivalent width yields $\sigma_1 = 0.531$ Wm^{-2} (see Table 1). The PSI-corrected ACRIM data correlate noticeably better with He 1083; the standard deviation decreases to $\sigma_1 = 0.478$ Wm^{-2}. Inclusion of the APSI-correction results in another clear improvement of the fit, diminishing the standard deviation to $\sigma_1 = 0.461$ Wm^{-2}. This improvement of the fit is about $\frac{1}{3}$ of the improvement achieved for the model applying no sunspot correction at all to ACRIM and the one applying the PSI correction ($\sigma_1 = 0.531$ and $\sigma_1 = 0.478$ Wm^{-2}, resp.). The mean absolute deviation between 81-day running means, d_{81}, decreases from approx. 0.178 Wm^{-2} (130 ppm) for ACRIM+PSI versus

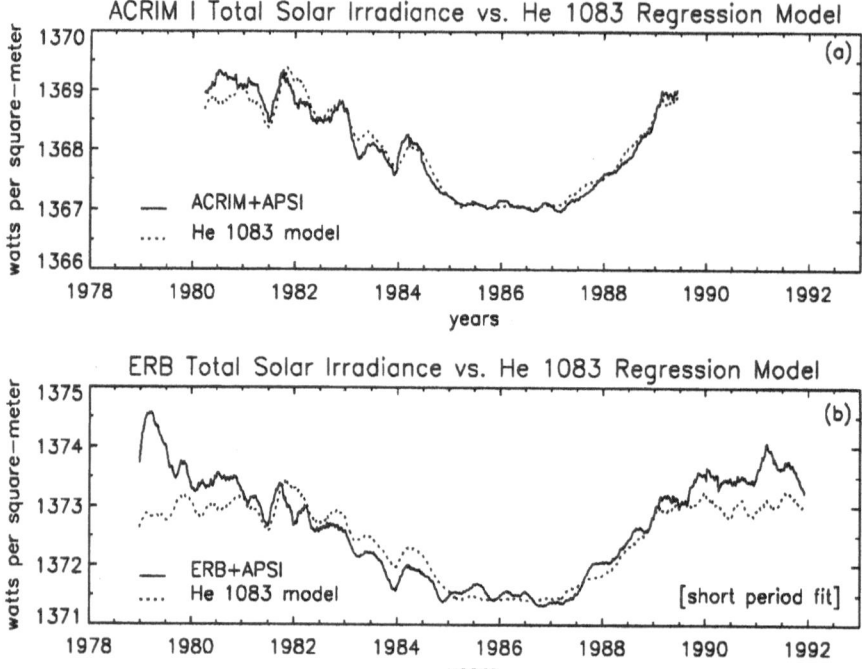

Fig. 1.   (a) 81-day running means of spot-deficit-corrected total solar irradiance
(ACRIM I) and He 1083 line equivalent width. The fit is good over the whole solar cycle,
with the exception of an irradiance excess in 1980 (end of maximum of solar cycle 21).
(b) 81-day running means of spot-deficit-corrected total solar irradiance (ERB) and
He 1083 line equivalent width. Note the breakdown of the correlation between observed
irradiance and irradiance model during solar maximum time periods.

He 1083 (Willson and Hudson, 1991; their d_{81} was estimated by us from
their Fig. 2) to approx. 0.124 Wm^{-2} (90 ppm) for ACRIM+APSI versus
He 1083 (cf. Figure 1a).

4.3. MODELLING HE 1083 NM VS. ERB (1980–89)

In order to compare the correlation of He 1083 vs. ERB with that vs.
ACRIM, we first investigate the short time period 1980-89. The results of
He 1083 vs. ACRIM are nearly reproduced, APSI again yields an improvement compared to PSI. It is, however, slightly less significant due to the
larger uncertainties inherent even in the revised ERB irradiance data used
here. If we extend the regression model calculated over the short period
into the cycle maxima, drastic discrepancies between the spot corrected irradiance and the He 1083 regression model become apparent during both
maximum periods. This reiterates the finding (e.g. Pap et al., 1992) that
there is a breakdown of the correlation between the observed total irradi-

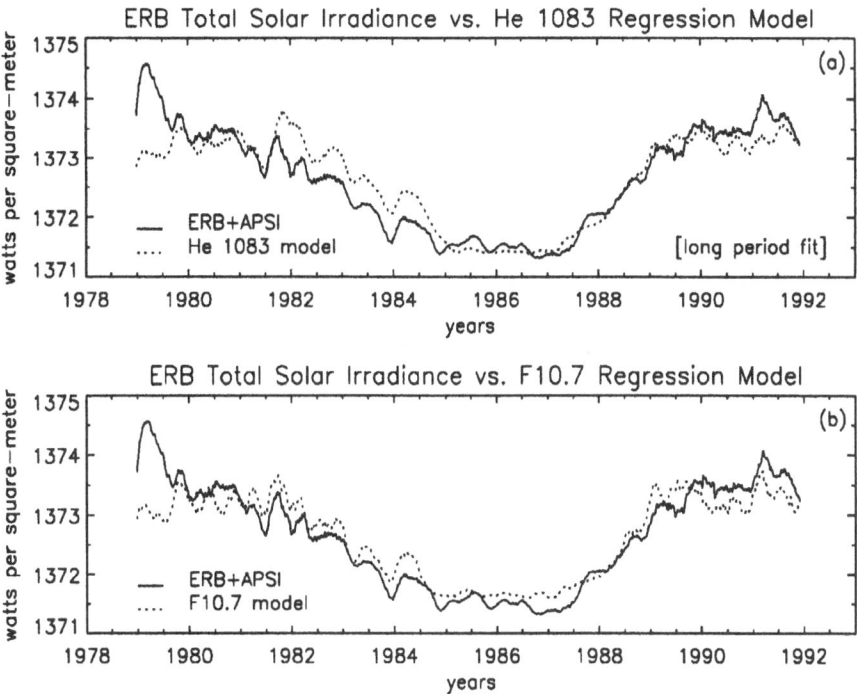

Fig. 2.   (a) 81-day running means of spot-deficit-corrected total solar irradiance (ERB)
and He 1083 line equivalent width. With the model calculated for the long time period
including the two maxima, the fit deteriorates also for the period between the cycle maxima, 1981–89.
(b) 81-day running means of spot-deficit-corrected total solar irradiance (ERB) and
10.7 cm radio flux.

ance and its model estimates during solar maximum (cf. Figure 1b).

4.4. MODELLING HE 1083 NM VS. ERB (1978–92)

In the next step, we calculate the He 1083 regression model for the long
period, 1978–92, in order to investigate whether a good fit between observed
irradiance and irradiance model can be achieved over the whole activity cycle
including the two maxima. It is evident that this is not the case. The d_{81}
values of the long period model increase by a factor of 1.6 compared to
the model based on the short period fit. However, from the calculated fit
parameters we infer that the APSI still yields a better fit than the standard
PSI (cf. Figure 2a). We note that there has been some discussion about
the reliability of the early ERB and of the 1980 ACRIM data (cf. Foukal
and Lean, 1988). But Hoyt et al. (1992) and Willson and Hudson (1991)
claim careful calibration of their data. Therefore, the discrepancies between
models and measurements occuring when the maxima of cycles 21 and 22

are included seem to indicate a general inadequacy of the models.

4.5. MODELLING F10.7 CM VS. ERB (1978–92)

Finally, we check the results of the previous step by calculating a 10.7 cm radio flux regression model. Again, a good fit over the whole solar cycle cannot be achieved. The fit parameters are not quite as good as the ones for the He 1083 model. Here, too, APSI results in a slightly better fit than PSI (cf. Figure 2b).

5. Conclusions

For both He 1083 nm line equivalent width and 10.7 cm radio flux irradiance models as well as ACRIM I and ERB irradiance data, the APSI yields an improved fit compared to the fit with the standard PSI. The improvement over PSI is approx. $\frac{1}{3}$ of the one gained through correcting the original irradiance data with the PSI. The area dependence of the sunspot bolometric contrast should be taken into account in future modelling efforts.

As a general result the models provide a good fit with the spot deficit corrected irradiance only during the period *between the maxima*. If both maxima are included the models show appreciably larger discrepancies to the spot-deficit-corrected measured irradiances; typically, the d_{81} values increase by a factor of 1.6. We find a general inadequacy of irradiance models calculated for the *entire* activity cycle during solar maximum.

Acknowledgements

We would like to thank R. Willson and D. Hoyt for providing the ACRIM I and Nimbus–7 data, and J. Harvey for providing the He 1083 data. Discussions with J. Pap have helped to clarify questions regarding sunspot data. NSO/Kitt Peak data used here are produced cooperatively by NSF/NOAO, NASA/GSFC, and NOAA/SEL.

References

Brandt, P.N., Schmidt, W., and Steinegger, M.: 1992, 'Photometry of Sunspots Observed at Tenerife' in R.F. Donnelly, ed(s)., *Proceedings of the Workshop on the Solar Electromagnetic Radiation Study for Solar Cycle 22*, NOAA/ERL/SEL: Boulder, 130.
Foukal, P. and Lean, J.: 1988, *Ap. J.* **328**, 347.
Hoyt, D. V., Kyle, H. L., Hickey, J.R., and Maschhoff, R.H.: 1992, *J.G.R.* **97**, 51.
Hudson, H.S.: 1988, *Adv. Space Res.* **8**, 15.
Hudson, H.S., Silva, S., Woodard, M., and Willson, R.C.: 1982, *Solar Phys.* **76**, 211.
Pap, J., Willson, R.C. and Donnelly, R.F.: 1992, 'Two-Parameter Model of Total Solar Irradiance Variation over the Solar Cycle' in K. Harvey, ed(s)., *Proceedings of the NSO/Sac Peak 12th Summer Workshop*, ASP Conference Series 27, 491.
Steinegger, M., Brandt, P.N., and Schmidt, W.: 1990, *Astrophys. Space Sci.* **170**, 127.
Willson, R.C. and Hudson, H.S.: 1991, *Nature* **351**, 42.

ON THE CAUSE OF TOTAL IRRADIANCE VARIATIONS OBSERVED BY THE CCD SOLAR SURFACE PHOTOMETER

JUN NISHIKAWA

Hiraiso Solar Terrestrial Research Center, Communications Research Laboratory, Isozaki 3601, Nakaminato, Ibaraki 311–12, Japan

Abstract. Spatially-resolved precise photometric observations of the whole Sun at wavelengths of 545nm (FWHM 40nm) were carried out by using the CCD solar surface photometer. Bright parts of photospheric network have contrast of several tenths of percent, and their contribution to the total irradiance is approximately half that of active region faculae. The solar irradiance variations estimated from sunspots, faculae and active network (contrast>0.3%) agreed with the ACRIM data. The quiet Sun irradiance used in the present results was different from the total irradiance at the solar minimum observed by the ACRIM, which indicates unmeasured components (contrast>0.1%) cause the 11-year cycle irradiance variation.

1. Introduction

The solar constant (the total solar irradiance, S) variations were detected by the active cavity radiometer irradiance monitor (ACRIM) experiment on board the *Solar Maximum Mission* (*SMM*) spacecraft (Willson and Hudson, 1981) and the Earth Radiation Budget (ERB) radiometer aboard *Nimbus 7* satellite (Hickey *et al.*, 1980). The influence of spots and faculae on S can be approximated by their areas and contrast values. Proxy methods (Bruning and LaBonte, 1983; Chapman and Boyden, 1986) roughly reproduced the variation of S with certain residuals relative to the ACRIM data. As expected by studies of LaBonte (1986), Foukal and Lean (1988), and Willson and Hudson (1988), network contributions to S was found to be present by precise broad-band solar surface photometry (Nishikawa, 1990, hereafter Paper I), in which the network contribution was shown at a half that of active region faculae. Photospheric network is really bright when detection limit of photometer becomes 0.1% in contrast (Paper I; Lin and Kuhn, 1993). It is now well known that S is modulated by sunspot, facular, and network, i.e. solar magnetic structures, activity. The details of the variation of S, however , is not unresolved yet. In order to understand the mechanism of the solar irradiance variations, spatially-resolved accurate surface photometric observations of the whole Sun with precise spacecraft observations of S are indispensable.

The CCD Solar Surface Photometer (CCDSSP), a precise surface photometric observing system for the whole Sun using two-dimensional CCD cameras, was developed in 1987 at the National Astronomical Observatory at Mitaka in Tokyo

Solar Physics 152: 125–130, 1994.

(Nishikawa, 1990, hereafter Paper II). The CCDSSP has a precision of 0.07% per pixel (7" square) in intensity measurements. The CCDSSP two-color observations showed several characteristics of network similar to faculae (Nishikawa and Hirayama, 1990 hereafter Paper III). In this paper, the cause of total irradiance variations are discussed based on results of Paper I, Paper II, and Paper III obtained from the CCDSSP observations.

2. Observations and Data Reduction

Surface photometry of the Sun was carried out using the CCDSSP for 28 days from December 1987 to May 1988 at the National Astronomical Observatory. The CCDSSP consists of dual optics using 2.5cm objectives (40cm focal length) with filters to make optimum brightness of solar images and with filter turrets to get flat field and dark frames. CCD video camera is put at each focal plane and the video signal is lead to an integrating video frame memory. Basically an averaged solar image was made from 44840 video frames of the Sun calibrating with 11210 dark and 11210 flat field frames obtained in a 236 min observation. The precision for each pixel (7" square) was 0.25% per minute and 0.07% per 236 min. The solar image data were treated as 160 macropixels square with about 14" resolution after binding 2 pixels square and regridding to make the solar position constant. A limb darkening subtraction with fourth-order polynomials was made for each solar image. Also a global pattern subtraction, which was made with median-filtered (a kind of low pass filter) residual image, was performed to get final residual intensity data. Areas outside 0.95 solar radius, which had significant deviations in brightness from the fitted limb darkening curve, were out of our considerations. Residual intensity is normalized by the total intensity of the quiet Sun. Detailed description of the CCDSSP and data reduction are in Paper II and Paper I.

3. Results

After all the calibration and reduction steps mentioned above and in Paper I and Paper II, we obtained residual intensity distributions (Fig.2 in Paper I). We can get each of sunspot, facular, and network contribution to solar irradiance by summing up the normalized residual intensity for dimmer elements, active region brighter parts, and total brighter points, separately (top of Figure 1). The active network contribution is the difference between the two brighter cases and is about half of the contributions from the

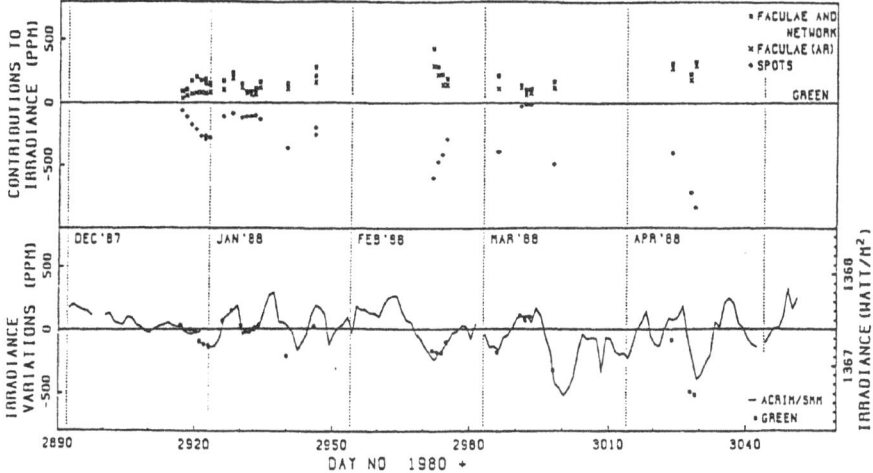

Fig. 1. The top diagram represents daily total contributions of negative (sunspots) and positive (facular plus network) components to the solar irradiance in green color. Active region facular contributions are separately plotted in the positive area. The bottom diagram shows the estimated irradiance variations in good agreement with the total irradiance values observed by the ACRIM (Willson, 1989), where the quiet sun irradiance value for the estimated variations is tuned to 1367.4 W m^{-2}.

active region faculae. Here points that have more than 0.3% contrast relative to the disk center intensity were used. The sum of the positive and negative contributions approximates the total irradiance variation and is plotted also in Figure 1 (bottom), along with the total irradiance measurements by ACRIM radiometer aboard the *SMM* (Willson, 1989). A variation of a few hundred ppm is seen in our daily estimated solar irradiance and this variation is in good agreement (r.m.s.= 20ppm for 20 data sets) with the variation of the observed irradiance value by the ACRIM. Here the offset level is assumed to be 1367.4 W m^{-2}.

4. Discussion

Patches of five–minute oscillations are seen over the whole Sun at an amplitude (or three times standard deviation) of about 0.5% in the green color (Nishikawa *et al.*, 1986; Paper I; Paper II). Since the size of oscillation patch, a few 10^4 km, is comparable to the network structure, we should pay attention to the intensity oscillation for studying the network contrast. The contamination of pressure mode oscillations should be reduced by averaging more than 100 image data taken over many periods of oscillations (or by integrating continuously more than 1.6 min which is derived under a sinusoidal assumption), when one want to get a precise solar intensity distribution with less than 0.05% residual amplitude of oscillation signal. Residual amplitude of intensity oscillation may be 0.03% for the best case in Paper I of a 236-image averaging, and 0.1% when 20 images are used for averaging. Smaller contrast pixels than the oscillation residuals after averaging have no reliability as steady intensity structures on the solar surface.

A quiet Sun level was determined by limb darkening subtraction with polynomials and by making a low-pass filtered intensity distribution. We cannot find any global intensity pattern of solar origin at 0.1% level from the quiet Sun image analysis. Scattered light measurement in two dimension and precise flat fielding may be needed to detect lower contrast global structures than 0.1%.

The irradiance variations shown in Figure 1 agreed with the ACRIM data. Here we have two free parameters. One is magnification of the variation, another is the quiet Sun irradiance value (about 1367 W m^{-2}) corresponding to unity for the surface photometric data.

The former parameter includes a bolometric correction factor, under estimation calibration factor caused by the extended point spread function, and so on. We are more interested in how the estimated variation is fit with the real total irradiance data by the ACRIM radiometer rather than the construction of the parameter. As in this case that the two variations are in good agreement, it is probable that the total irradiance variations in the observed period are caused by the counted elements which have larger contrast than 0.3% in green color, i.e. sunspots, faculae and network. If the counted elements are different from the real area of the cause of the total irradiance variations, estimated variations are probable to show different shapes. We may conclude that the total irradiance variations in a few months is driven by sunspots, faculae, and photospheric network which areas have larger contrast than 0.3% in green color.

The latter parameter, the quiet Sun total irradiance (constant), for the present observations was determined as 1367.4 W m^{-2} compared with the ACRIM data (Willson, 1989), which differs from the mean irradiance at the solar minimum of 1367.19 W m^{-2} (Willson and Hudson, 1988). This result indicates the 11-year cycle variation of S, which was shown to be about 0.1% peak-to-peak amplitude (Willson and Hudson, 1988), cannot be explained by the elements with more than 0.3% contrast. This offset discrepancy is about 0.015% of the total irradiance. Indeed there were many bright areas, network and surroundings of faculae, with less than 0.3% contrast (see Figure 5 in Paper I). Bright network elements with 0%–0.1% (average of 0.05%) contrast, however, can produce only 0.01% increase of irradiance, even if they occupy all the network area (assuming one fifth of the solar disk), which contrast is considered to be too low to contribute to the irradiance variations. Areas with 0.1%–0.3% (0.2% average) contrast can explain the offset discrepancy of 0.015% irradiance when they have 7.5% area of the solar disk. But the 0.2% contrast elements can radiate only 0.04% of S with 20% filling factor, which becomes only 40% of 11–year cycle increase of S. Actually we can find bright areas with 0.1%–0.4% contrast have strong contribution to S (see Figure 4 in Paper I). And we can read out mean contribution of 0.1%–0.3% elements in the observation period, however, is about 0.008% of S, which covers roughly half of the offset discrepancy. Quiet Sun subtraction error under low (about 14") spatial resolution may affect offset discrepancy. To the offset problem, under–estimation of facular emission may contribute, which caused by a difficulty to determine the quiet Sun level at large facular areas. According to the considerations above and in Paper I and Paper II, the cause of 11–year cycle variation of S might be not one but plural. Possibilities are in 0.1%–0.3% contrast network (may be more than half of the cause), large facular areas, global high temperature zone (or islands) (e.g. Kuhn, Libbrecht, and Dicke, 1988; Yoshimura, 1985), and a long–term radial pulsation of the Sun. Notice that the unmeasured contribution from facular regions outside 0.95 solar radius might affect the offset.

Network is composed of small facular bright points whose size is about 0."2 (e.g., Muller and Keil, 1983). Observations with 0."1 spatial resolution are not so easy, and the real structure of facular bright points, however, is not resolved yet. Resolved facular bright points in network and faculae were showed to have similar characteristics (Paper III). Whether facular and network areas are in self–balanced of radiation energy, i.e. have surrounding cool photosphere, or not is important for the estimation of irradiance variations. If the cool photosphere is extended, we make bright elements contributions to S over–estimated. More detailed observations are needed for facular and network energy balance problems.

Acknowledgements

The author is grateful to Professors T. Hirayama for helpful discussions and constant encouragements. Professor H. S. Hudson kindly provided the data of the ACRIM radiometer to the author, which were compared with the estimated irradiance variations. The data analysis was made by using FACOM M780 computer of the Astronomical Data Analysis and Calculation Center at the National Astronomical Observatory. Travel costs were partially defrayed by U.S.National Science Foundation Grant ATM-9224968.

References

Bruning, D. H. and Labonte, B. J.: 1983, *Astrophys. J.* **271**, 853.

Chapman, G. A. and Boyden, J.: 1986, *Astrophys. J. (Letters)* **302**, L71.

Foukal, P. and Lean, J. 1988,: *Astrophys. J.* **328** 347.

Hickey, J. R., Stove, L. L., Jacobwitz, H., Pellegrino, P., Maschoff, R. H., House, F., and Vonder Haar, T. H.: 1980a, *Science* **208**, 281.

Hirayama, T., Hamana, S., and Mizugaki, K.: 1985, *Solar Phys.* **99**, 43.

Kuhn, J. R., Libbrecht, K. G., and Dicke, R. H.: 1988, *Science* **242**, 908.

LaBonte, B. L.: 1986, *Astrophys. J. Suppl. Series* **62**, 229.

Lin, H. and Kuhn, J. R.: 1993, *Solar Phys.* **141**, 1.

Muller, R. and Keil, S.: 1983, *Solar Phys.* **87**, 243.

Nishikawa, J.: 1990a, *Astrophys. J.* **359**, 253. (Paper I)

Nishikawa, J.: 1990b, *Astrophys. J. Suppl. Series* **74**, 315. (Paper II)

Nishikawa, J., Hamana, S., Mizugaki, K., and Hirayama, T.: 1986, *Publ. Astron. Soc. Japan* **38**, 277.

Nishikawa, J. and Hirayama, T.: 1990, *Solar Phys.* **127**, 211. (Paper III)

Willson, R. C.: 1989, private communication.

Willson, R. C. and Hudson, H. S.: 1981, *Astrophys. J. (Letters)* **254**, L185.

Willson, R. C. and Hudson, H. S.: 1988, *Nature* **332**, 810.

Yoshimura, H.: 1985, *Publ. Astron. Soc. Japan* **37**, 171.

Latitude and Cycle Variations of the Photospheric Network

R. Muller and Th. Roudier

*Observatoire Midi-Pyrénées, URA 1281 Pic du Midi,
65200 Bagnères de Bigorre*

Abstract

New measurements of the number density of the Network Bright Points confirm the variation of the density of the photospheric network at the centre of the disk during the solar cycle and that in the period 1983-1985, this number density (i.e. the magnetic flux in the quiet sun) was maximum, both at the poles and at the equator.

1. Introduction

The irradiance of the Sun varies with time scales of days, months, and 11-years, directly related to active regions, active complexes, and the solar cycle of activity (Livingston et al., 1992 and references therein). Irradiance variability, as well as the solar cycle, are due to the interaction between convection and magnetic fields. Consequently, in order to understand this variability, it is important to know how the solar magnetic flux varies. The magnetic field manifests itself on the solar surface, at the photospheric level, as dark sunspots and bright faculae in active regions, and elsewhere as a cellular network, called the photospheric network.

Most observed irradiance variation of short period can be explained satisfactorily by active region modulation. It is, however, not clear "whether the additional solar-cycle variation is caused by global solar changes or by simply the enhanced emission from the bright regions" (the photospheric network), as questioned by Livingston et al., 1992.

It thus appears important to know the variation of the photospheric network for two reasons. First, for its contribution to the variation of the solar irradiance; second, for the amount of flux contained in the quiet sun and its variation, in order to understand the interaction between convection and magnetic field which is responsible of the solar cycle and the solar variability.

Solar Physics **152**: 131–137, 1994.
© 1994 *Kluwer Academic Publishers.*

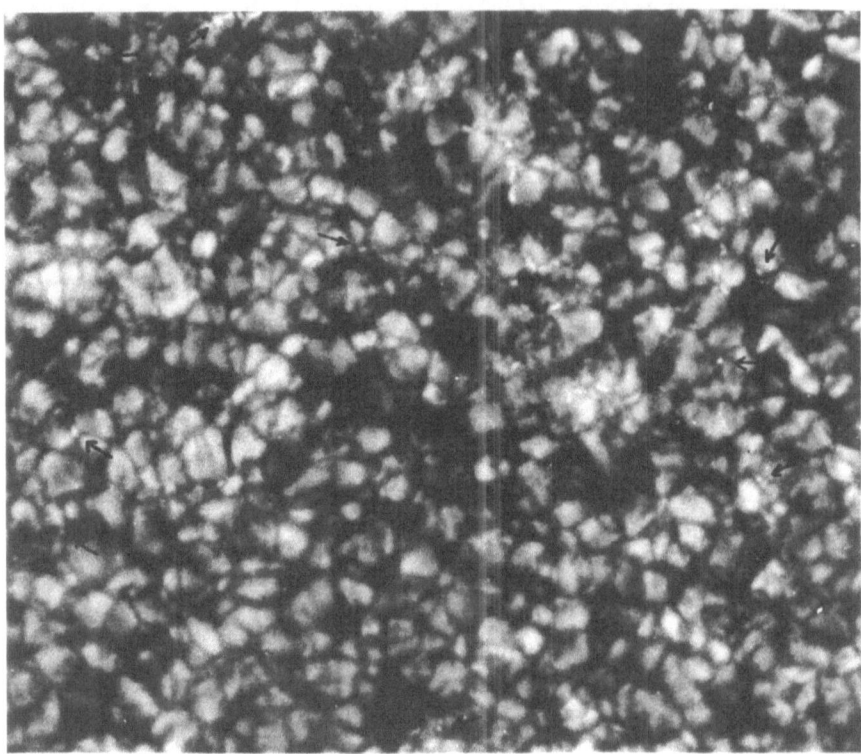

Figure 1 : Network Bright Points embedded in the granular pattern, observed through a 10 Å interference filter, centered at 4308 A, with the 50 cm refractor of the Pic du Midi Observatory. Many NBPs are clustered in a supergranule boundary, while others appear as isolated points (some shown by an arrows).

Magnetic flux is measured with full disk magnetographs, like those of Mount Wilson, Stanford or Kitt Peak. They provide daily magnetograms from which it is possible to derive the distribution of the flux over the solar surface, or the flux integrated over the whole Sun, and their variations with time (Rabin et al., 1992 and references therein). The spatial resolution of the observations is rather poor, not better than 1". In addition, they are integrated over a much larger area, when they are presented as synoptic maps (1° square, for example, in the case of the Kitt Peak synoptic magnetograms, Rabin et al. 1992). This spatial integration is required to improve sensitivity. The main disadvantage of such a magnetogram processing, is that the smallest magnetic elements are lost, especially in the quiet Sun, and some magnetic flux of opposite polarity is lost, too. Another weakness of magnetograms is that they do not provide a measure of the brightness of the magnetic features.

Magnetic features can also be observed on filtergrams, where they appear as dark (sunspots and pores) or bright (faculae and photospheric network) features. In the quiet Sun, the photospheric network is formed from very small (<0"5) bright points called Network Bright Points (NBPs). For this reason, high resolution observations are required. NBPs are visible either in white light filtergrams or in spectral line filtergrams, where they are easier to identify because they are much brighter than in white light. Our observations are made in the CN band at 4308 A, through an interference filter, 10 A wide (Figure 1). We use NBPs as indicators of magnetic elements, and because they have a small range of sizes and magnetic field strengths, we can assume that each of them carries a quantum of magnetic flux (2.5 x 10^{17} Mx for flux tubes of diameter 150 km and field strength 1500). Thus, counting the number of NBPs per surface unit is an indirect measurement of the solar magnetic flux (Muller and Roudier, 1984). Of course, in such a way, it is not possible to map the whole Sun. The observations are restricted to the disk centre and to various positions along the central meridian, and also along the equator for comparison. They are useful for the study of meridian and cycle variation. The contribution of the network to the solar luminosity and variance can also be derived. Although the resolution of our filtergrams is as high as 0"25, it must be noted that all the magnetic elements present at the surface of the Sun are not detected : those which are singificantly smaller than this limit and those which are darker than the average photosphere. The drawback to our method is that we have no direct measurement of magnetic fluxes and that we cannot map the whole Sun.

In our previous papers (Muller and Roudier, 1984; Muller, 1989) we have reported the following results :

- at the centre of the disk, the number of NBPs (that means the magnetic flux in the quiet Sun) varies in antiphase with the sunspot number, in the period 1975-1987, by a factor 3 or so ; surprisingly this variation is opposite to that shown in Livingston et al., 1992 ;

- the number density of NBPs (the magnetic flux) is not uniform along the central meridian: in 1983, it was maximum at the equator and at the poles.

In this paper we extend the observed variation at the disk centre, by a new measurement made in 1988 and we present a centre-to-limb variation for 1985, which will be compared to that of 1983. NBPs identification and correction for centre-to-limb loss of visibility has been described in Muller and Roudier 1984.

2. Variation at the Disk Centre for the Period 1975-1988

The measurements of 1988 confirm the decrease of the number of NBPs observed since 1983 (Figure 2). This figure shows that the number (it is equivalent to, say, the magnetic flux of the photospheric network) varies almost in antiphase with the sunspot number; it was a minimum in 1979, at the time of sunspot maximum, and a maximum in 1984, in between the Sunspot maximum of 1979 and the Sunspot minimum of 1986.

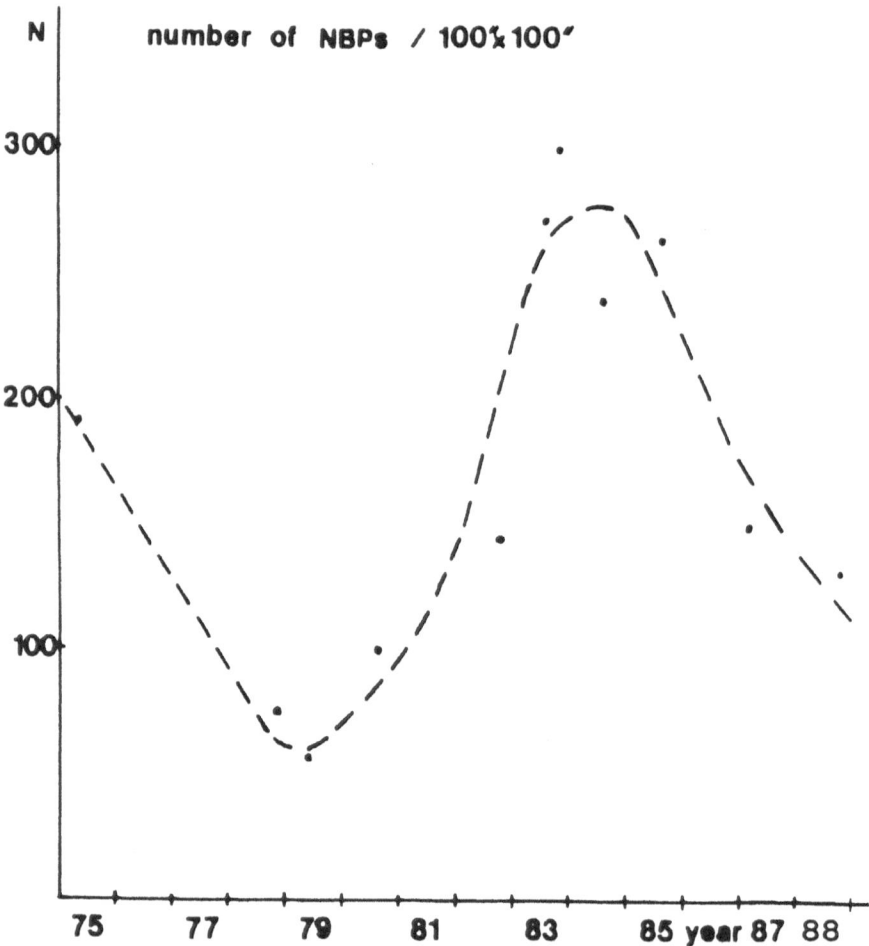

Figure 2 : Time variation of the number of Network Bright Points (per surface unit of 100" x 100") at the solar disk centre.

3. Latitude Distribution of NBPs : Comparison of 1983 and 1985

The latitude distributions show that NBPs are not uniformly distributed over the surface of the Sun. In 1983 and in 1985, the number was maximum at the centre of the disk and at the pole. In 1983, the number at the pole was very high, in agreement with Sheeley (1964) who reported that the number of polar faculae reaches a maximum value about 4 years after the maximum. It is interesting to note that the maximum of NBPs at the centre of the disk is also reached 4 years after the sunspot maximum (Figure 2).

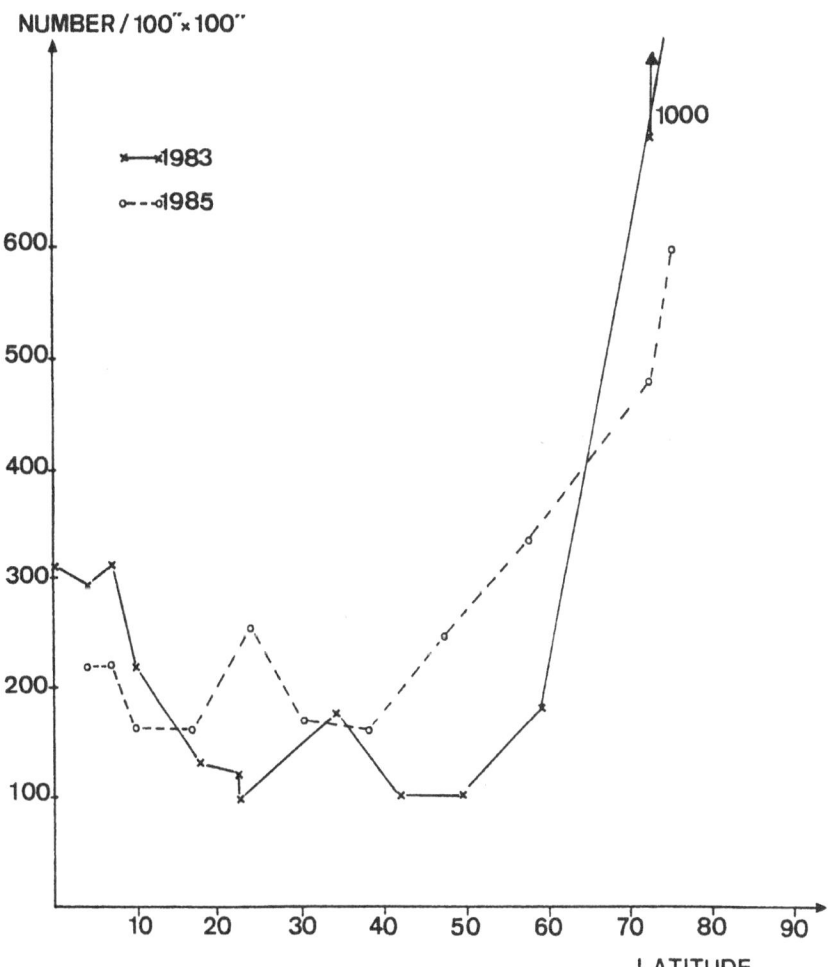

Figure 3 : Meridional variations of the number of Network Bright Points (per surface unit of 100" x 100").

The lower number in 1985, rather than in 1983, is also in agreement with the variation observed by Sheeley. Both distributions exhibit a small maximum at median latitudes (25° in 1985, 35° in 1983), just in between the sunspot latitude belt and the polar active latitude. This maximum must be confirmed, however.

4. Discussion

Our high resolution filtergrams allow us to observe the distribution of the magnetic elements at the surface of the Sun, away from active regions. As we cannot map the whole Sun, our observations are made along the central meridian; in the following discussion we assume that the distribution is homogenous in longitude. This is certainly true if we avoid active regions in the quiet Sun and the concentrations of magnetic elements found in remnants of old plages. The longitude homogeneity is verified by our observations along the equator (Muller and Roudier, 1984). Because of the very high resolution of the filtergrams (0"2), many more magnetic elements are identified than on classical magnetograms. Only the magnetic features smaller than the resolution and those which do not show up as bright features (Keller, 1992 ; Title et al., 1992) are not detected.

We have found that the magnetic flux of the quiet Sun was a maximum at the equator and at the poles from 1983 through 1985. It was a minimum at a latitude close to the sunspot belt. This minimum shifts from 35° to 25° during this period. A small secondary maximum visible in between the sunspot and high latitude belts must still be confirmed. The accumulation of magnetic flux both at the poles and at the equator, a few year after the solar maximum, certainly results from the diffusion of flux emerged in active regions in the sunspot belts. The accumulation at the poles correspond to the transformation of the toroidal magnetic field into a poloidal field.

For the moment we have only analysed the observations obtained during these two years, although we have observed the Sun regularly from 1983 until now. The complete results will be given in subsequent papers. At the centre of the disk we are able to present the analysis of a much more extended set of observations, from 1975 to 1988, covering more than one solar cycle. The result is a variation of large amplitude, almost in antiphase with the sunspot number. The maximum number of NBPs is at least three times larger than the minimum number, observed 4 years after a solar maximum. On the contrary, the Kitt Peak magnetograms show a variation of the magnetic flux of the quiet Sun, in phase with the sunspot number (Livingston et al., 1992 ; Harvey, this volume), while the Mt

Wilson magnetograms did not show any variation (La Bonte and Howard, 1982). The discrepency between the two kinds of variations can be understood if we keep in mind that, with our high resolution filtergrams we have observed the weakest network at the disk centre, excluding remnants of activity, while the Kitt Peak and Mt Wilson magnetic flux measurements map the whole Sun and include some remnants of active regions.

At the centre of the disk, there are about 150 points per surface unit 100" x 100" (Figure 3). Their average observed size is 0"33 = 230 km, and their average brightness is 1.08, relative to the mean photosphere. The filling factor is about 0.1 % (150 NBPs of diameter 0"33 pers surface unit of 100" x 100" correspond to a filling factor of 150 x $\pi/4$ 0"33 x 0"33 / 100" x 100" $\approx$ 0.1 %). If we assume, as a first approximation, that their distribution and brightness are uniform on the Sun, the excess of brightness of the photospheric network is 0.01 % of the luminosity of the Sun. The amplitude of the variation is about 100 points around the average value of 150, which means a variation of luminosity of about 0.01 % too. This is one order of magnitude smaller than the observed variation of irradiance measured by NIMBUS and ACRIM. In addition, the variation of the network excess is opposite to the variation of the solar irradiance, which is in phase with the activity cycle.

The results presented in this paper confirm that high resolution filtergrams provide a powerful tool for investigating the variability of the quiet Sun magnetic flux on the one hand, and the contribution of the photospheric network to the cycle variation of the solar irradiance, on the other hand.

References

Harvey, K. : 1994, in "The Sun as a Variable Star", IAU coll. no. 143, ed. J. Pap.

Keller, C.U. : 1992, Nature **359**, 307.

La Bonte, B.J. and Howard, R. : 1982, Solar Phys. **80**, 15.

Livingston, W., Donnelly, R.F., Grigoriev, V.G., Demidov, M.L., Lean, J., Steffen, M., White, D.R., and Willson, R.L. : 1992, in : "Solar Interior and Atmosphere", eds. A.N. Cox, W.C. Livingston, M.S. Matthews, p.1109.

Muller, R. : 1989, Adv. Space Res. Vol.8, N°7, p.159.

Muller, R. and Roudier, Th. : 1984, Solar Phys. **94**, 33.

Rabin, D.M., De Vore, C.R., Sheeley, N.R., Harvey, K.L., and Hoeksema, J.T. : 1992, in : "Solar Interior and Atmosphere", eds. A.N. Cox, W.C. Livingston, M.S. Matthews, p. 781.

Sheeley, N.R. : 1964, Astrophys. J. **140**, 731.

Title, A.M., Topka, K.P., Tarbell, T.D., Schmidt, W., Balke, C., and Scharmer, G. : 1992, Astrophys. J. **393**, 782.

Variability of the Solar Chromospheric Network Over the Solar Cycle

R.Kariyappa and K.R.Sivaraman

Indian Institute of Astrophysics, Bangalore 560 034, India

Abstract. From a large sample of the Kodaikanal spectroheliograms in the CaII K line we have studied the variations in the intensity of the network elements over two solar cycles and have estimated their contribution to the overall variability seen in the disc-averaged K line profiles. The relative contribution of the network elements and the bright points to the K-emission are of the order of 25% and 15% respectively. We have shown that the area of the network elements is anti-correlated with the solar activity, and it increases by about 24% during the solar minimum compared to the maximum period.

1. Introduction

The H and K lines of ionized Calcium have been recognized as useful indicators for identifying regions of chromospheric activity on the solar surface since the time when Hale and Ellerman (1904) and Deslanders (1910) first observed the bright reversals in these lines. A two-dimensional image of the Sun in the CaII H or K line under high spatial resolution shows that the three agencies responsible for CaII emission are :

i. Plages, which are most conspicuous by virtue of the emission that far exceeds the emission from other features and represent the active regions on the sun,

ii. the network elements which are co-spatial with the boundaries of supergranular cells in the underlying photospheric levels, and

iii. the bright points with dimensions of 1-2 arc sec that populate the interior of the supergranular cells.

One important point is that the spatial correlation between areas of chromospheric emission and the photospheric magnetic fields is well established, and this relation holds good right from plages down to features as small as a few arc-secs across (Babcock and Babcock, 1955; Leighton, 1959; Skumanich, Smythe and Frazier, 1975; Sivaraman and Livingston, 1982). From minimum to maximum of the 11-year solar cycle, the fraction of the solar disc covered by the plages increases, and there is a corresponding increase in the observed K emission from the full solar disc, i.e. for the sun observed as a star (White and Livingston, 1978; 1981; Sivaraman, Singh, Bagare and Gupta, 1987). During the peak of solar activity, plages occupy about 10% of the solar surface. Although this is only a fraction of the area occupied by all of the network elements put together, because of the higher intensity

Solar Physics **152**: 139–144, 1994.

---

contrast of the plages, their contribution to the total K-flux is very large compared with those of the remaining contributors. The area occupied by the plages over the visible solar surface varies in time and exhibits modulation of two different time scales. The first one is the 27-day modulation caused by solar rotation (Bappu and Sivaraman, 1971), and the second one is the well-known 11-year cycle modulation (Kuriyan, 1967). There is a less evident variation related to the smaller-scale chromospheric activity that occurs in the network structure bordering supergranule cells. Skumanich, Smythe and Frazier (1984) have derived a three-component model of chromospheric activity variation over the solar cycle using the observed data of White and Livingston (1981). Their model includes the flux from a cell component, a network component, and a plage component. They have estimated the fractional area covered by the cell points and their contrast and concluded that they cannot be important contributors to the K-flux. Pap (1992) has shown that the 72% of the solar-activity-related changes in Lyman alpha irradiance arise from plages and network elements and the network contribution is estimated by the correlation analysis to be about 19%, which is in good agreement with the photometric results of Pap, Marquette and Donnelly (1990b). The main contribution of the emission from the network elements to the K-flux, its temporal evolution in the intensity enhancements with the solar cycle, and the behaviour of the area of the CaII bright network with the solar activity are less documented and remain unanswered.

In this paper we made an attempt to answer these problems using the CaII K-spectroheliograms covering the period from 1957 to 1983.

2. The observations and the data analysis procedures

We have used in this study the Kodaikanal collection of CaII K-spectro heliograms obtained using the spectroheliograph. The K-spectroheliograph is a two-prism instrument with a dispersion of 7Å/mm near λ3934Å. The spectroheliograms employ a 60 mm image with a spectral window of 0.5 Å centered on K_{232}. The image scale is about 33 arc sec/mm. The span of years included is from 1957 to 1983, with plates clustering around dates of sunspot maxima, the intermediate and minima during this span. The main criteria used for selection of spectroheliograms were that the seeing be good to excellent; further that the region around the disc centre up to $\mu=0.8$ be free from active regions.

We have singled out the chromospheric network elements of dimension 10 mm x 5 mm at the disc centre on each plate and which correspond to 330 arc sec x 160 arc sec on the sun. Then we made the two dimensional scans of this region on all the plates using the PDS Microphotometer of Indian Institute of Astrophysics. The scanning aperture used is 50μ x 200μ, which corresponds to 1.5 arc sec x 6 arc sec on the sun. The digitized density values have been converted into relative intensities via the calibration curve following the standard photometric reduction procedure. We have derived the intensity plots for all the plates for the scanned regions. We have marked the network elements on the intensity plots (keeping the background intensity as a lower threshold value) after projecting the scanned region of the plate to the same size as the plots using a Carl Zeiss enlarger cum projector. From these plots we have estimated the residual intensity and the area of the network elements.

3. Results and Discussion

3.1. THE VARIATION OF INTENSITY ENHANCEMENT AT THE NETWORK ELEMENTS WITH THE SOLAR CYCLE AND THEIR CONTRIBUTION TO K-EMISSION FLUX

We have shown in Figure 1 the variation of the residual intensity of the network elements with time. The year 1975/1976 was the period of the lowest emission and can, therefore, be taken as the reference over which the excess in other years can be measured. Between 1957/58 and 1963/1964 the averaged network emission decreased by 64%, and remember that 1957/1958 was a most powerful solar maximum. Between 1969/1970 and 1975/1976, the averaged network emission decreased by 23%, while in 1979/1980 the increase to solar maximum was 16%. It is very clear from Figure 1 that the variation in the intensity enhancement at the network elements is related to solar activity. White and Livingston (1981) have measured the intensity variation of the quiet CaII network and supergranular cell centers observed near the equator and shown that the network elements do not participate in the solar cycle variation. In the present study we have measured the intensity enhancement only for the network elements without including the supergranular cell centers in it, and the network elements were selected for the measurements at the center of the solar disc on a quiet region. We find from Figure 1 that the network elements do participate in the solar cycle variation.

Pap et al. (1991) have shown that there is a good relationship between the variation of the Lyman alpha irradiance and the other solar indices like, for example, Ca plage index. Pap (1992) has also shown that the plage contribution to Lyman alpha irradiance is about 50%. Using this figure and also assuming that the remaining 50% contribution from the network elements, the inner network bright points and the weak emission seen between the bright points (background emission), we have derived the relative contribution from these structures from our network intensity plots to the total K-emission flux. The estimated quantities of the relative contribution to K-emission flux from various chromospheric structures for the years of the

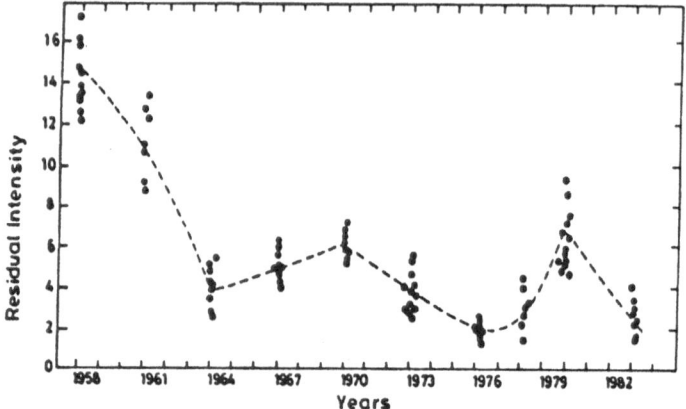

Fig. 1.   The cycle variation of residual intensity of network elements.

TABLE I
The relative contribution from the network elements, the bright points and the background emission to the K-flux

Year	Network (%)	Bright point (%)	Background (%)
1958	30	12	8
1964	23	18	9
1970	22	17	11
1976	23	15	12
1980	25	15	10

solar maximum and minimum are presented in Table I. We conclude from
Table I that the network elements and the bright points will contribute, respectively, an average of 25% and 15% to the K-emission flux. This network
contribution is in good agreement with the value (19%) estimated in Lyman
alpha by Pap (1992), and it also shows solar cycle related changes.

3.2. THE VARIATION OF THE AREA OF THE NETWORK ELEMENTS WITH THE SOLAR CYCLE

We have measured the area of the CaII bright network from the intensity
plots of the network using a planimeter. We have estimated the error in
the area measurements by repeating several times, and it comes to about
± 0.542 cm^2. Figure 2 shows the variation of the area of the CaII bright

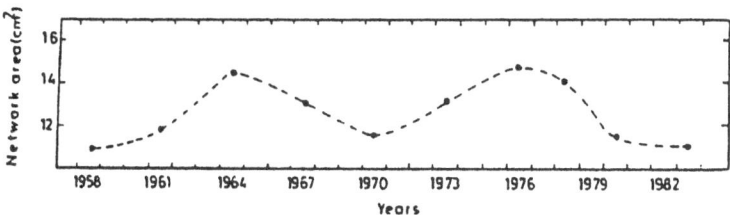

Fig. 2.   The cycle variation of area of the network elements. The image scale on the intensity plot is about 0.65 arc sec /mm.

network with time. It is clear that the area occupied by the network elements decreases with the increase of solar activity but it is not a constant quantity as shown by Raghavan (1983). Singh and Bappu (1981) have measured the supergranular cell size defined by the locus of peak intensity as a function of epoch, along with bright network, by the auto-correlation technique. They find that the cell size decreases from 22000km at solar minimum to 20000km at solar maximum. But Raghavan (1983) has shown that the average interior network cell size decreases with decreasing activity. In the present study, we have shown that the area of the network elements is anti-correlated with solar activity, and also we find that (Figure 2) there is an increase of 24% in the area of the CaII bright network during the solar minimum (1964) compared to the solar maximum (1958). Figure 2 shows that the area occupied by the network elements is larger at sunspot minima than during maxima. Since the magnetic enhancements in the upper photosphere are co-spatial with the emission seen in the network elements (Skumanich, Smythe and Frazier, 1975), this would imply that the magnetic fields occupying a larger area of the solar disc during sunspot minimum.

4. Conclusions

We find that the variation in the intensity enhancement at the network elements is related with solar activity. The estimated relative contributions from the network elements and the bright points to the K-emission flux are of the order of 25% and 15% respectively. We find that the area of the CaII bright network is anti-correlated with solar activity and increases by about 24% during solar minimum compared to solar maximum.

5. Acknowledgements

One of us (R. K.) wishes to express with pleasure his grateful thanks to Prof. Ramanath Cowsik, Director, Indian Institute of Astrophysics and to

Drs. Judit M. Pap & Richard F. Donnelly for providing the financial support to attend and present this paper at the IAU Colloquium No.143: The Sun as Variable Star, Solar and Stellar Irradiance Variations held at Boulder during June 20-25, 1993. R. K. also received support from NSF Grant ATM - 9224968. We thank an unknown referee for useful comments and suggestions.

References

Babcock, H.W. and Babcock, H.D.: 1955, *Astrophys.J.* 121, 349.
Bappu, M.K.V. and Sivaraman, K.R.: 1971, *Solar Phys.* 17, 316.
Deslandres, H.: 1910, *Ann.Obs.Astr.Phys.Paris* 4, 1.
Hale, G.E. and Ellerman, F.: 1904, *Astrophys.J.* 19, 41.
Kuriyan, P.P.: 1967, *Kodaikanal Obs.Bull.* No.172.
Leighton, R.B.: 1959, *Astrophys.J.* 130, 366.
Pap, J., Marquette, H.W. and Donnelly, R.F.: 1990b. *Adv.Space Res.* 11 (5), 271.
Pap, J., London, J. and Rottman, G.J.: 1991, *Astron.Astrophys.* 245, 648.
Pap, J.: 1992, *Astron.Astrophys.* 264, 249.
Raghavan, N.: 1983, *Solar Phys.* 89, 35.
Sivaraman, K.R. and Livingston, W.C.: 1982, *Solar Phys.* 80, 227.
Sivaraman, K.R., Singh, J., Bagare, S.P.and Gupta, S.S.: 1987, *Astrophys.J.* 313, 456.
Singh, J. and Bappu, M.K.V.: 1981, *Solar Phys.* 71, 161.
Skumanich, A., Smythe, C. and Frazier, E.N.: 1975, *Astrophys.J.* 200, 747.
White, O.R. and Livingston, W.C.: 1981, *Astrophys.J.* 249, 798.

A YOHKOH SEARCH FOR "BLACK-LIGHT FLARES"

LIDIA VAN DRIEL-GESZTELYI
Kiso Observatory, Institute of Astronomy, University of Tokyo, Japan

HUGH S. HUDSON
Institute for Astronomy, University of Hawaii, USA

BACHTIAR ANWAR*
Kwasan and Hida Observatories, University of Kyoto, Japan

and

EIJIRO HIEI
Meisei University, National Astronomical Observatory, Japan

Abstract. Calculations which predict that a phenomenon analogous to stellar negative pre–flares could also exist on the Sun were published by Hénoux *et al.* (1990), and
Aboudarham *et al.*, (1990), who showed that at the beginning of a solar white–light flare
(WLF) event an electron beam can cause a transient darkening before the WLF emission
starts, under certain conditions. They named this event a "black light flare" (BLF). Such
a BLF event should appear as diffuse dark patches lasting for about 20 seconds preceding
the WLF emission, which would coincide with intense and impulsive hard X-ray bursts.
The BLF location would be at (or in the vicinity of) the forthcoming bright patches.
Their predicted contrast depends on the position of the flare on the solar disc and on the
wavelength band of the observation.

The *Yohkoh* satellite provided white–light data from the aspect camera of the SXT
instrument (Tsuneta *et al.*, 1991), at 431 nm and with a typical image interval of 10–12 s.
We have studied nine white–light flares observed with this instrument, with X–ray class
larger than M6. We have found a few interesting episodes, but no unambiguous example
of the predicted BLF event. This study, although the best survey to date, was not ideal
from the observational point of view. We therefore encourage further searches. Successful
observations of this phenomenon on the Sun would greatly strengthen our knowledge of
the lower solar atmosphere and its effects on solar luminosity variations.

1. Introduction

In the last two decades there were several reports on flares of dMe stars which
displayed a kind of pre-flare event: short-lasting (tens of seconds up to 2–3
minutes) brightenings (so-called precursors) or diminutions (dips or negative
pre-flares) (Rodonò *et al.*, 1979; Cristaldi *et al.*, 1980). Sometimes they were
observed simultaneously: a brightening in the U-band and a dip at the red
end of the spectrum (Flesch and Olivier, 1974). Statistical investigations
showed that dips are more common in the near IR wavelengths (70 % of the
cases according to Bruevich *et al.*, 1980; about 40 % of the cases observed
at λ_{eff}=0.8 μ according to Grinin, 1983). On the other hand in the B-band,
precursors were found to be 2-3 times more common than dips (Cristaldi *et
al.*, 1980).

---

* Permanent address: National Institute of Aeronautics and Space, Bandung, Indonesia

Solar Physics **152**: 145–151, 1994.
© 1994 *Kluwer Academic Publishers.*

The most successful explanation of the pre-flare events was published by Grinin (1973, 1983), who explained the negative pre-flare events by an increase of H^- continuum absorption resulting from impulsive heating of the chromosphere by a downflow of material, which causes the optical flare when it reaches the photosphere.

In the solar case, negative flare (in integrated sunlight) have only been reported, as far as we know, in microwaves. At 10 cm the "negative burst" is a well-known phenomenon (e.g. Kundu, 1964) which is usually explained in terms of absorption by overlying material.

TABLE I
List of *Yohkoh* white–light flares

YYMMDD	Begin UT	Max UT	End UT	X Ray Class	Opt Imp	Loca- tion	NOAA Reg.
911024	2224	2241	2251	M9.8	1N	S12E46	6891
911027	0537	0548	0712	X6.1	3B	S13E15	6891
911110	2004	2013	2033	M7.9	1N	S15E43	6919
911115	2233	2239	2254	X1.5	3B	S13W19	6919
911203	1631	1639	1724	X2.2	2B	N17E72	6952
920126	1521	1533	1606	X1.0	3B	S16W66	7012
920214	2304	2310	2342	M7.0	2B	S13E02	7056
920708	0942	0950	1026	X1.2	1B	S11E46	7220
920716	1653	1700	1712	M6.8	2B	S10W61	7222

Hudson (1972) and Aboudarham and Hénoux (1986a,b; 1987) proposed that an electron beam can cause nonthermal ionization of hydrogen in the low chromosphere. Hénoux *et al.* (1990) and Aboudarham *et al.* (1990) proposed that this electron bombardment increases the electron number density, and through this the H^- number density. The enhanced electron number density produces an increase in emission by recombination, particularly in the upper layers; on the other hand the enhanced H^- number density increases the opacity particularly in the lower chromosphere. If the temperature decreases with height, this results in a decrease of the continuum intensity that will depend upon the limb center distance and the wavelength. During this "beam–ionized stage" of the flare, the theory shows that an electron energy flux (above 20 keV) of $F_1 = 10^{12}$ erg cm^{-2} s^{-1}, at a spectral index $\delta = 4$, produces a negative contrast at almost all wavelengths except a narrow ≈ 40 nm band at the Paschen limit (820.4 nm).

There seem to be similarities with the pre-flare event observations on dMe flare stars. The pre-flare dips are frequently observed in the near infrared (Bruevich *et al.*, 1980; Grinin, 1983) and occasionally simultaneous U-band

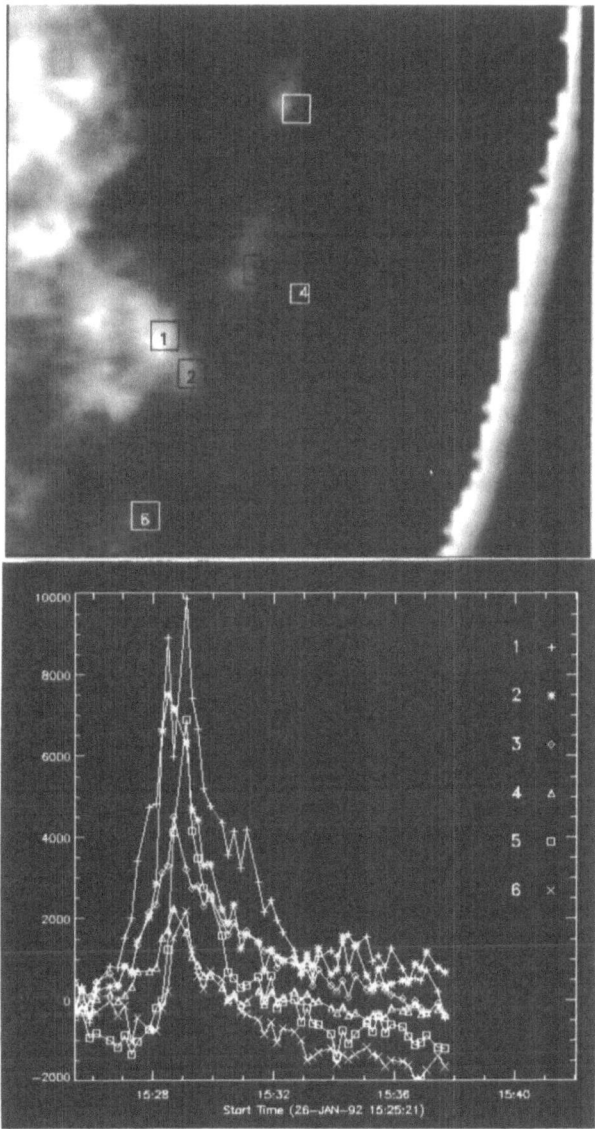

Fig. 1. The *Yohkoh* "spotless" white–light flare of 26 Jan. 1992. SXT aspect camera image
identifying the patches (upper panel) and their light curves. The latter exhibit impulsive
features that coincide precisely with different spikes of the hard X-ray burst (see Fig. 2),
and the suggestion of a BLF effect in patches 5 and 6. The maximum excess brightness of
the WLF was about 50% of the local photospheric brightness.

precursors and red negative preflares are observed (Flesch and Oliver, 1974). However, we note that the negative contrasts observed in stellar flares may exceed that possible with the BLF mechanism discussed here. According to Hénoux et al. (1990) and Aboudarham et al. (1990), the effect of the electron beam depends on its flux. An electron beam with a flux $F_1 = 10^{12}$ erg cm^{-2} s^{-1} producing a flare continuum observed at $\lambda=500$ nm and at disk centre will have a contrast of -5%, while for $F_1 = 10^{10}$ erg cm^{-2} s^{-1} the effect of the beam is negligible. The negative contrast will be reduced even in case of a beam with large flux if some heating occurs in the bombarded atmosphere (i.e. in the case of a slowly–rising electron flux).

BLF events are expected to be associated with intense and impulsive hard X-ray bursts. The BLF event could be observable only for the first 20 sec or so following the onset of the hard X-ray burst in the same place where the subsequent WLF emission appears.

2. Search for Black–Light Flares

It is not surprising that BLFs have remained undiscovered. Even WLFs, which last several minutes and have large contrasts (sometimes doubling the solar surface brightness) are often poorly recorded.Even WLFs, which last several minutes and have large contrasts (sometimes doubling the solar surface brightness) are often poorly recorded. This is even more true for the observations of the beginning stage of WLFs. In the WLF catalog by Neidig and Cliver (1983), among the listed 57 events only 26 were observed at onset.

The reasons for the poor observations of WLFs (and therefore BLFs) in-clude the occurrence of WLFs in regions of large image contrast (sunspots and faculae), the non–linear response of film recording, and of course see-ing. The Yohkoh data correct many of these problems. For example, we find that we can make very flat difference images, using time–wise cancellation (Uchida and Hudson, 1971) to reduce the effects of the background bright-ness variations.

The Yohkoh data come from a CCD camera viewing the Sun through an interference filter 431 nm wavelength, 3 nm bandpass, with 2.46 arcs pixel size. For each of the flares studied we have made a sequence of differ-ence images using a pre–flare image of the active region as a reference, and registering the individual images to this reference to a precision probably exceeding one arc s. This procedure eliminates the effects of sunspots from most of the images in the time series. In these difference time series, we have searched for BLF emission by plotting light curves of individual pixels or groupings of pixels.

We have analyzed nine white-light flares observed by Yohkoh (Table 1) in this way. In most cases we did not find any sign of pre–flare darkening. The

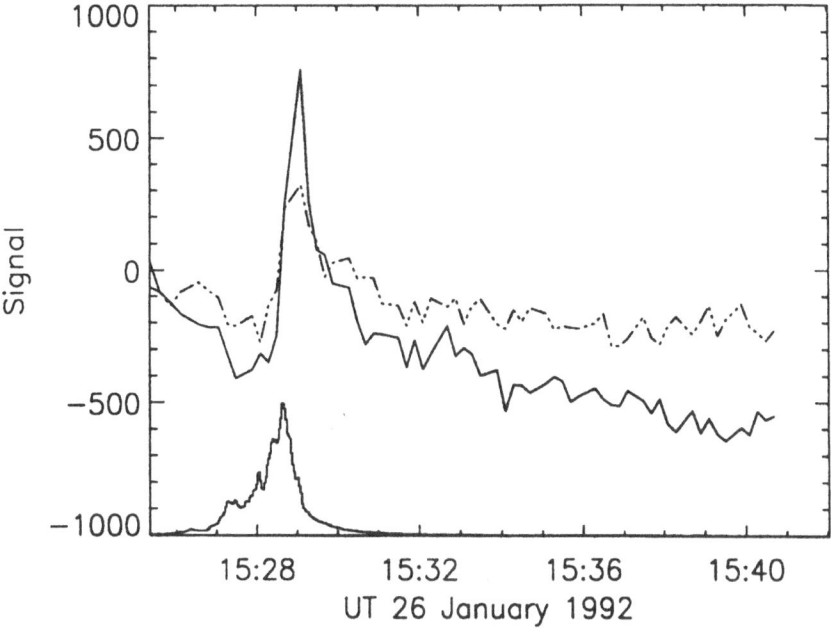

Fig. 2. Light curves for individual pixels in the most northerly (No. 5, cf. Fig. 1 - full line) and southerly (No. 6 - dashed line) footpoints of the "spotless" white-light flare of 26 Jan. 1992. The N footpoint is the clearest case we have seen of a possible negative signature, and matches reasonably well with the theory. The hard X-ray light curve (29–29 keV) shown (lowest curve) is for the entire flare; the bulk of the hard X-ray emission is associated with other white–light flare patches in the region (cf. Fig. 1). As noted in the text, however, this is not yet a definitive observation of the BLF phenomenon.

white–light intensity increased practically simultaneously with the hard X-ray emission (also observed by *Yohkoh*; see Hudson *et al.*, 1992). There was one flare which showed some tendency for pre–flare dips: the first "spotless" WLF of 26 Jan. 1992 (Hudson *et al.*, 1994). It was an X-class flare and it was observed close to the limb.

The best example we have found is shown in Figure 2, which shows the light curves of individual pixels at the most northern and southern patches of the 26 Jan. 1992 flare shown in Figure 1. In the best pixel the (negative) contrast exceeds 10%, compared with an rms fluctuation of about 5% and a peak excess (WLF) signal of about 50%. We note that these contrasts are exaggerated relative to ordinary WLF contrasts because we are observing at 431 nm and also near the limb. Incidentally, the separation of these two footpoints was at least 10^5 km, and the simultaneity of the two footpoints was closer than about 6 s (half the image interval); this implies an exciter speed in excess of 2×10^4 km/sec, explicable most easily with non-thermal electron energy transport.

3. Conclusions

We have used the *Yohkoh* white-light data, the first obtained from outside the Earth's atmosphere, to study white-light flares and in particular, to search for the predicted "black-light flares". In other respects, the *Yohkoh* data confirm the relationship between white-light continuum emission and the presence of high-energy electrons in the impulsive phase of a flare (Hudson *et al.*, 1992). We have found no unmistakable examples of the black-light flare phenomenon. Although Figure 1 and 2 are suggestive, we prefer to describe it in the words of Carrington (1859): "one swallow does not make a summer". The search continues as we develop better data-analysis tools and examine additional flares. Why have we not succeeded so far? Assuming the correctness of the theory, we suggest the following observational limitations:

(i) The data do not have high enough time resolution (typically 10-12 sec) to detect such a brief event.

(ii) The data do not have high enough spatial resolution (the *Yohkoh* aperture is 5 cm diameter, the pixel size 2.46 arc sec).

(iii) Other limiting noise sources exist, uncorrectable pointing jitter that converts spatial patterns into time-series noise, and background solar variability (*e.g.* p-modes). We note that the noise levels in the Figures, as inferred from the point-to-point fluctuation, greatly exceed the counting statistics, so that these systematic noise sources, (e.g. p-modes), dominate the search.

(iv) The *Yohkoh* flare trigger, which initiates the high time resolution, may occur late enough in the flare that pre-heating has occurred. According to the theory, this can quench the black-light flare phenomenon.

(v) In particular, the events observed by *Yohkoh* were not among the most energetic flares (the X6.1 of 27 October 1991 flare was poorly observed by *Yohkoh* from the point of view of data coverage, the high time-sequence started when the flare was in progress already).

We note that many of these limitations can be overcome with ground-based observations, which may indeed have better angular resolution and sampling than the *Yohkoh* data. The best searches will reduce the noise level to the photon statistics limit.

Acknowledgements

The authors would like to express their thanks to the SXT group, and especially to L. Acton and S. Tsuneta for their support of observations with the SXT aspect camera. The work of Hudson has been supported by NASA under contract NAS8-37334.

References

Aboudarham, J., Hénoux, J.C.: 1986, *Astron. Astrophys.* **168**, 301.

Aboudarham, J. and Hénoux, J.C.: 1987, *Astron.Astrophys.* **174**, 270.

Aboudarham, J. and Hénoux, J.C., Brown, J.C., van den Oord, G.H.J., van Driel-Gesztelyi, L., and Gerlei, O.: 1990, *Solar Phys.* **130**, 577.

Bruevich, V.V. et al.: 1980, *Izv. Krim. Astrofiz. Obs.* **61**, 90.

Carrington, R.C.: 1859, *Monthly Not. Roy. Astron. Soc.* **20**, 13.

Cristaldi, S., Gershberg, R.E., Rodonò, M.: 1980, *Astron. Astrophys.* **89**, 123.

Flesch, T.R. and Oliver, J.P.: 1974, *Astrophys. J.* **189**, L127.

Grinin, V.P.: 1973, *Izv. Krim. Astrofiz. Obs.* **60**, 179.

Grinin, V.P.: 1983, '' in in P.B. Byrne and M. Rodonò (eds.), ed(s)., *Activity in Red Dwarf Stars*, Reidel, Dordrecht, 613.

Hénoux, J.C., Aboudarham, J., Brown, J.C., van den Oord, G.H.J., van Driel-Gesztelyi, L., and Gerlei, O.: 1990, *Astron. Astrophys.* **233**, 577.

Hudson, H.S.: 1972, *Solar Phys.* **24**, 414.

Hudson, H.S., Acton, L.W., Hirayama, T., and Uchida, Y.: 1992, *Publ. Astron. Soc. Japan* **44**, L77.

Hudson, H.S., Strong, K.T., Bennis, B.R., Zarro, D., Inda, M., Kosugi, T., and Sakao, T.: 1993, *Astrophys. J. L.* , (submitted).

Hudson, H.S., van Driel-Gesztelyi, L., and Neidig, D.F.: 1994, , (in preparation).

Kundu, M.R.: 1964, *Solar Radio Astronomy*, New York: Interscience.

Neidig, D.F. and Cliver, E.W.: 1983, *AFGL-TR-83-0257 Enviromental Research Papers* , No. 856.

Rodonò, M., Pucillo, M., Sedmak, G., de Biase, G.A.: 1979, *Astron. Astrophys.* **76**, 242.

Uchida, Y., and Hudson, H.S.: 1971, *Solar Phys.* **26**, 414.

CORONAL INDEX OF SOLAR ACTIVITY: YEARS 1939-1963

M. RYBANSKÝ, V. RUŠIN, M. MINAROVJECH AND P. GAŠPAR

Astronomical Institute, Slovak Academy of Sciences
059 60 Tatranská Lomnica, The Slovak Republic

Abstract. A coronal index of solar activity (CI), based upon the total irradiance of the coronal 530.3 nm green line, has been constructed for the period 1939-1963 from observations at five stations worldwide. The monthly average CI exhibits a cyclic pattern with time, with succesive peaks monotonically increasing since solar cycle 18. Potential uses of the CI are discussed.

1. Introduction

The origin of solar activity is one of the basic problems of solar physics. In order to solve the problem we need to introduce a measure of solar activity - some index. One approach toward understanding solar activity involves studying the temporal relationship between different forms of activity; to do this, indices of the different types of solar activity would be useful.

A second reason for developing a solar activity index is its potential value in studying the solar influence on the interplanetary and terrestrial environment. This direction of research was stressed last, because some proof exists of a connection between solar activity level and the total irradiance of the Sun, e.g. Pap et al. (1991).

The basis for the establishment of different solar activity indices are observations of individual features of solar activity, e.g. sunspots, flares, prominences, radio flux, etc. Also the fluxes (emissions) from regions (lines) of the electromagnetic spectrum, from X-ray to radio flux, are introduced.

Each of such indices has its own advantage or disadvantage. As examples, first, the longest data set, the Wolf sunspot number (R), is determined from visible wavelength observations of the solar surface. The disadvantage of this index is that we do not know the substance of the sunspots; therefore, we do not know how to derive the physical influences of the index. Second the disadvantage of indices derived from measurements of radio flux of the Sun is the anisotropy of irradiance and conditions of its propagation.

A most natural index of solar activity– especially for solar-terrestrial study–would seem to be the flux of emissions which arises at some regions (lines) of the X or EUV part of the spectrum. However, these emissions do not pass through the Earth's atmosphere, and they may only be observed from satellites. At present, these observations may only be made over short time intervals. The hardest emission and the highest variation (amplitude) of flux is generally valid for these indices.

Some X-ray emissions from the solar corona have variable fluxes that are caused by the same processes which are responsible for the increase of coronal emission line intensities. This was the basic idea for creating a coronal index of solar activity (CI). The index is derived from ground-based observations (around the solar limb) of the 530.3 nm coronal line. This green solar corona is emitted by Fe XIV ions and is observed presently at only four coronal stations, according to international agreement. Interrupted data for the green corona have been available since 1939.

CI represents the total irradiance of the green corona into one sterradian towards Earth. An exact definition of CI and the method of its calculation are described by Rybanský (1975), Rybanský et al. (1993 and references therein). We note that CI data were published

Solar Physics **152**: 153–159, 1994.
© 1994 *Kluwer Academic Publishers.*

over the period 1964-1991, which means for the period of the existence of the Lomnický Peak coronal station.

In this paper the method of calculation and results for 1939-1963 are presented. The basic data were taken from Waldmeier (1951) and The Quarterly Bulletin on Solar Activity (1947-1963).

In the second section of the paper a method of establishing data homogeneity from different coronal stations is described and CI data are given for 1939-1991.

In the third section we discuss the final results of CI, its reliability, advantages and disadvantages, and possibilities for its use.

2. Data Homogeneity and Results

Green coronal intensities (Fe XIV, 530.3 nm) obtained from different coronal stations vary greatly among themselves. The reasons for such coronal data behavior are different and they were discussed in the paper by Rybanský et al. (1993).

In our former papers all the data were transformed to the Lomnický Peak photometric scale where every observation is calibrated to the Sun's center. We can then easily check all lags of observations and computations of CI.

Originally, in order to homogenize former data from different coronal stations to one photometric scale, we used the following method: with the aid of intensity comparisons from common observational days to transform data from individual coronal stations to the Pic du Midi photometric scale, because it is one of the oldest stations in the world. Then, data obtained in this manner can easily be transformed to the Lomnický Peak photometric scale using the relation

$$LS = 0.54 \times PM \tag{1}$$

where PM are intensities derived at Pic du Midi.

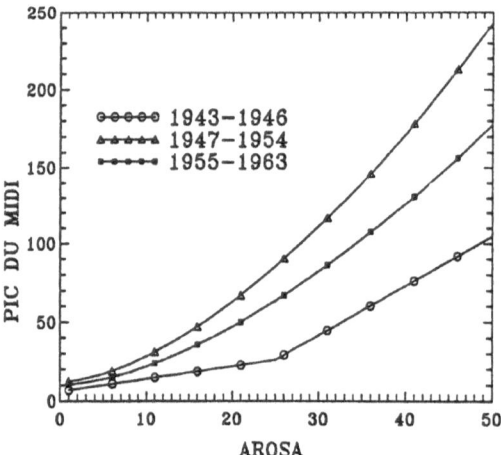

Fig. 1. The relationship between the Arosa and Pic du Midi coronal intensities for 1943-1963.

The relation (1) was obtained from a comparison of common observational days from

1965-1974 (Rybanský, 1979; Rybanský and Rušin, 1983), and we also intend to use it before the year 1964 for continuity.

This procedure, the comparison of data from common observational days, was done for every day and for each coronal station: Arosa (AR), Wendelstein (WS), Kanzelhoehe (KH) and Norikura (NO).

Apart from Norikura, where a big dispersion was observed, this procedure gave good results for the period 1955-1963. A big jump in the calibration is observed in the Pic du Midi data between measurements for the years 1954 and 1955. This claim was confirmed from the comparison between individual stations and Pic du Midi. A similar jump in the Pic du Midi photometry occurred again between measurements for the years 1946 and 1947. To prove that calibration again changed at the Pic du Midi is problematic because we have had for our disposal only data from two stations: Pic du Midi and Arosa.

Because the most stable results over this period were found at the Arosa coronal station, we decided to transform data from all other coronal stations for the period 1939-1954 to the Arosa photometric scale, and then transform that data to the Lomnický Peak photometric scale, according to the relation

$$LS = 5.4 + 0.143 \times AR^{1.65}. \tag{2}$$

This relation was obtained by comparison of data from common observational days between Arosa and Pic du Midi for the period 1955-1963 and the relation (1). Results of the comparison, together with approximate curves are shown in Figure 1.

Our methods are supported by comparison of our relation curve with that of Waldmeier (1951) which transformed relative units into absolute coronal units, and we obtained nearly the same values.

Using this process, a homogeneous data set was developed for the period 1939-1963, and missing data for days without observations were interpolated. Number of observational days is shown in Table 1. There are indicated interpolated data (IP) in Table 1.

The course of CI monthly means over the entire period (1939-1991) is shown in Figure 2. As follows from Table 1, the CI daily data over the period 1939-1944 have mostly only a formal character and meaning because of the small number of observational days. Nevertheless, some of the monthly means, where observations were available (1939: February, March; 1940: January, March, July and October; 1941: January - April, and July-September; 1942: January -April, and July - November; 1943: January - March, and August - October; 1944: January - March, May, and September - October), are true.

3. Discussion and Conclusion

One may see in Figure 2 that the temporal variations of CI have standard characteristics of a solar activity index. The 11-year, long-term cycle, intermediate-term (approximately semi-annual) and 27-daily periodicity due to the solar rotation are clearly seen. Other intermediate-term variabilities in the period 1939-1963 have not been studied, if they even exist.

The results obtained are given in units of W/sr. More frequently, especially in emissions detected from satellites, the emitted output is expressed in [W/m²], which means power counted in 1 m² at the distance of 1 AU, or in the number of photons crossing 1 cm² per second [photons/cm²/s] at the distance of 1 AU. To convert CI data into similar units,

M. RYBANSKÝ ET AL.

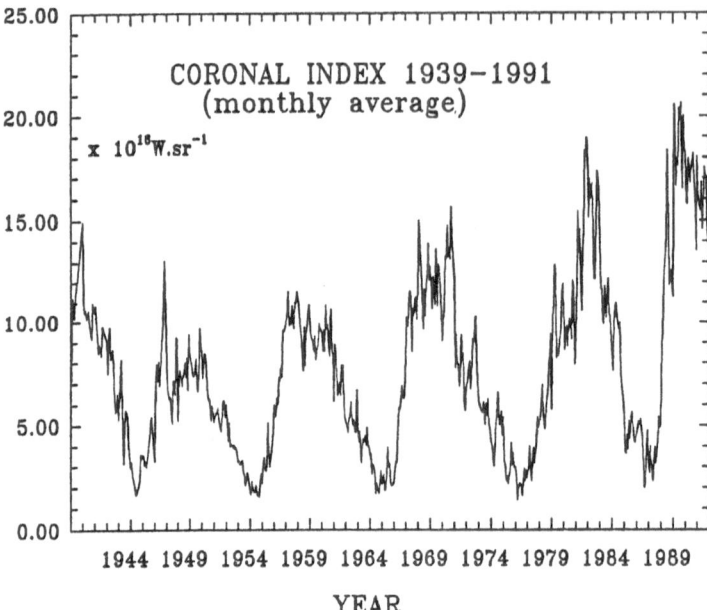

Fig. 2.   The course of CI monthly mean data from 1939-1991.

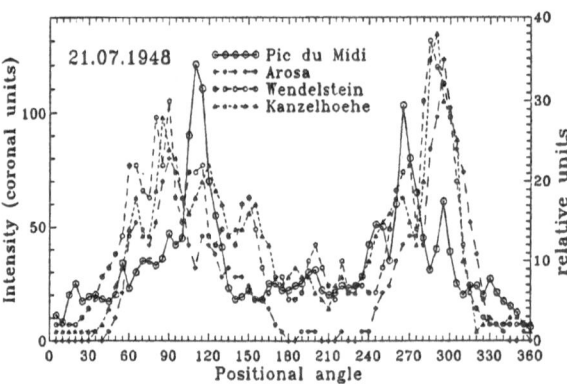

Fig. 3.   Comparison of daily green corona variations between individual coronal stations.

the following relation should be used: 1×10^{16} W/sr $\equiv 4.5 \times 10^{-7}$ W/m$^2 \equiv 1.2 \times 10^8$ photons/cm^2/s.

The amplitude of CI varies roughly from 1×10^{16} to 20×10^{16} W/sr, which means from 1.2×10^8 to 2.4×10^9 photons/cm^2/s. We note that according to Barth et al. (1990), the amplitude of Lyman-alpha (121.6 nm) irradiance varies from 2.5×10^{11} to 4.0×10^{11} photons/cm^2/s over the period 1982 - 1988, this means that the amplitude of change for

TABLE I
Numbers of observational days used for CI computation

year/Station	AR	PM	WS	KH	NO	IP
1939	24					341
1940	29					337
1941	60					300
1942	80					285
1943	82	15				263
1944	54	12				300
1945	91	34				240
1946	88	40				238
1947	44	66	61			194
1948	75	85	57	24		125
1949	78	89	52	23		123
1950	48	119	53	21		124
1951	42	93	59	12	45	114
1952	31	73	51	8	59	144
1953	41	113	45	4	39	123
1954	28	96	29	3	45	126
1955	32	97	47	10	53	126
1956	26	152	37	18	39	94
1957	27	129	30	31	50	98
1958	30	116	40	36	44	99
1959	32	119	48	34	52	80
1960	27	91	25	29	57	137
1961	18	136	52	24	22	113
1962	33	110	34	20	41	127
1963	24	101	40	5	31	164

CI is 12.5 times higher than for Lyman-alpha.

The trustworthiness of the CI data depends on the accuracy of the intensity measurements of the green corona, which we are only able to estimate. From 1965-1990, when a photographic method of photometry was used, one may estimate a 20 % error. Since 1991, when photoelectric measurements were introduced at the Lomnický Peak coronal station (Minarovjech and Rybanský, 1992), the accuracy is higher, within 5 % . To estimate the accuracy before 1965 is very difficult. This was discussed in Section 2. From Figure 3 one may estimate that errors should reach 80-100 %.

It is quite possible, that the disagreement between CI, Wolf number and the 2800 MHz radio flux (importance of their values) in cycle 19 could be explained by this effect. On the other hand, such an explanation must be rejected, because the transform relation for the homogeneous data are the same for both cycles 19 and 20. High, unknown variability between solar activity indices was found in cycles 19 and 20 between faculae area and the 2800 MHz radio flux or plage index (Lean, 1987).

As stated previously, CI data may be used to study the influence of solar activity in the solar corona, and/or possible influence of this activity on terrestrial processes.

The advantages of CI are:

a) length of the time sequence (53 years to date)

b) relationship with the X-ray flux (in regards to the mechanism of their origin)

c) possibility to study variability of solar activity not only around the solar equator, but around the entire Sun.

In connection with the last conclusion, it is quite possible, using the same assumptions, to derive a picture of intensity distribution in front of the Sun (from limb observations). Here a comparison between the green coronal chart and X-rays from the Sun (Vaiana et al., 1969) on May 20, 1966 is shown in Figure 4. The agreement between the images is good, on the large scale. Clearly, no rapid changes have occurred between the time of the image and the period of limb data acquisition that altered the large scale coronal forms then present.

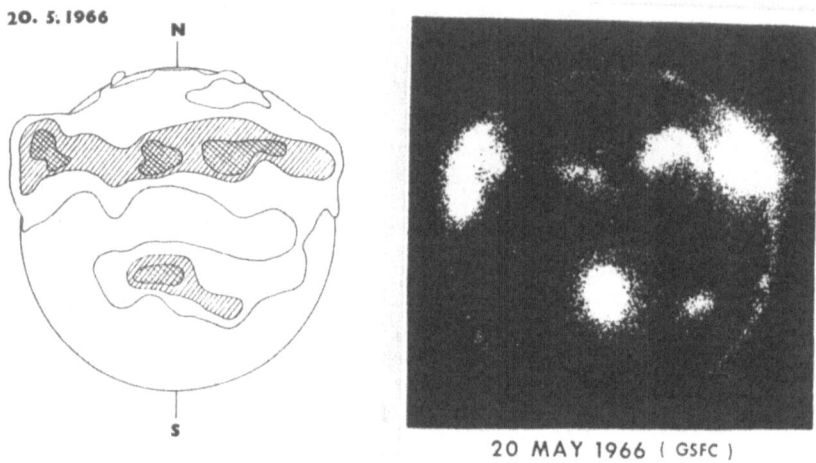

Fig. 4.   The green corona emissivity distribution in front of the Sun (left) and the X-ray image of the Sun (right), May 20, 1966.

The disadvantages of CI are:

a) uncertainity at limb data extrapolation onto the solar disk

b) long-term gaps in data, especially at the beginning of the period searched

c) errors which result from different methods of observation at different coronal stations.

Acknowledgements

The authors are grateful to the Slovak Grant Agency for Science (Grant No. 59/1993) for partial support of this work. Travel costs (V.R.) were partially defrayed by U. S. National Science Foundation Grant ATM-9224968 and SOLERS 22 officials. The authors wish to acknowledge all known and unknown observers at individual coronal stations around the world for their data without which this paper would not have been possible. We are grateful also to our collegaues Mr. R. Mačura, Mr. K. Maník and Mr. L. Scheirich for preparing the input data for CI computation and to the unknown referee for improving this paper.

References

Barth, C. A., Tobiska, W. K. and Rottman, G. J.: 1990, *Geophys. Res. Lett.* **17**, 751.

Lean, J.: 1965, *J. Geophys. Res.* **92**, 839.
Pap, J., London, J. and Rottman, G. J.: 1984, *Astron. Astropys.* **245**, 648.
Minarovjech, M. and Rybanský, M.: 1992, *Solar Phys.* **139**, 1.
Rybanský, M.: 1975, *Bull. Astron. Inst. Czech.* **26**, 367.
Rybanský, M.: 1975, *Bull. Astron. Inst. Czech.* **26**, 374.
Rybanský, M., Rušin, V., Gašpar, P. and Altrock, R.C.: 1993, *Solar Phys.* in press.,
Vaiana, G. S. and Giaconi, R.: 1991, 'Plasma Instabilities in Astrophysics' in D. G. Wentzel and D. A. Tidman, ed(s)., *Gordon and Breach, Science Publishers*, New York, 91.
Waldmeier, M.: 1951, *Die Sonnenkorona*, Verlag: Birkhauser, Basel.

ROTATIONAL CHARACTERISTICS OF THE GREEN

SOLAR CORONA : 1964-1989

J. RYBÁK

Astronomical Institute, Slovak Academy of Sciences
059 60 Tatranská Lomnica, The Slovak Republic

Abstract. Fe XIV 5303 Å coronal emission line observations have been used for the estimation of the rotation behaviour of the green solar corona. A homogeneous data set, created from measurements carried out within the framework of the world-wide coronagraphic network, has been examined with a correlation analysis to reveal the averaged synodic rotation period as a function of latitude and time over the epoch from 1964 to 1989.

The values of the synodic rotation period obtained for the epoch 1964-1989 for the whole range of latitudes and for a latitude band $\pm 30^0$ are 28.18±0.12 days and 27.65±0.13 days, respectively. The differential rotation of the green solar corona was confirmed, together with local maxima of the rotation period at latitudes 45^0 and -60^0 and a minimum at the equator, but no clear cyclic variation of the rotation has been found for the epoch examined.

1. Introduction

The differential rotation of different solar atmospheric layers and features has been demostrated both by spectroscopic and tracer methods. This has been done also for other parts of the solar corona and for some coronal structures. Also the solar emission-line corona has been studied with the aim of determining the synodic rotation period on the basis of Fe XIV 5303 Å line measurements from patrol observations.

Although we know some limitations of the technique used in this study and a criticism of this type of data for the rotation period determinations (Newkirk, 1967), we have decided to continue in this field, opened by the paper of Antonucci and Svalgaard (1974) dealing with the epoch 1947-1970, and then greatly broadened by the paper of Sime, Fisher and Altrock (1989), which covers the period from 1974 to 1985 (paper SFA).

The most important aim of this paper is to extend the epoch investigated in paper SFA with the same quality of rotation period results obtained there. In addition, the overlapping period of SFA data and our data is used to check the possible effects on the synodic rotation period determination due to the data homogenisation from different observing instruments to a common photometric scale.

2. Data

'The Homogeneous Data Set of Coronal Green Line Intensities over the Period 1964-1990', published by Rybanský and Rušin (1992), has been used in the present study. The homogeneous data set (HDS) was created following the method of Rybanský (1975), on the basis of measurements from different coronal stations, which have been reduced to the Lomnický Peak photometric scale and a height of 40" above Sun's limb.

The data are calibrated on an absolute scale and expressed in absolute coronal units (1 ACU = an equivalent width of 10^{-3} mÅ of the solar disk center continuum at nearly the same wavelength as the green line).

The data over the period 1964-1989 used for the rotation period determination represent 36 years of daily records, each of 72 data points with a position angle lag of 5°, where days with no available measurements have been interpolated. Values are integer numbers, very rarely exceeding the level of 250.

3. Methodology

Due to our intention to evaluate the effects of the data homogenisation on the rotation period determination, we have decided, as a first step of the period estimation computation, to follow *precisely* the method used in the SFA paper. Many possible influences could take place in the homogenisation procedure — different coronal stations have used not only different photometric procedures but also different heights of measurements over the Sun's limb, e.g. Lomnický Peak - 40" and Sacramento Peak - 140". This is the main reason why we have selected the same calculation procedure and not any other numerical method, which would add also its own effects.

We have followed the following main steps of the data evaluation in order to estimate the synodic rotation period in selected time and latitude intervals of the green coronal line measurements :

− Only the east limb data have been used.
− The data have been divided into yearly intervals and averaged into latitude bins of 15° degrees extention, centred on latitudes ±75°, ±60°, ±45°, ±30°, ±15° and 0°.
− For each yearly latitude bin an extention of Pearson's correlation coefficient (Press et al., 1986) - the autocorrelation function (ACF) $r(k)$ - was computed as

$$r(k) = \frac{\sum_{i=1}^{N}(x_i - \bar{x})(x_{i+k} - \bar{x})}{\sqrt{\sum_{i=1}^{N}(x_i - \bar{x})^2} \cdot \sqrt{\sum_{i=1}^{N}(x_{i+k} - \bar{x})^2}} \tag{1}$$

where $\bar{x} = \sum_{i=1}^{N} x_i/N$ is the mean of the selected coronal line data, x, of length N=365 (366) days. Parameter k is the lag of the data and ACF was calculated for intervals from 0 to 90 days.

− The local maximum of each ACF was selected in the interval from 25 to 35 days and then the center of mass of the ACF was estimated in this interval.
− Following the papers of Hansen, Hansen and Loomis (1969), Fisher and Sime (1984) and the SFA paper, the above specified estimation of the ACF center of mass was taken to be the synodic rotation period, which is based on the assumption that in all data sets of every year and latitude bin the green coronal emission line structures 'live' longer than at least 28-32 days. This has then a consequence of a local maximum in the ACF $f(k)$.
− The rotation period latitude and temporal averages together with their standard deviations have been computed for the whole epoch as well as for the common epoch of the SFA paper and the IIDS data : 1975-1985.

4. Results

Using the IIDS data, the average synodic rotation period of the green solar corona has been calculated for the epoch 1964-1989 for the whole range of latitudes and the latitude band (±30°), and they are 28.18±0.12 days and 27.65±0.13 days, respectively, with errors of one standard deviation quoted.

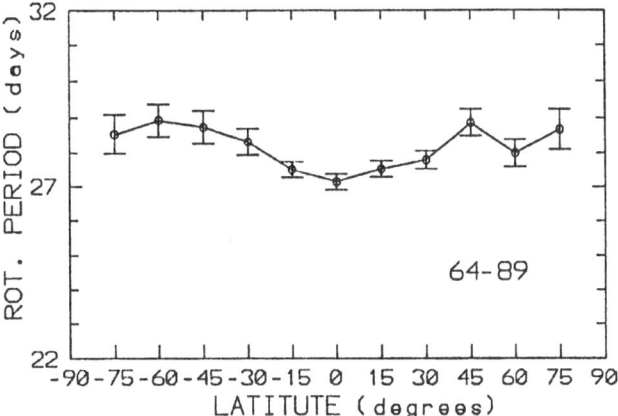

Fig. 1.   The time-averaged synodic rotation period behaviour for the epoch 1964-1989 as a function of latitude. The error bars displayed are one standard deviations of the mean.

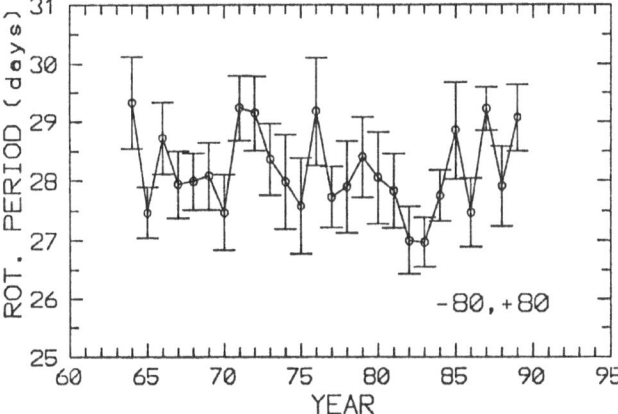

Fig. 2.   The values of the yearly averages of the synodic rotation period over the whole interval of latitudes for the time interval 1964-1989 with the corresponding errors.

All other results, obtained from the calculated periods are presented in form of figures, where error bars indicate one standard deviation from the mean.

'4.1. DIFFERENTIAL ROTATION

Figure 1 shows the synodic rotation period separately for latitude bands from -75° to 75° (but they contain information from the range ±80°), when the time-averaging over the whole epoch is applied. The curve is relatively very flat compared to those obtained from other solar atmosphere layers, but the differential rotation of the green corona is clearly shown. The lowest value of thr period is at the equator (just over 27 days) and the highest values are at -60° and at 45° (less than 29 days). Only one value of the rotation period for the latitude bin at 75° does not follow the typical behaviour.

J. RYBÁK

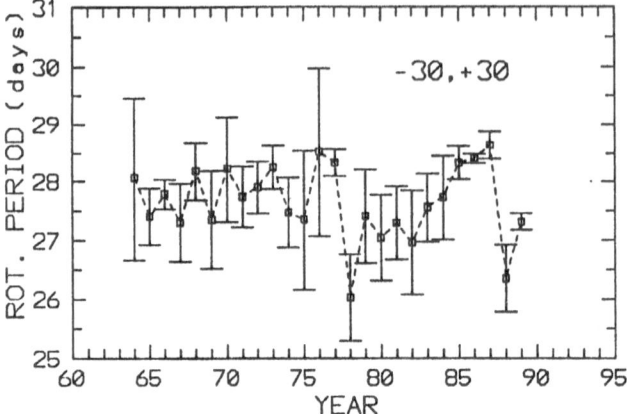

Fig. 3. The values of the yearly averages of the synodic rotation period for the equatorial latitude interval
(latitudes ±30°) for the time interval 1964-1989 with the corresponding errors.

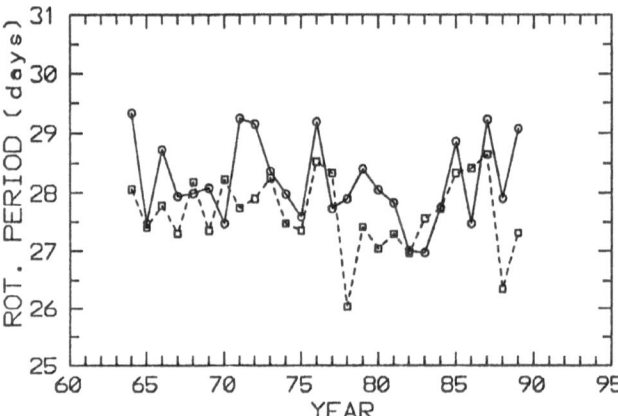

Fig. 4. The behaviour of the yearly averages of the synodic rotation period for the whole east-limb interval
(solid line) and for the equatorial interval of latitudes ±30° (dashed line) for the time interval 1964-1989
without error bars from Figures 2 and 3.

4.2. TEMPORAL VARIATIONS

The temporal behaviour of the rotation period from 1964 to 1989 is presented in the following figures, where averaged values over all latitude intervals are plotted in Figure 2
and averaged values from the latitude band ±30° are shown in Figure 3. In Figure 4 the
behaviour of the above mentioned averages is repeated, but without error bars.

Changes of amplitude are at a level of at about 2 days for both data sets. The approxi-
matively very similar behaviour of the results in both intervals is seen to be different only
in the years 1971-1972 and 1978.

A comparision of the data in Figure 4 shows that the equatorial band is rotating with
a very slightly lower value of the synodic period than the all-latitude average from almost

the whole limb from pole to pole (including the equatorial band).

As for any cyclic variation of the rotation period, it seems that in the case of the errors obtained and the scatter of the values, it is impossible to point out any clear variation of the rotation period with the solar cycle on the basis of the data used. Neverless, other estimations with the help of other methods should be used in the future.

5. Comparision with SFA Results

There is a long data overlap between the IIDS, which covers the epoch 1964-1989, and the Sacramento Peak data, presented in the SFA paper from 1974 to 1985, of the whole volume of the second ones. Even the constuction of the latitude bands 15° wide can be done correctly for both data sets.

A direct comparision of corresponding values of the synodic rotation period for all yearly averages in each latitude band has been done, and the following results have been obtained :
- A relatively larger interval of IISD results was studied than in the SFA results.
- The time-averaged synodic rotation period behaviour as a function of latitude from the SFA paper and from HDS are of the same shape with systematically greater periods of the latter one (approximately a 0.4 day difference).
- 10 from 11 above-mentioned, time-averaged periods show overlapping intervals of the corresponding 1σ errors with SFA and our results.
- Values of the yearly averages of the synodic rotation period over the whole interval of latitudes from the SFA paper and computed from HDS are very similar, with only 2 yearly averages (1976 and 1985) without overlapping of the corresponding 1σ errors.

6. Conclusions

On the basis of this first attempt to use the 'Homogeneous Data Set of Coronal Green Line Intensities over the Period 1964-1990' for the purpose of the synodic rotation period determination, it can be concluded that this data set can be used for such types of work.

The results obtained show good consistency with periods obtained from 'implicitly' homogeneous measurements from one instument, which increases the weight of our results for the periods 1964-1974 and 1986-1989.

Nevertheless, no clear cyclic variation of the synodic rotation period has been revealed. But, on the other hand, the behaviour and the amplitude of the differential rotation can be seen very well.

For more precise determinations of the rotation period values it would be helpful to test the numerical stability of the results obtained with modifications of the ACF method, and also to check the values from the ACF method with some other methods.

Acknowledgements

The author wishes to express his deepest thanks to the SOC of the colloquium for an exemption to present this paper as a 'missing author'. The author is thankful to Dr. V. Rušin for his comments and suggestions, as well as for the presentation of this paper at the colloquium. Thanks are due to dr. R. Howard for his kind and very helpful editorial work.

Finally, the author would like to thank all who collected, reduced and typed the huge data volume at all coronal stations around the world over the period of many years, which has been used in this study.

This work has been supported under grant GA 506/1993 by the Slovak Academy of Sciences.

References

Antonucci, E. and Svalgaard, L.: 1974, *Solar Phys.* **34**, 3.
Fisher, R. and Sime, D.G.: 1984, *Astropys.J.* **287**, 959.
Hansen, R.T., Hansen, S.F. and Loomis, H.G.: 1969, *Solar Phys.* **10**, 135.
Newkirk, G.: 1967, *Ann. Rev. Astron. Astrophys.* **5**, 213.
Press W.H., Flannery B.P., Teukolsky S.A. and Vetterling W.T.: 1986, *Numerical Recipes : The Art of Scientific Computing*, Cambridge University Press, Cambridge, 484.
Rybanský, M.: 1975, *Bull.Astron.Inst.Czech.* **26**, 374.
Rybanský. M. and Rušin, V.: 1992, *Contrib.Astron.Obs.Skalnaté Pleso* **22**, 229.
Sime, D.G., Fisher, R.R. and Altrock, R.C.: 1989, *Astrophys.J.* **336**, 454 (paper SFA).

SOLAR CYCLE VARIATION OF THE MICROWAVE
SPECTRUM AND TOTAL IRRADIANCE

E.J. SCHMAHL AND M.R. KUNDU

Astronomy Department

Univ. of Maryland

Abstract. We have extended the proxy relationship between irradiance and microwaves by using the daily solar fluxes from Toyokawa Observatory at 1000, 2000, 3750 and 9400 MHz in addition to the Ottawa 2800 MHz flux for the years 1980-1989. It turns out that the flux at 1000 MHz is better correlated with irradiance than the flux at higher frequencies–an unexpected result. We have also found that the spectrum of the flux shows shape changes that are related to the number and type of active regions. Because of this the five-frequency spectral measurements of microwave flux allow one to separate the sunspot and coronal features, providing an improved proxy of solar variability.

1. Introduction

The Active Cavity Radiometer Irradiance Monitor (ACRIM) aboard the Solar Maximum Mission (SMM) operated almost continuously from February 1980 to November 1989 and obtained irradiance data with absolute accuracy of the order of 0.2 percent. These observations have demonstrated that the solar irradiance varies on time scales of days to years, with opposite contributions from spots and faculae, and possible contributions from the magnetic network (Fröhlich *et al.*, 1991).

It is well known that the daily 2800 MHz microwave flux is correlated with the sunspot number and other measures of solar activity, and with the solar irradiance. This correlation arises simply because the microwave emission is generated in the same chromospheric and coronal magnetic flux tubes which are rooted in the sunspots, plage and faculae seen in visible light (Kundu, 1965; Tapping and deTracy, 1990). Since the spectra vary among these various magnetic structures, different microwave frequencies will produce different correlations with optical measures of activity. This has motivated us to use the solar flux data at 1000, 2000, 3750 and 9400 MHz from Toyokawa Observatory, along with the 2800 MHz flux data from Ottawa, to make correlations with ACRIM irradiance measurements. We have also computed linear regressions with the irradiance residuals after corrections for sunspot blocking using sunspot area indices, and with the Photometric Sunspot Index (PSI) itself (courtesy of Pap and Hudson). These correlations reveal several new aspects of the microwave Sun proxy relationships.

Solar Physics **152**: 167–173, 1994.
© 1994 *Kluwer Academic Publishers.*

2. Microwave-ACRIM Correlation

After reducing the Toyokawa fluxes to a solar distance of 1 AU, and making an interpolation of one missing single-frequency flux value, we have performed a detailed comparison of the Toyokawa and Ottawa data with the ACRIM irradiances (S), both corrected and uncorrected for sunspot blocking (PSI). In an earlier study (Schmahl and Kundu, 1991) we found that the correlation of microwave flux and irradiance peaks at a frequency near 1000 MHz, and that the correlation coefficient declines toward higher frequencies. This suggested that the spectrum changes in response to sunspot and plage variations. We have investigated this response in several ways.

Figure 1a shows the microwave spectrum as a function of time for the years 1980-1989. Each of the 10 columns of the figure is a contour plot of frequency vs day number for the year. Each column covers the frequency range 1000-9400 MHz (left-right). Intermediate frequency fluxes were computed by a spline fit to the Toyokawa and Ottawa flux data. This representation clearly shows both enhancements in flux and changes in spectral slope. Comparison with figure 1b for the PSI (courtesy of J. Pap and H. Hudson), shows that every case of enhancement in the sunspot index corresponds to a spectral flattening.

2.1. Correlation with the Uncorrected ACRIM Irradiance

We have cross-correlated the microwave data with the ACRIM irradiance time series (S), using low-pass and high-pass filters to determine the effects of the short- and long-term variations. We have also computed the best-fit linear combination of the five fluxes ("linear regression fit"), for the same cases. For no smoothing (length=1 day), the linear regression fit has a correlation of 77% with the irradiance. The correlations of the individual frequencies are significantly smaller (39%-56% for 9400-1000 MHz). As the smoothing length increases, so do the correlations because smoothing eliminates the sunspot-blocking dips in irradiance, which are anti-correlated with microwave enhancements. When virtually all short-term variations are smoothed out (length > 600 days), all the correlations increase to better than 95%.

High-pass filtering of the microwave data eliminates the long-term variations. ACRIM/Microwave correlations at the individual frequencies are found to be negative for periods less than 40 days, with minimum correlation values when the maximum period passed is about 16-20 days. For these shorter periods, the sunspot-blocking dips in the irradiance signal dominate the correlation. We find, however, that the multiple linear regression correlation is never negative. The spectral information available in the five frequencies provides enough information about the sunspot component, that it is partially subtracted out, even for periods shorter than 20 days.

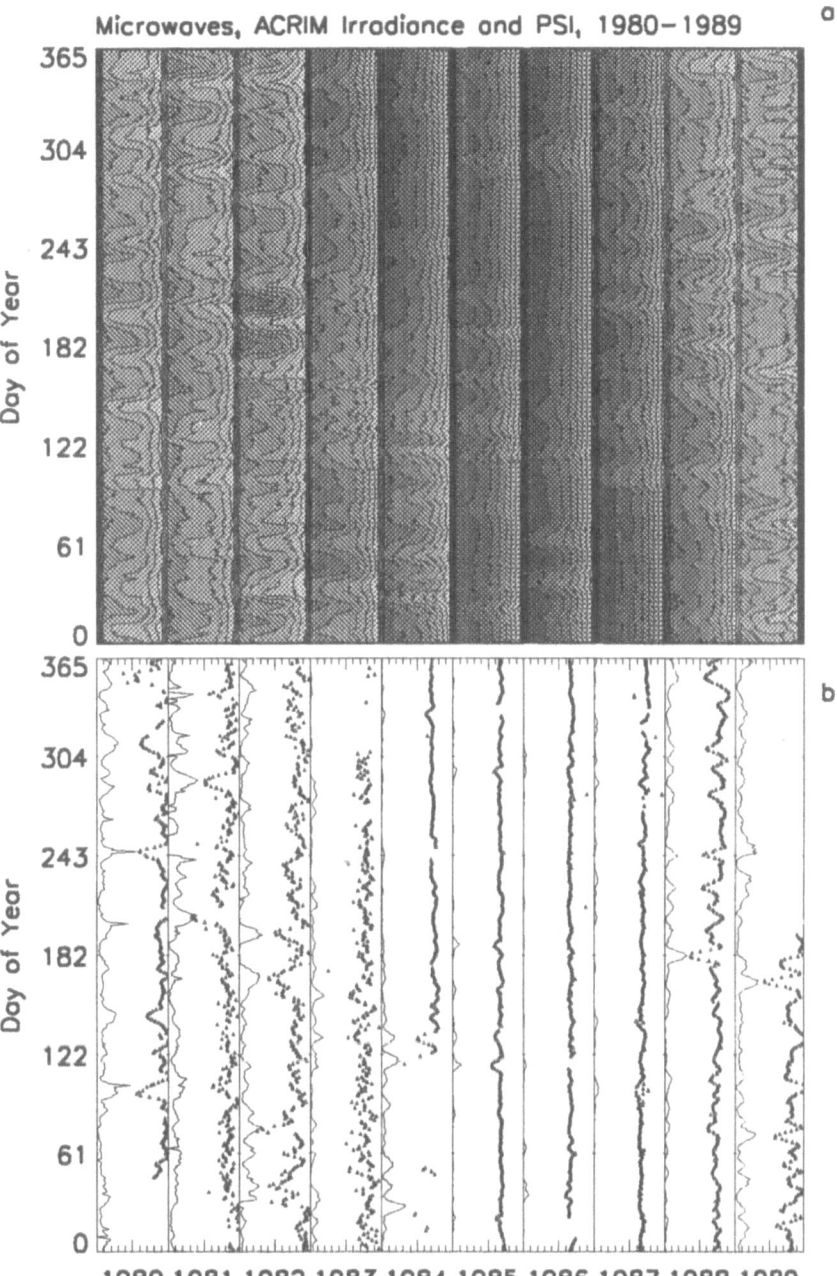

Fig. 1. (a) The microwave spectrum as a function of time for the years 1980-1989. Each of the 10 columns of the figure is a contour plot of frequency vs day number for the year. Each column covers the frequency range 1000-9400 MHz (1000 MHz on the column's left, 9400 MHz on the right). Intermediate frequency fluxes were interpolated using cubic splines. (b) Combined plot of the Photometric Spot Index and the ACRIM Irradiance in a format similar to Fig. 1a. The PSI (solid curve) is shown along the left abscissa for each year Figure 1c, while the irradiance (dotted curve) is shown just to the right of the PSI. Note that the pronounced "dips" in the irradiance correspond to the increases in PSI. (The PSI were provided by J. Pap and H. Hudson.)

2.2. Correlation with the ACRIM Irradiance Residuals

As might be expected from previous work using the 2800 MHz flux (Hudson and Willson, 1982; Lean 1988), the primary effect of adding the PSI to the ACRIM irradiance is to raise the irradiance/microwave correlation dramatically. We have correlated the irradiance residuals with the microwave fluxes using high and low-pass filters to determine the effects of short and long term variations.

For no smoothing (smoothing length = 1 day), the correlation values are 83, 86, 87, 89 and 91% for the five frequencies (9400, 3750, 2800, 2000, and 1000 MHz, respectively) and the linear regression fit has a correlation coefficient of 92%. All of the single-frequency correlation values also increase as a function of smoothing length up to ~ 100 days of smoothing, at which point the correlations no longer increase substantially, each reaching its own asymptotic limit (96-98.6% for 9400-1000 MHz). This suggests that sunspots or other components with time scales less than 100 days, once smoothed over, no longer counter the correlation. However, this flattening of the correlation is not found in the linear regression fit, which steadily approaches 100% as the smoothing length increases up to 600 days. This is the result of continuous, small changes in the spectrum on timescales of 3 months or more.

When periods longer than ~ 20 days are eliminated using a high-pass filter, the correlation values are all below 10%. When the cutoff is changed to reject periods longer than ~ 100 days, the 1000 and 2000 MHz correlations go from the lowest pair to the highest of the five frequencies. That is, the lower frequencies are better correlated with long-term evolution, while the higher frequencies correlate better with short-term changes. All the correlations rise rapidly as the filter passes longer periods, reaching their maximum values when periods $\leq$ a few hundred days are included.

3. Modeling the Sources of the Correlation

It has been shown that the 2800 MHz flux correlates best with irradiance after correction for sunspot blocking (Hudson and Willson, 1982; Lean, 1988). This removal of the sunspot component using optical data is required because of the significant fraction of the 2800 MHz flux which arises in spots. Since optical observations are normally required to determine the sunspot-associated fraction, it is important to show that multifrequency microwave observations by themselves can provide an independent measure of the sunspot component.

3.1. The Sunspot Component

In our earlier work, we discovered a strong sunspot signature in the five-point Toyokawa-Ottawa spectrum (Schmahl and Kundu, 1991). The spec-

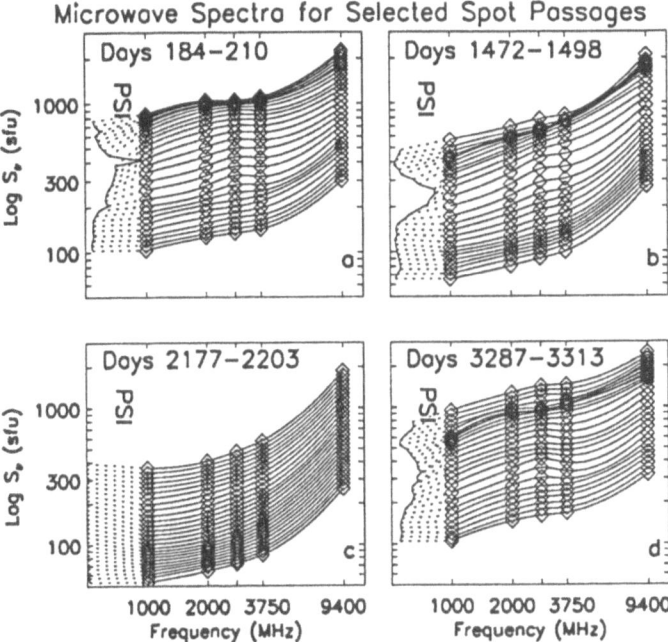

Fig. 2. Examples of the day-to-day evolution of the 1000-9400 MHz spectrum (smoothed with a 3-day boxcar and interpolated using splines) during the disk passage of large spot groups. Each curve, starting from the bottom, has been shifted up by a constant factor (1.08) from the preceding curve. (a) The microwave spectrum on successive days during the disk passage of a large spot group in June-July 1980. (b) Similar plots for solar rotations in January 1984, during a secondary sunspot maximum in the solar cycle. (c) Spectra during November-December 1986, at sunspot minimum. The PSI values are so small they appear as zeroes on this plot. The shape of the sunspot-minimum spectrum may be taken as a base reference for the other curves. (d) Representative spectra in January 1989 during the rise towards towards the maximum of cycle 22.

trum shows distinctive changes in shape on day-to-day time scales when large sunspots cross the disk. Figure 2 shows examples of the day-to-day evolution of the 1000-9400 MHz spectrum (smoothed with a 3-day boxcar and interpolated using splines) during the disk passage of large spot groups. (Each curve, starting from the bottom, has been shifted up by a constant factor of 1.08 from the previous curve.) Against the left vertical axis we plotted the daily PSI, with a dotted line connecting each PSI ordinate value to the corresponding microwave spectrum.

Figure 2a shows the microwave spectrum on successive days during the disk passage of a large spot group in June-July 1980. The daily PSI alongside the curves shows the direct association between the appearance of a "bump" in the spectrum and the central meridian passage of a spot group. Figures 2b-d show similar plots for solar rotations in January 1984, during a secondary sunspot maximum, November-December 1986, at sunspot minimum, and January 1989 during the rise towards the next solar maximum. The shape of the sunspot-minimum spectrum (Figure 2c) may be seen as the basic minimum spectrum underlying the other curves, by which the sunspot component of the spectrum may be "subtracted out".

It is important to note that these spectral bumps cannot be caused by free-free emission, which must have a positive spectral slope, hence they cannot be due to plage. The only plausible mechanism is cyclotron emission from sunspots, where the coronal magnetic field is strong (i.e. $> 10^2 G$). An analysis of these spectral shapes is beyond the scope of this paper.

3.2. THE LINEAR REGRESSION PROXY

We find that the linear regression of the daily 5-frequency fluxes with the uncorrected 1980-1989 values of S gives a correlation of 83%, which compares favorably with the 86% value found for the correlation of the 2800 MHz flux with S+PSI. For the particular case in which daily microwave fluxes are used on all days where ACRIM data exists from 1980 to 1989, the best linear combination of the fluxes is:

$$S = 0.0401\, F_{1000} + 0.0154\, F_{2000} - 0.0036\, F_{2800} - 0.0342\, F_{3750} + 0.0018\, F_{9400} + S_0$$

where S is in watt m^{-2}, $S_0 = 1366.7651$, and the fluxes $F_{1000} - F_{9400}$ are in sfu. The coefficients depend on the time period chosen, but vary only slightly for intervals greater than 5 years. The salient features of the linear combination are the negative signs of the coefficients for 2800 and 3750 MHz, and the dominating coefficient for 1000 MHz. The 1000 MHz flux has the lowest correlation with PSI of the five frequencies (possibly because the low corona is faint above sunspots), hence it is least affected by the sunspot-blocking effect. As noted above, the 2800 and 3750 MHz fluxes have a significant fraction of their origin in sunspot cyclotron emission. Hence they provide a measure of sunspot area and mimic PSI in their behavior. A more detailed analysis of the regression combined with other measures of temporal and spectral variation (beyond the scope of this paper) may establish the fraction of cyclotron emission more definitively.

3.3. LONG TERM SPECTRAL CHANGES

We have also found that the shape changes in a well-defined manner on multi-year time scales. As the solar cycle progresses from maximum to minimum, the spectrum becomes (1) smoother, (2) reduced at all frequencies, and (3) steeper. The first of these changes results from reduced magnetic activity and therefore reduced cyclotron emission, which often produces maxima in the spectrum around 2800-3750 MHz (cf. figure 2). The second results from the reduced areal coverage of sources (plage and associated coronal structures) and reduced density in those sources. The third type of change is very likely due to the enhanced reponsiveness of optically-thin coronal emission (at 1000 MHz) to density changes, and relatively diminished response of the optically-thick chromospheric emission (9400 MHz) to similar changes.

4. Conclusions

We have shown that the microwave fluxes from 1000 to 9400 MHz are all highly correlated with the ACRIM irradiance over the 10-year period 1980-1989. Unexpectedly, the 1000 MHz flux shows the highest correlation; we speculate that this is due to the absence of X-ray emission from the atmosphere above sunspots. For short time intervals (less than 100 days), the microwave fluxes are all anti-correlated with the irradiance due to the close association of the microwave emission with sunspots which block the irradiance. A linear regression of the five frequencies with the ACRIM irradiance (not corrected for sunspot blocking) provides a good proxy for the irradiance over all time scales in the range 100 days to 10 years. Even for shorter time intervals, the linear regression proxy does not have negative correlation with the irradiance. Inspection of the regression coefficients computed for intervals > 100 days shows that the 2800 and 3750 MHz fluxes have negative coefficients, indicating that the cyclotron emission seen at these frequencies provides the information needed to determine a measure of the sunspot-blocking component of the irradiance function. The day-to-day changes in shape of the spectrum reveals a signature of the cyclotron emission from sunspots, and the "bumps" in the spectrum provide a fairly accurate measure of the sunspot blocking factor in irradiance.

5. Acknowledgements

We thank Drs. Pap and Hudson for providing their tables of the Photometric Sunspot Index. We also thank Mrs. Betty Jo Stevenson for digitizing the Toyokawa fluxes from published tables. Travel costs were partially defrayed by U.S. National Science Foundation Grant ATM-9224968.

6. References

Fröhlich, C., Foukal, P.V., Hickey, J.R., Hudson, H.S., and Willson, R.C.: 1991, 'Solar Irradiance Variability from Modern Measurements', C.P. Sonett, M.S. Giampapa, M.S. Matthews, (eds.) *The Sun in Time*, University of Arizona: Tucson, 11.

Hudson, H.S., and Willson, R.: 1982, 'Sunspots and Solar Variability', in L. Cram and J. Thomas, (eds.) *Physics of Sunspots*, Sacramento Peak Obs.: Sunspot, 434.

Kundu, M.R.: 1965, *Solar Radio Astronomy*, Interscience: New York, 185.

Lean, J.: 1988, *Adv. Space Res.*, 8, 85.

Schmahl, E.J., and Kundu, M.R.: 1991, *EOS* 22, 219.

Tapping, K.F., and DeTracey, B.: 1990, *Solar Phys.*, 127, 321.

SOLAR BRIGHTNESS DISTRIBUTION AND ITS VARIABILITY AT 3 MILLIMETER WAVELENGTH

V.G. NAGNIBEDA, V.V. PIOTROVITCH

Astronomical Institute St. Petersburg University, St. Petersburg, 198904, Russia

ABSTRACT. Solar mapping at 3.4 millimeter wavelength has shown the existence of various spatial structures in brightness distribution. The most prominent structure is presented by the local sources of the slowly varying component of the solar radio emission. Usually they coincide with active regions. Some sources have no corresponding optical counterparts. Using synoptic radio maps, latitudinal belts of enhanced brightness were detected at the north and south hemispheres. These belts seem to coincide with sunspot zones, but the enhanced emission exists independently of sunspot group appearance. Comparison of our maps with XUV images shows a noticeable resemblance. Sources above active regions and latitudinal belts of enhanced brightness are seen in both ranges as well as coronal holes.

1. Introduction

The upper solar atmosphere radiates mainly in two spectral ranges — radio and XUV. Therefore it is worthwhile to compare solar images at both ranges. Non-burst solar millimeter-wave emission has a thermal nature and is generated in local thermodynamic equilibrium. Thus it provides a simple diagnostic of physical conditions in the chromosphere and the lower part of the transition region. Unfortunately, the strong influence of the terrestrial atmosphere and the shortage of observation time at large radio telescopes leads to the absence of long sets of solar observations at millimeter wavelength range.

Since 1987 quasi-regular solar observations have been carried out with the RT-7.5 Moscow State Technical University radio telescope. Using the long set of solar maps at 3.4 mm wavelength with 2.4 arc min spatial resolution, we can distinguish various spatial components in solar brightness distribution. They are: the emission of the quiet Sun as a whole, the local sources of the slowly varying component of the solar radio emission, and latitudinal belts of enhanced brightness in "activity zones". Comparison of the radio map with XUV images obtained at space probe Phobos–1 reveals noticeable resemblance between images.

Solar Physics **152**: 175–180, 1994.

2. Observations

Solar mapping is carried out by means of raster or radial scanning. The radio radius is measured using the set of angular distances between the extrema of the scan derivative. Polar or equatorial radio radii are measured by raster scanning along or across the central meridian, respectively. A detailed description of the telescope and observations are in Solovyov *et al.* (1992).

3. Brightness Distribution at 3 MM Wavelength

The average value of the solar radio radius was found to be equal to 1.011 ± 0.003 of the optical one. Polar and equatorial radii are the same. The brightness temperature of the quiet Sun was measured using the new moon as a calibrating standard. The narrow beam helps us to exclude the input from the slowly varying component. The ratio of central brightness temperatures is 43.6 ± 2.5. Using the brightness temperature of the new moon from (Krotikov & Pelushenko, 1987) we obtain the brightness temperature of the quiet Sun : $6900 \pm 400 \, K$. Assuming a flat brightness distribution, we estimate the quiet Sun flux as 11300 ± 680 s.f.u.

The local sources (LS) of the slowly varying component are seen on maps as compact areas of enhanced brightness. Ordinary enhancements are about several percent above the quiet Sun level. As a rule they are associated with active regions and coincide with plages in size and shape. The contribution from LS into the total solar flux is about 1–3%.

Some sources have no corresponding active regions. An example of such a source is shown in Figure 1 and discussed by Piotrovitch & Solovyov (1989). It is interesting to note that there is no active region at the source location during the preceeding and subsequent solar rotations. Using 1.35 cm observations on the same day with the radio telescope of the Crimean Astrophysical Observatory, we have measured the slope of the source spectrum as $\sim \lambda^{-2}$. So the source is optically dense, and therefore it has its origin in the chromosphere. The stanford magnetogram shows the hill of the large scale magnetic field coinciding with the source (Solar Geophysical Data, 1987).

Maps obtained with RT-7.5 demonstrate a permanent depression along the solar equator (Nagnibeda *et al.*, 1989). To show this phenomenon more clearly the synoptic map was constructed using daily maps (Figure 2). Evidently this depression appears due to enhanced emission from so-called "zones of activity" clearly distinguished between heliolatitudes of 10° and 50° in both hemispheres. LS appear

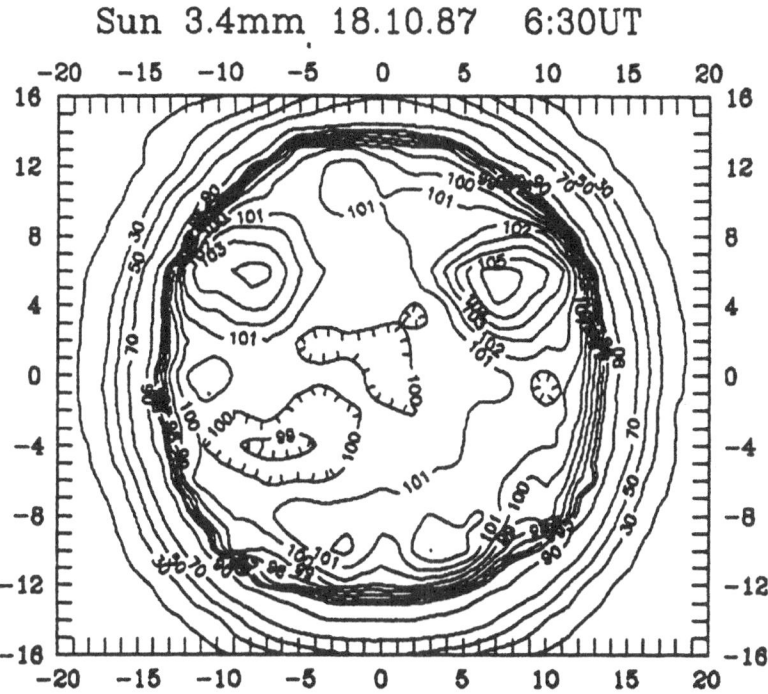

Fig. 1.   Contour map Oct. 18, 1987.   The north pole is above. The source without an active region is in NE quadrant.

from time to time at these zones, but enhanced brightness exists all time. The brightness temperature difference between equatorial and latitudinal belts is equal to about 100 K. The corresponding flux is about 1% of the total solar one. Apparently the enhanced emission of activity zones is the basic component of solar radio emission at millimeter wavelengths and defines the solar cycle dependence of the quiet Sun level (Nagnibeda & Piotrovitch, 1990).

The quasi-simultaneity of our mapping of the Sun with XUV observations from space probe Phobos–1 (time difference $< 1^d$) gave us an opportunity to compare solar images in both spectral ranges. Figure 3 presents a solar map obtained with RT-7.5. Comparison of the map with published solar images at 175 Å (Fe IX–XI lines with corresponding *temperature* $\simeq 10^6$ K) and 304 Å (He *II*, $3 \times 10^4 - 10^5$ K) (Sobel'man *et al.*, 1990) show an interesting similarity. Local sources above active regions and latitudinal belts of enhanced brightness are

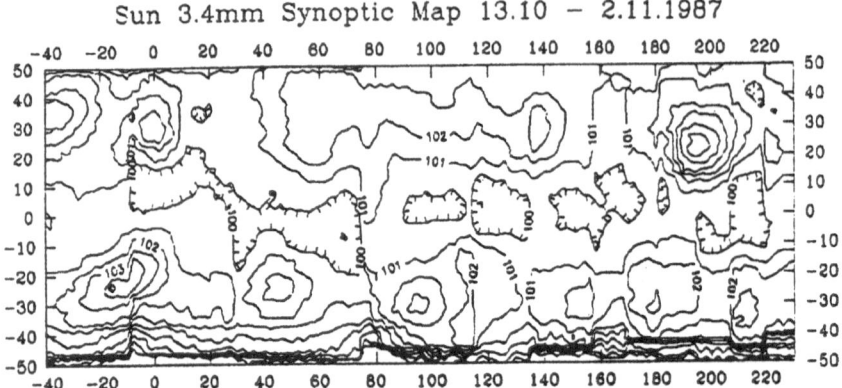

Fig. 2. Synoptic map. Contours are drawn from 95% through 1% from the quiet Sun level. Carrington rotation No. 1794.

clearly seen in both ranges. Polar and equatorial coronal holes are seen as 1% depressions from the quiet Sun level at 3.4 mm wavelength.

4. Discussion

It is interesting to estimate the contribution to millimeter-wave brightness from the layer containing He II. Ivanov-Kholodny & Nikolsky (1962) introduce the generalized emission measure $GEM \equiv \int N_e^2 \times T_e^{-3/2} dh$, which is proportional to the XUV line intensity and to the optical depth (τ) of thermal bremsstrahlung at millimeter waves. Using the observed contrast from Phobos–1 images (Sobelman *et al.*, 1988) we have obtained GEM for different solar regions. An estimation of the optical depth shows $\tau \ll 1$, thus we can calculate He II layer contribution $T_b = T \times \tau$ to the brightness temperature at 3.4 mm wavelength (Nagnibeda & Piotrovitch, 1991). He II ($3 \times 10^4 < T < 10^5 K$) layer input into millimeter brightness enhancements constitutes less than a half of the observed ones. The contribution from the corona is negligible at millimeter waves due to higher transparency. It seems that the bulk of the millimeter-wave excess in latitudinal belts is generated below the He II layer, possibly, in the regions of enhanced magnetic field in and out of plages, as in the case with LS without active regions mentioned above. Large coronal holes are seen at millimeter waves as low depression regions. Some brightenings found in coronal holes at 8 mm wavelength (Kosugi *et al.*, 1986) seem to be associated with local enhancements of

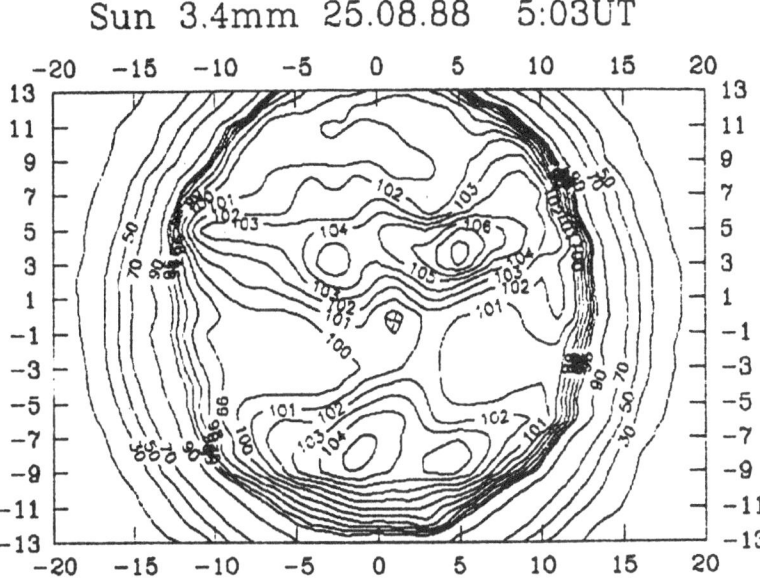

Fig. 3. Contour map Aug. 25, 1988. Brightness distribution correlates with XUV image from Phobos–1 (Sobel'man *et al.*, 1988).

magnetic field. Evidently the similarity of millimeter-wave and XUV images is the manifestation of the large-scale structure of the solar atmosphere from the upper chromosphere to corona.

Acknowledgements

The authors thank Dr. G.P. Apushkinsky for providing the solar data from RT-22 radio telescope of the Crimean Astrophysical Observatory.

References

Ivanov-Kholodny, G.S., Nikolsky, G.M.: 1962, *Astron. Zh.* **39**, No.5, 777 (*in Russian*).

Kosugi, T., Ishiguro, M., Shibasaki, K.: 1986, *Publ. Astron. Soc. Japan* **38**, No.1, 1.

Krotikov, V.D., Pelushenko, S.A.: 1987, *Astron. Zh.* **64**, No.2, 417 (*in Russian*).

Nagnibeda, V.G., Piotrovitch, V.V., Solovyov, G.N.: 1989, in Teplitskaya, R.B. (ed.), *'Solar Magnetic Fields and Corona'*, Nauka, Novosibirsk, **2**, 243.

Nagnibeda, V.G., Piotrovitch, V.V.: 1990, *Astron. Nachr.* **311**, 413.

Nagnibeda, V.G., Piotrovitch, V.V.: 1991, in Martirossian R.M. (ed.), *Proc. IVth Sov.-Finn. Symp. on Radio Astron.*, Ashtarack, 41.

Piotrovitch, V.V., Solov'ev, G.N.: 1989, in Teplitskaya, R.B. (ed.), *'Solar Magnetic Fields and Corona'*, Nauka, Novosibirsk, **1**, 296.

Sobel'man, I.I., Zhitnik, I.A., Valnicek, B., Rybansky, M. *et al.*: 1988, *Preprint FIAN* No. 241, Moscow (*in Russian*).

Sobel'man, I.I., Zhitnik, I.A., Valnicek, B., Rybansky, M. *et al.*: 1990, *Pis'ma Astron. Zh.* **16**, 323 (*in Russian*).

Solovyov, G.N., Rozanov, B.A., Ivanov, V.N., Nagnibeda, V.G., Piotrovitch, V.V.: 1992, in Fedorov, I. (ed.), *'Topics in Radioelectronics and Laser System Design'*, CRC Press, Inc., Boca Raton, 167.

LATITUDINAL VARIABILITY OF LARGE-SCALE CORONAL TEMPERATURE AND ITS ASSOCIATION WITH THE DENSITY AND THE GLOBAL MAGNETIC FIELD

M. GUHATHAKURTA * and R.R FISHER
NASA/GSFC
Greenbelt, MD, 20771

Abstract. In this paper we utilize the latitude distribution of the coronal temperature during the period 1984-1992 that was derived in a paper by Guhathakurta et al, 1993, utilizing ground-based intensity observations of the green (5303Å Fe XIV) and red (6374Å Fe X) coronal forbidden lines from the National Solar Observatory at Sacramento Peak, and estabish its association with the global magnetic field and the density distributions in the corona. A determination of plasma temperature, T, was estimated from the intensity ratio Fe X/Fe XIV (where T is inversely proportional to the ratio), since both emission lines come from ionized states of Fe, and the ratio is only weakly dependent on density. We observe that there is a large-scale organization of the inferred coronal temperature distribution that is associated with the large-scale, weak magnetic field structures and bright coronal features; this organization tends to persist through most of the magnetic activity cycle. These high-temperature structures exhibit time-space characteristics which are similar to those of the polar crown filaments. This distribution differs in spatial and temporal characterization from the traditional picture of sunspot and active region evolution over the range of the sunspot cycle, which are manifestations of the small-scale, strong magnetic field regions.

1. Introduction

The Solar cycle has been studied in its various forms and manifestations over a century now. However until recently no one had looked at the large-scale solar cycle variabllty of the coronal temperature (Guhathakurta et al., 1993 (GFA); Guhathakurta et al., 1992; Guhathakurta and Altrock, 1992; Guhathakurta and Fisher, 1992). Recent study bv GFA show that the latitude distribution of the coronal temperature follows a spatio-temporal evolutionary pattern that is characteristically different from the traditional picture of sunspot and active region evolution over the range of a sunspot cycle. The picture that emerged from this study was that the coronal temperature could be seen as the sum of two components on the solar surface at a height of 1.15 $R_\odot$; the high latitude component around ±50-60° which represents high-temperature in the corona and the low latitude component which is the conventional sunspot activity belt confined to latitudes ±40 °. This study considered only the largest scale structure of the solar corona.

* affiliated to USRA

Solar Physics **152**: 181–188, 1994.

The photometric data were sampled at a daily rate, and in the creation of the synoptic charts of distributions a sample size of 6° of solar latitude was used. This means that this study was band limited to features with a spatial scale not less than 12° of latitude in width as seen at the limb and at least two day's rotation ($\sim$ 26°) in longitude. Scale sizes set by these limits are approximately 0.2 and 0.4 $R_\odot$ respectively, and include helmet streamer, streamer belt, and larger scale coronal hole structures.

In this study we establish a correlation between the global magnetic field and the density (polarized brightness) distributions with the inferred temperature structure in the corona. Such studies ultimately should put constraints on the theoretical modeling of the dynamo, which will have to reproduce these two components which are spatially and temporally out of phase with each other.

2. Coronal Data 1984 -1992

Guhathakurta et al. (1993) have presented four data sets and the estimates of the coronal temperature as inferred from the Fe X/Fe XIV line ratio in Figure 2 of their paper. For a 8-year period, 1984-1992, they displayed the synoptic data sets from top to bottom for:

(1) the polarized brightness distribution of the K-corona, (HAO/ MAUNA LOA)

(2) the photospheric longitudinal magnetic field, (NSO/KITT PEAK)

(3) the distribution of coronal temperature as inferred from the Fe X and Fe XIV line ratio data,

(4) the Fe X (6374 Å) emission line, (NSO/SACRAMENTO PEAK)

(5) and the Fe XIV (5303 Å) emission line (NSO/SACRAMENTO PEAK).

To eliminate high-frequency noise associated with rapid evolution and transient activity, they have added together data sets for two rotations so as to form a two-rotation average synoptic distribution for the observed quantities.

These two-rotation synoptic averages are plotted adjacent to each other so that the y-axis is solar latitude and the x-axis represents both solar longitude (in the given 54 day sample period) and time ($\approx$ 13.4/2$\approx$6.7 samples per year, where it is assumed that there are 13.4 Carrington rotations in a year) as a function of time so that the data are given as both a function of solar latitude (y-axis) and a mixed coordinate which has both spatial and temporal information (x-axis). On the scale of a single, two-rotation average, the observed quantity is displayed as a function of solar latitude and longitude, assuming a latitude invariant Carrington rotation rate. Thus, there are two physical time-scales visualized as the eye ranges over the x-axis. Rotational modulation of the observed signal occurs at a frequency of

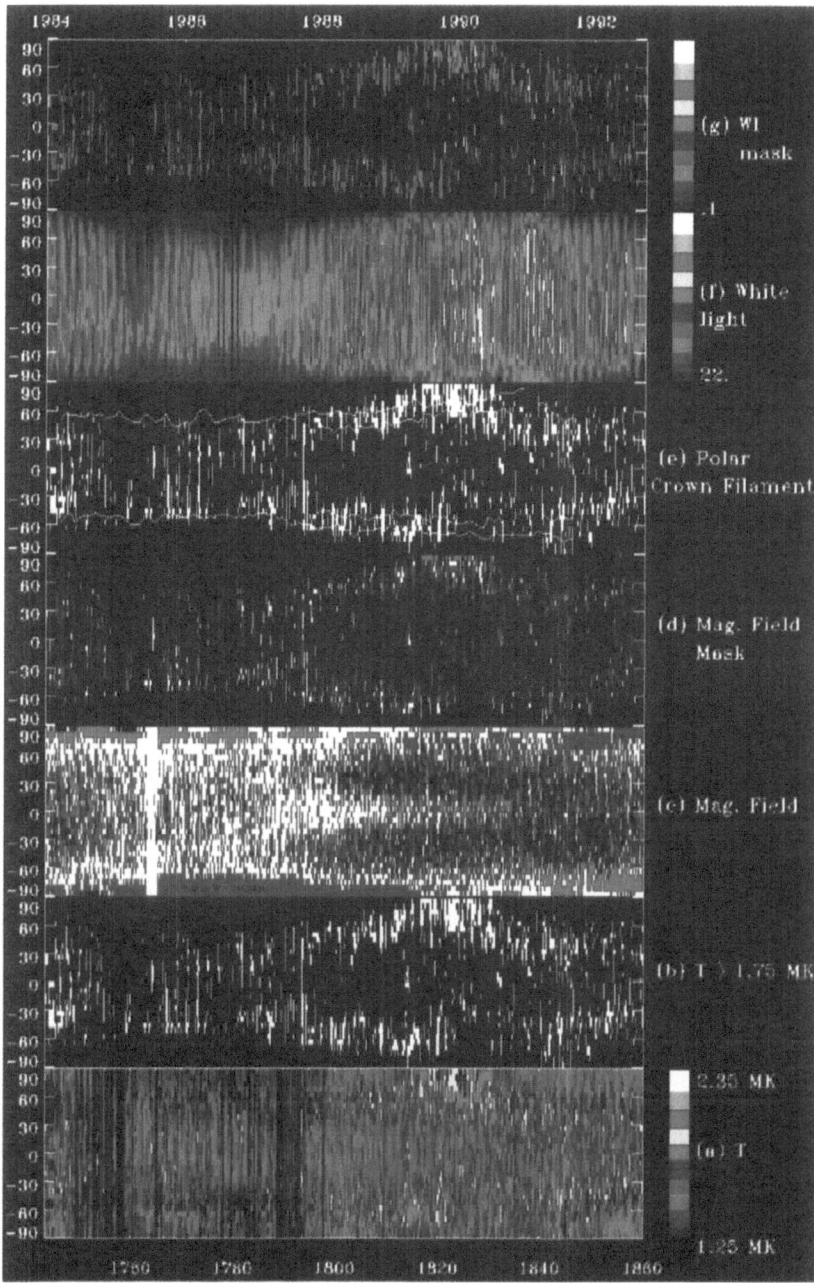

Fig. 1. a) The temperature estimate (in MK) from line ratio of the inner corona at 1.15 $R_\odot$, b) Temperature mask where all temperature greater than 1.7-1.8 MK (see text) is colored white c) the photospheric magnetic field from Kitt Peak where white represents magnetic field between $\pm$ .8-1G (see text), blue/red represents +/- field below 4G during minimum and 8G during the rest of the cycle, and black represents the high field strength, d) magnetic field through the high temperature mask, e) polar crown filament (lines) over plotted on the high temperature ban, f) the coronal white-light at 1.3 $R_\odot$ from Mauna Loa Hawaii in units of 10^{-8} $I_\odot$, and g) white light through the high temperature mask for Carrington Rotations 1743-1860 (1984-1992) for heliographic latitudes -90° to 90°.

about 6.7 cycles/year and magnetic activity cycle modulation occurs at the rate of 1/11 per year.

In order to explore the association of the high-temperature region with the large-scale magnetic field and density distributions, we have adopted the same data, data format and choice of color table as given in Figure 2 of GFA for our Figure 1 of this paper.

3. Location and Evolution of Elevated Temperature Zones

The inferred FeX/Fe XIV coronal temperature was presented in the middle strip of data in Figure 2 of GFA (for convenience to readers the same plot is presented as Figure 1a in this paper). The method used to estimate this quantity is fully described by Guhathakurta, et al.,(1992) and GFA. In that figure relatively cool regions in the corona were colored green ($\sim$ 1.25-1.35 MK) the intermediate regions ($\sim$ 1.45-1.6 MK) were colored beige/blue, and the pink, yellow and red regions were areas where the inferred temperature was higher (near 1.8-2.1 MK). The most prominent features of the data display were the two bands of higher temperature (pink, yellow and red) material which are seen at relatively high latitudes on either side of the solar equator. The average temperature in these bands was found to be 1.85 $\pm$ .10 MK which is about 500,000 K hotter than the average temperature (1.32 $\pm$ .07 MK) at the poles during solar minimum and 300,000 K hotter than the mean equatorial temperature (1.57 $\pm$ .11 MK) determined for the period of this study.

At the time near solar minimum, at 1986, there is relatively little variation of latitude as a function of time of the high-temperature regions. Near the maximum of the magnetic activity cycle there is a drift of the poleward boundary toward higher latitudes, first seen in the southern hemisphere and then about half a year later, a similar enlargement of the high-temperature region of the corona is seen to migrate toward the north polar region.

The relationship between the high-temperature and its association with the magnetic field and the density distributions can be demonstrated with a masking technique. In Figure 1b of this paper, a mask is constructed which allows inspection of the distribution of polarized brightness of the corona or photospheric magnetic field through a mask which is set to zero (black) for all values of temperature below a certain threshold and which is one (transparent) for all values of the temperature greater than the selected threshold. We used a threshold value of 1.75 MK during the descending and ascending phase of the solar cycle, 1.7 MK during the minimum phase of the solar cycle and a value of 1.8 MK during the maximum phase of the solar cycle.

4. Correlation Between Regions of Elevated Coronal Temperatures and the Photospheric Magnetic Field Distribution

In Figure 1c we have presented the longitudinal component of the photospheric magnetic field. The data have been colored so as to simply encode strong fields (positive and negative) as black. Weaker fields have been encoded with either blue (+ field) or red (-field) as conventional longitudinal magnetograms while neutral lines (or close to zero magnetic field) have been encoded white. During the minimum phase of the cycle (1985-1987) anything above the absolute value of 4 G was considered strong field and anything between ±.8 G was considered neutral. During the rest of the cycle anything above the absolute value of 8 G was considered strong field and anything between ±1. G was considered neutral.

The relationship between locations of elevated coronal temperature and the photospheric magnetic field can be viewed by using the spatial mask as presented in Figure 1b and superposing that on Figure 1c. The resultant image is plotted as Figure 1d.

Inspection of Figure 1d indicates that regions of higher temperature tend to be located over positions where weak photospheric fields of opposite sign are in the process of being transported toward the pole of rotation. Occasionally, particularly in the declining phase of the magnetic activity cycle, a lower latitude region of enhanced temperature is detected in the corona. These areas are also located over areas where large-scale weak magnetic fields of opposite polarity are detected, usually associated with the interface between well defined boundaries separating magnetic active regions.

4.1. COMPARISON WITH POLAR CROWN FILAMENT DATA

From the compilation of H_α synoptic charts McIntosh (1979) uses the dark disk filaments (prominences seen against the bright solar disk) in mapping the complete patterns of large-scale solar magnetic polarity. These long, persistent filaments which mark most of the patterns, reveal lines of polarity reversal also known as 'neutral' lines (NLs). Prominences, regions of cool, dense material at coronal heights in the vicinity of these magnetic neutral lines, are thought to be supported against solar gravity by a surrounding magnetic field structure with a spatial scale comparable to the prominence length ($\sim$ .3 $R_\odot$ Allen, 1973). The NLs that underlie the polar crown of filaments are the longest on the solar surface, interconnecting the two hemispheres. The so-called 'rush to the poles' by the polar crown filaments is part of a process of poleward migration by a series of large-scale patterns that merge into a single feature in the year or two before maximum. This merger creates the nearly-continuous chain of filaments known as the polar crown (McIntosh, 1979). We observe a striking similarity (Figure 1e) in the latitudinal distributions of these polar crown filaments (McIntosh, 1992; McIntosh, 1993) during 1984-1991, and the inferred high-temperature coronal regions. We observe that as soon as the polar crown begins to move poleward (1988-1990), filaments appear on the next NL equatorward. These define a secondary polar crown equatorward of the 'true' polar crown (with a maximum latitude of around 40° around this time of the cycle) which is continuously present and becomes the polar crown of the next sunspot cycle. The latitude of both the primary and secondary polar crown filament falls on the observed high-temperature band during this period.

5. Correlation Between Coronal Density and Temperature Distributions

An indication of the total content of mass in the corona is given by the plot of the distribution of polarized brightness as detected by the HAO K-coronameter in Figure 2 from GFA (same plot is presented as Figure 1f in this paper). The polarized brightness distribution in Figure 2, to first order, reflects the total electron content of the corona independent of temperature.

We have used sixteen levels of color to represent white-light data ranging from .1 - 22 $\times 10^{-8}$ $I_\odot$ (disk center brightness). Here black represents any data point below 1.38 $\times 10^{-8}$ $I_\odot$, the range of colors from dark to light green is 1.38-4.13 $\times 10^{-8}$ $I_\odot$, beige is 4.13-4.5 $\times 10^{-8}$ $I_\odot$, light blue to dark blue is 4.5-9.63 $\times 10^{-8}$ $I_\odot$, purple to pink is 9.63-13.75 $\times 10^{-8}$ $I_\odot$, yellow to red is 13.75-19.25 $\times 10^{-8}$ $I_\odot$ and grey to white is 19.25-22. 10^{-8} $I_\odot$. The total intensity and latitude extent of the white-light corona is modulated at the frequency of the solar magnetic cycle. In the white-light plot, it is seen that as the cycle declines in 1984 toward the minimum at 1986, the white-light pB drops in amplitude to a minimum and then towards the middle of 1987 begins to increase toward maximum values of pB which are displayed in pink/yellow/red near the maximum of this current cycle 1989-1990.

The relationship between the density and temperature distributions can be demonstrated with the same spatial mask as described in Figure 1b and superposing that on Figure 1f we obtain Figure 1g.

From Figure 1g we observe that the coronal polarized brightness distributions (density) in the high-temperature band show the presence of bright features with a distinct local maximum. These bright features usually correspond to the centers of helmet streamers/prominences.

At the minimum in the magnetic activity cycle a large fraction of the white light corona is associated with the location of high-temperature material. In fact, the density and temperature distributions are quite well correlated in terms of spatial location during the 'period around the minimum of the activity cycle from 1985 to 1987'. The corona tends to be bright in those regions correlated with high latitude filament locations. At the onset of the new cycle the contribution to the total electron content of the corona by structures over strong magnetic field regions is increased and the correlation between the temperature distribution and the coronal density distribution is less clearly demonstrated.

6. CONCLUSIONS

The following conclusions were drawn:

(1) There is an organization of the large-scale coronal temperature distribution which is associated with the large-scale structures in the solar magnetic field. Generally this organization takes the form of two zonal bands each about 20 degrees of solar latitude in width where the low latitude boundaries are located around 50N and 50S.

(2) These structures tend to persist through most of the solar activity cycle. Near the maximum of the sunspot cycle there was a poleward expansion of these zones while small-scale manifestations of activity such as sunspots and active regions were slowly drifting towards the equator.

(3) Figure 1d shows that the high-temperature zones tend to lie over regions where magnetograph observations indicate a change of polarity of weak large-scale magnetic fields.

(4) Figure 1g shows that the high-temperature zones tend to lie over bright coronal features where there is a local maxima in the coronal polarization brightness. These bright features usually correspond to the centers of coronal helmet streamers.

(5) Finally, the latitude distributions of the small-scale aspects of the solar activity as observed in sunspots and active regions and reflected in the green and red line data (Figure 2 of GFA) and the longitudinal component of the magnetic field in Figure 1c (the so-called 'butter-fly' diagram) are significantly different from the distributions of the large-scale features as observed in white-light data and the inferred temperature structures. The latitudinal distributions of the large and small-scale structures change in very different ways during the rise to maximum phase (1987-1990) of the solar cycle. During this phase the active region latitudes drift equatorward whereas bright coronal features in white-light and the inferred temperature structures drift poleward.

Acknowledgements

Research at GSFC was supported in part under a NASA grant . Observations at NSO/SP were obtained by Evans Facility observers under the supervision of Lou B. Gilliam, Chief Observer and were kindly provided by Richard Altrock. K-coronameter data were supplied to us by D. Sime of the HAO. Some data reduction support was provided by Timothy W. Henry

(NSO/SP) and Vic Tisone (NCAR/HAO). We would like to thank J.W. Harvey (NSO/KP) for providing the synoptic magnetogram data.

References

Allen, C.W.: 1973, *Astrophysical Quantities*, 3d ed., Athlone: London.

Guhathakurta, M., Fisher, R.R., and R. Altrock: 1993, *Astrophys. J. Lett* 414, 145.

Guhathakurta, M., and Fisher, R.R., 1992, *Bull. Amer. Astron. Soc.*, 24, 4,1254.

Guhathakurta, M., and Altrock, R.C.: 1992, *Astr. Soc. of the Pacf. Conf. Ser.* 27, 395.

Guhathakurta, M., Rottman, G.J., Fisher, R.R., Orrall, F.Q. and R. Altrock: 1992, *Astrophys. J.* 388, 633.

Hundhausen, A.J.: 1993, *J. Geophys. Res.* , in press.

McIntosh, P.S., 1979, *UAG Reort 70*, World Data Center A for Solar Terrestrial Physics, Boulder, Co.

McIntosh, P.S.: 1992, *Astr. Soc. of the Pacf. Conf. Ser.* 27, 14.

Mcintosh, P.S., 1993 *Private communication.*

ACTIVE REGION EVOLUTION AND

SOLAR FLUX VARIATIONS

T. P. HARTSELL

LASP/APAS Department, University of Colorado, Boulder, CO, 80309-0391, USA

P. L. BORNMANN

*National Oceanic and Atmospheric Administration, Space Environment Laboratory,
Boulder, CO, 80303, USA*

Abstract. The goal of this study is to relate the changes in the solar radiative output to the growth and decay of magnetic active regions. We will test the assumption that each index of radiation variability is a convolution of an active-region magnetic driving function and a response function. The first step has been to identify the appropriate driving function. This driving function was assumed to have been data from the magnetic active regions derived from the Mount Wilson daily magnetograms (Howard, 1989). The daily magnetic reports were sorted to give active-region sequences. To estimate the magnetic flux of active regions outside the observing window, (i.e., behind the limb) we fit the data to a growing-and-decaying exponential function, which permits independent growth and decay. This double exponential gives reasonable fits to the observed temporal evolution of active-region magnetic flux.

1. Introduction

Solar indices representing temporal variation seen in various wavelength regions are often correlated although not identical. For a given change in the magnetic flux at the Sun's surface, each index responds in a slightly different manner. These different responses make it difficult to predict variation in one index on the basis of variation observed in another index. Therefore, an attempt has been made to associate the temporal variation of these indices to the magnetic flux variations.

Ultimately the magnetic flux variations will be used to quantify the differences in the radiation index responses. The assumption was made that each index, T, is a convolution of magnetic driving function, F, and a response function, R, as illustrated in Figure 1. This convolution method has several advantages over the usual flux comparison methods (Skumanich et al., 1984; Tobiska and Barth, 1990). A convolution is a smearing process, in which the response to a driver can begin at a delayed time and the effects of the driver in the resulting time series continue long after the driver has ceased. In a prior effort to achieve this smearing effect, Hinteregger et al. (1981) created a model of EUV temporal variation that depended on both the instantaneous F10.7 cm radio flux and an 81-day average of F10.7. Similarly, a convolution will give less

Solar Physics **152**: 189–194, 1994.
© 1994 *Kluwer Academic Publishers.*

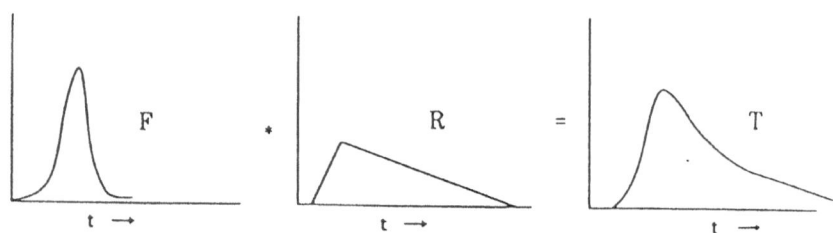

Figure 1.   The convolution of a driving function F with the response R to make the observed time sequence T is represented as F*R=T.

importance to instantaneous correlations and more to the cumulative effects of previous activity as a sort of memory of prior events.

This cumulative response has been suggested by other studies in this field. For example, the conventional two-component model for temporal variation of the solar spectrum (Cook et al., 1980) consisted of a quiet-Sun basal component with an active-region component that varied with solar activity. Lean et al. (1982) subsequently recommended that this two-component model be replaced by a three-component model, which added the emission from the active network as the third component. This active network was postulated to consist of magnetic remnants of decayed active regions. We therefore speculate that the delayed response represents the active network's contribution to index variation.

Because we are considering a delayed response to activity, the driving function must contain information about the entire development of active regions, not just when an active region is visible. For example, any solar feature within about 30 degrees of the limb will be severely foreshortened. This makes accurate measurement of active region's magnetic field, location, and area difficult. Therefore, only active regions within 60° of the central meridian were recorded, giving a 120° coverage of the entire solar surface. In addition, the locations of active regions on the Sun are needed to correct for limb darkening.

2. Data

The magnetic active-region data from the Mount Wilson daily magnetograms, extracted by Howard (1989), will be used as the driving function. These data were recorded as a chronological daily list of active regions. For each of these reports, the list included the date, the flux-weighted position, and the region's total magnetic flux in Mx. The data from January 1974 through December 1989 were used to develop the driving function.

This temporally ordered list had to be converted to a list ordered by active-region. The active-region reports of the first day in the chronological list were used to establish the first active-region sequences. The differential rotation curve determined from this data set by Howard (1990) was used to predict the location for the active regions on subsequent days. This predicted location served as the center of a target region. The size of this target increased with elapsed time between successive observations of an established active region to allow for increased uncertainty in the motion of the active regions. When the location of a new report fell within the target area about an established active region, that new report was added to the established active region sequence and became the basis for the next predicted target location. If a new report's location failed to fall within any of the target areas, then that new report was used to establish a new active-region sequence.

This algorithm reported 2423 different active-regions sequences when applied to the data from January 1974 through December 1989. Of these sequences, most covered only one transit of the solar disk, but approximately one third were observed during more than one solar rotation. For each active region, this algorithm listed the location and magnetic flux for each report, the elapsed time between reports, the difference in CMD and latitude between the region and its predicted target location, the photospheric magnetic flux, and the one-sigma uncertainty in the photospheric flux.

The plots shown in Figures 2a and 2b show the location and magnetic flux history of one active region. Figure 2a shows the location of the active region in CMD and latitude and each point is labeled with the day number, where day 0 is January 1, 1965. Disk transits of an active region are shown by sequences of crosses that progress from right to left in the plot. Figure 2b shows the magnetic flux of the active region as a function of time. The error bars represent one-sigma uncertainties, which include the uncertainty in the magnetic flux measurement and the uncertainty in the position of the active region. The cross-hatched area indicates the time when the active region was located more than 60° from the central meridian. The small diamonds near the bottom of this figure show the minimum flux that would be observable at the heliocentric angle for the location of the active region for that date; they are set to zero if no data were taken on that day.

3. Curve Fitting

The active region's magnetic flux was reported for active regions within 60° of central meridian to minimize errors due to the foreshortening effect. Therefore, to estimate the magnetic flux when the regions were beyond 60° CMD

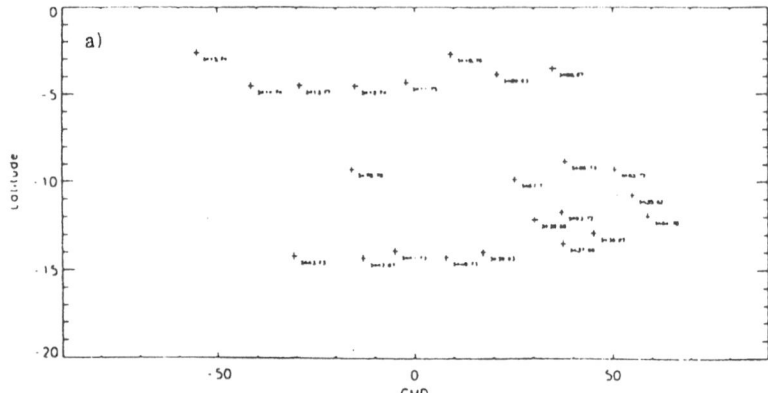

Figure 2a). Data for an active region that was first observed on May 2, 1974. The location of the active region as a function of CMD and latitude, is indicated by crosses labeled with the day number.

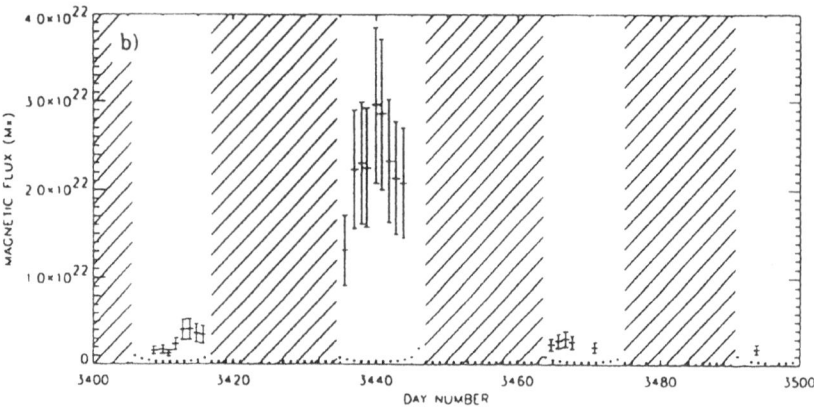

Figure 2b). Data for an active region that was first observed on May 2, 1974. The magnetic flux as a function of time is shown with one-sigma error bars. The cross-hatched area covers the time when the active region is unobservable; located at more than 60° CMD. The small diamonds near the bottom of this figure show the minimum observable flux; if no data were taken for a given day, the diamonds are set to zero.

and to extrapolate across data gaps, the data points were fit to a double exponential of the form

$$F(t) = \frac{1}{e^{Ct-A} + e^{D-Et}} + f_{min}, \qquad (1)$$

where A, C, D, and E are free parameters and f_{min} is the minimum flux of the active region. A least-squares fit to this nonlinear function was determined using

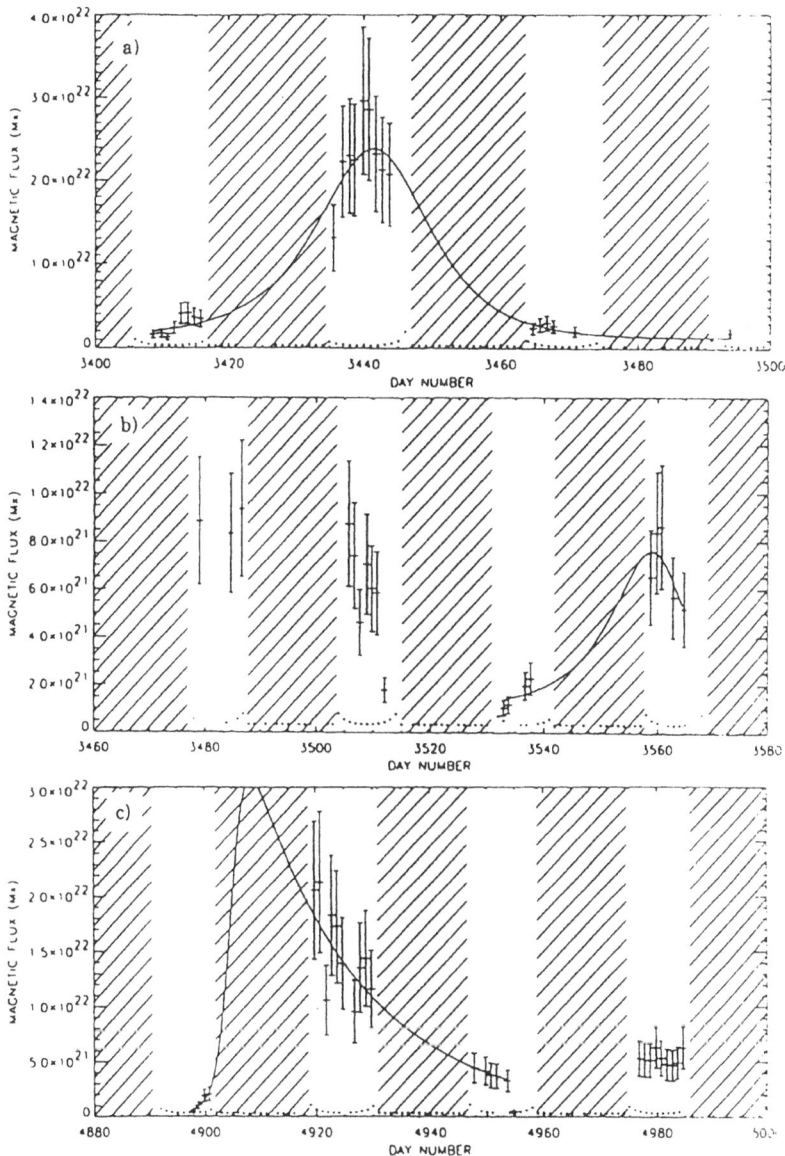

Figure 3: Three examples of the fit to the data using the double-exponential function: a) an active-region beginning May 2, 1974, b) an active-region beginning July 11, 1974, c) an active-region beginning January 5, 1977.

the gradient-expansion method based on Bevington's (1969) CURFIT. Three examples of these fits are shown in Figure 3. As can been seen, the functional form in equation (1) can fit the evolutionary trend of the data points reasonably well. The small-scale deviations from the smooth fit may be due to intermittent injections of additional magnetic flux.

4. Conclusions

Magnetic active regions have been successfully sorted from a chronological listing derived from Mount Wilson daily magnetograms. The process has been automated to guarantee consistency in the decisions involved with the sorting. The resulting magnetic active-region sequences contain information about the evolution of the region's magnetic flux and also reflect any systematic changes in the rotation of the active region over its lifetime.

Initial results from the curve-fitting analysis show promise in the parameterization of the magnetic flux of active regions over a lifetime of several solar rotations. These parameters may show systematic trends that could provide insight into general evolution of magnetic flux in active regions.

The magnetic active regions provide an impulse function to be used in convolution analysis of solar flux variations. We anticipate that the convolution approach will produce a better understanding of the connection between magnetic activity on the sun, as typified by magnetic active regions, and variations in different solar indices.

References

Bevington, P. R.: 1969, *Data Reduction and Error Analysis for the Physical Sciences*, McGraw Hill
Cook, J. W., Brueckner, G. E., VanHoosier, M. E.: 1980, *J. Geophys. Res.*, **85**, 2257
Hinteregger, H. E., Fukui, K., Gilson, B. R.: 1981, *Geophys. Res. Letters*, **8**, 1147
Howard, R.: 1990, *Solar Phys*, **126**, 299
Howard, R.: 1989, *Solar Phys*, **123**, 271
Lean, J. L., White, O. R., Livingston, W. C., Heath, D. F., Donnelly, R. F., Skumanich, A.: 1982, *J Geophys. Res.*, **87**, 10307
Skumanich, A., Lean, J. L., White, O. R., Livingston, W. C.: 1984, *Astrophys. J.*, **282**, 776
Tobiska, W. K., Barth, C.A.: 1990, *J. Geophys. Res.*, **95**, 8243

ON THE VARIABILITY OF SOME CHARACTERISTICS
OF SOLAR RADIATIVE FLUX

E.A. MAKAROVA
Sternberg State Astronomy Institute.Moscow State University
Universitetsky prospekt 13, 19899, Moscow, RUSSIA

T.V. KAZACHEVSKAYA
Institute of Applied Geophysics
Rostokinskaya street 9, 129226, Moscow, RUSSIA

A.V. KHARITONOV
Astrophysical Institute, Academy of Sciences of Kazakhstan
Kamenskoe plato, Alma-Ata, 480069, KAZAKHSTAN

ABSTRACT. We present estimates of the variability of the total solar flux and its spectral components of different wavelength intervals in the region from 1 nm to 240 mm with a time-scale to 11-years. We used the data published in the monograph "Solar radiation flux"(by E.A.Makarova et al.,1991) as well as new available data.

1. INTRODUCTION

We discuss some results collected in our monograph "Solar radiation flux" (in Russian "Potok solnechnogo izluchenia", Moscow, Nauka, 1991, by Makarova E.A., Kharitonov A.V. and Kazachevskaya T.V.). This book considers the principal solar spectrophotometric characteristics observed in a broad range of wavelengths from X-rays to the short-wave boundary of the radioband, i.e. the spectral energy distribution in the center of the solar disc, $I_\lambda(0)$, spectral irradiance, S_λ ,the total irradiance, S_0, and their variability. In this report we study possible variation in two spectral ranges: the "visible" 300 nm - 2400 nm and EUV ranges, 1 nm - 300 nm. In the "visible" where variability remains undetectable we wish to determine which is the more accurate: the average of the most reliable data in a given range or the best individual measurement. We will show the range of solar cycle variability in the UV range where the changes have been measured.

2. THE "VISIBLE" SOLAR ENERGY SPECTRAL DISTRIBUTION
IN THE RANGE 300 - 2400 NM

The energy in the solar spectrum in region 300-2400 nm contains 95% of the total energy emitted from the Sun. As is well known today, the total radiative energy emitted by the Sun, the solar irradiance S_0 ,can vary within the limits of 0.15- 0.4% during the cycle or during active region development. Therefore, the variability of the spectral components of S_λ ,i.e. spectral energy distribution $I_\lambda(0)$ or S_λ should be very much less

Solar Physics 152: 195–200, 1994.
© 1994 *Kluwer Academic Publishers.*

than S_0, i.e. less than 0.1% (except for strong lines formed in the upper photosphere and above).

The energy distributions given by different observers are not in good agreement. About 15 years ago, observed energy distributions had differences of as much as 15-20 % even near the maximum of the distribution and were supposed to be equivalent, (White, 1977, see also Makarova and Kharitonov 1972, Fig 44). Probably, the differences appear not only in the process of the absolutization, but also there are some distortions in the relative scale. Displacements of the energy maximum are observed not only in amplitudes but in the positions, for each series of observations. Now the best recent measurements of $I_\lambda(0)$ and S_λ are more precise; the interval errors are within the limits of 2-3 %, but apparently there are systematic errors in each series, see Figure 1, where we plot the deviation from mean of five measurements.

We believe that the most reliable and most accurate data on $I_\lambda(0)$ and S_λ in the range of 300 - 2400 nm can be obtained only by weighted averages of the best observational series, and this necessarily gives a "time" average as well. Then possible systematic errors of each series emerge as random errors and are cancelled out by the averaging. The selection of " the best" series of energy distributions is always a difficult problem. Earlier Makarova and Kharitonov, 1972, took into consideration rather subjective estimations: the accuracy of measurements, the method of the absolutization, and so on. It is necessary to stress that even with this subjective approach, the mean weighted distribution of the energy in the solar spectrum was very close to the recent, more reliable measurements.

2.1. Energy Distributions and Multi-Color Indices.

In our book we applied Hardrop's (1980) method for determination of the reliability of each measured energy distribution by using multi-color photometry. Each series is examined from the point of view of whether there are or are not any "deformations" in the distribution.

We compared synthetic colour indices of the Sun that have been calculated on the basis of the given energy distribution with colour indices of the Sun as a star belonging to the G2V class. Colour indices of a G2V star are deduced for solar-like stars and by direct determination from solar observations.

Sometimes synthetic colour indices calculated on the basis of a given energy distribution are unacceptable for the Sun in the selected multi-color system. For example, the energy distribution of the Sun adopted in the USA in 1974 (Anon., ASTM Standart, 1974) gives values of B-V and U-B that are too small. The energy distribution of the star, Stair and Johnson (1956), gives U-B = -0.02. This corresponds to a star belonging to a class earlier than Vega. A star with such UBV colour indexes scarcely exists. We conclude that the UBV inadequately samples the solar energy distribution. So we chose the Arizona thirteen-colour system used by Johnson and Mitchell (1975). It takes a wider region of spectra from 337.1 to 1198.4 nm and has more acceptable widths of spectral bands.

We consider eleven series of the solar energy distribution in absolute units and over a sufficiently wide spectral region and apply the scheme proposed by Hardrop. These series begin from Abbot's 1920 (1932) data and end with that of Neckel and Labs (1984). However, only three among these series have sufficiently small systematic errors to meet the Hardrop criteria. These data are as follows: Arvesen et al.(1969), Neckel and Labs (1984) and the photometer data 1979-1987, see Makarova et al.(1991 tab.9.2). These three energy distributions of the Sun are used as the mean data of S_λ and I_λ (0) In Tables 9.2 (page 327) and on IX (page 362) in the book by Makarova et.al.(1991).

2.2. Two New Solar Energy Distributions for 1992 and Preliminary New Mean Data.

Great success was achieved in 1992 in absolute solar energy measurements. There are two new distributions of energy in the spectrum of the Sun in absolute units in the visible region. After nearly a quarter of a century, two new recent results of the observations have been published: Kiev astronomers Burlov-Vasiljev, Gurtovenko and Matvejev (1992) carried out the observations at the high mountain station; Lockwood, Tug and White (1992) have compared the spectrum of the Sun directly with the spectrum of Vega. There is no doubt that both series are good enough application of the Hardrop colour-index criterion.

We made a preliminary mean energy distribution of the Sun, S_λ , using two new series in the spectral region up to 680 nm covered by all five measurements. Figure 1 shows these five distributions plus two mean distribution : one Makarova et al.(1991) and one new, obtained from these five. All values of each distribution are divided by the corresponding values of mean distribution from Makarova et al.(1991). As one can see, the correspondence between distributions of the mean data from 1991 and the mean of all five distributions is better than between either two series of observations. Each series of observations seems to have its own systematic errors.

So the question arises: when should we use the mean data and when should we use a specific measured distribution. As we expect, no one can decide which of these five distributions in Figure 1 is the best because we have neither arguments nor criteria to make such a choice. The differences between individual series are as much as 6-8%, or larger. For the mean data, differences are up 1-2%. It seems to us that if we want to select a more reliable energy distribution for the construction of solar models or to calibrate an atlas of solar spectra like the Kitt Peak atlas, then we should use the mean data because we would make smaller systematic errors.

If, on the contrary, we want to study the variability of any detail of the spectrum or some part of the spectrum, then we should use a particular distribution.

3. VARIATION OF IRRADIANCE IN THE UV SPECTRAL REGION

The variety of processes producing the emission from solar structures between 10 to 300 nm determines the complexity of the solar spectrum and possibly the ambiguity in its variation with wavelength. At the present time, the short-wave radiation shows temporal variations over the entire wavelength range daily, monthly and over solar activity cycle time scales. Long-duration measurements from the ground and satellites in recent years allows us to state the range of variability more accurately. We present only some characteristics of measurements in the range of $\lambda < 130$ nm obtained in our country and possibly less well known.

EUV time series show rotation modulation effects at the following periods: 32; 28.4; 13.5 days in the wavelength ranges of 175- 190 nm, 208-240 nm and L_α. There is also a 27-day variation of the emission which is modulated in a random fashion by the active region variation and the active longitude shifts. The observations show that the greater the wavelength, the less is the radiation flux variability Lean (1987) presents radiation variations as a function of wavelength, that has been determined from the observational data from satellites Solar Mesosphere Explorer (SME) and Nimbus 7: 6-9 % for 200- 170 nm; 15-20% for L_α and greater than 400% for $\lambda < 10$ nm.

Now we turn to short-wave radiation variations in the 11-year cycle of solar activity. The measurements in the short-wave spectrum region have been carried out for nearly 30 years. However, the differences in instruments and in calibrations do not permit us to draw a conclusion about amplitude of the 11-year variation. The variations of the radiation flux in the 11-year cycle have a strong dependence on the wavelength; these variations diminish with increasing wavelength.

Table 1 presents the data on the short-wave radiation variation in the maximum and in minimum of solar activity. The group of Prof. Hinteregger (1981) from USA carried out investigations using the Atmospheric Explorer satellite AE-E from 1976-1979, i.e., from the minimum to the maximum of the 21st cycle. We carried out measurements on Russian

Table 1. Variations of short-wave radiation in the 11-year cycle of the solar activity.

Range of λ, nm	Years	Intensity of radiation ergs cm^{-2} s^{-1}		Reference
		max	min	
5-105	1976-79	7.2	2.1	Hinteregger(1981), Torr(1979)
5-120	1978-85	4.6	2.4	Kazachevskaya et al.(1986)
5-120	1988-89	4.5-5		Kazachevskaya et al.(1991)
1-6	1970-86	$5 \cdot 10^{-2}$	$1 \cdot 10^{-2}$	Zhitnik et al. (1989)
$\lambda < 2$	1970-86	$(1-2) \cdot 10^{-2}$	$1 \cdot 10^{-4}$	Zhitnik et al. (1989)
$\lambda < 1$	1970-86	$(2-3) \cdot 10^{-3}$	$(1-2) \cdot 10^{-5}$	Zhitnik et al. (1989)

satellites from the "Prognoz" series from 1979-1985 using the same type of instruments using the thermoluminescent phosphorus receiver of radiation 1979-1985, i.e., from the maximum to the minimum of the 21st cycle. In 1988-1989 the measurements onboard "Phobos" were carried out with the same instrument (there is a report by Kazachevskaya,

Lomovsky and Nusinov devoted to this subject). We also list X-ray data obtained by Zhitnik (1989). These data are associated with the radiation of the "quiet Sun", i.e., without flares.

Approximately 90% of the entire energy of flare radiation is concentrated within the short-wave region. The intensity variations are especially strong in spectral lines. The measurements show that the increase in L_α irradiance is different for flares of different brightness and average characteristic as follows: for class 3 about 20%, i.e., $2.8 \cdot 10^{27}$ erg/s; for class 2 about 6-8%, i.e., $1\ 10^{27}$ erg/s; for class 1 about 2%, i.e., $2.8 \cdot 10^{26}$ erg/s, when we set the L_α irradiance equal to $1.4 \cdot 10^{28}$ erg/s for the "quiet Sun". In the X-ray region the radiation intensity increases by 2-4 orders of magnitude, i.e., by a factor up to 10,000 X.

4. DISCUSSION

We see that two ranges of spectrum solar energy distribution discussed here are remarkably different in the limits of variability. In the "visible" range the variability is too weak to be measured in existing observations. We conclude that the mean data of the best "visible" distributions selected by using Hardrop's colour index criterion are the most reliable now. On the other hand, in the short wavelength range, 1-130 nm, only the data for a given phase of the 11-year cycle give us the best estimate of the spectral energy distribution. because the variability is so large.

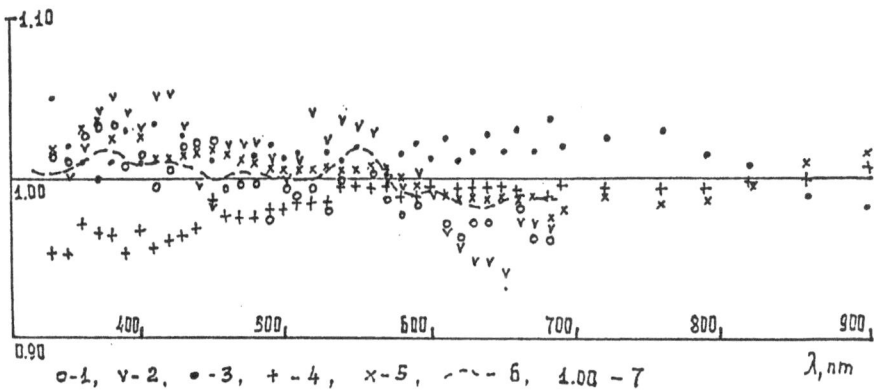

Fig.1. The deviations of a separate solar distribution, S_λ , from the mean data of Table 9.2 (Makarova et al.,1991). 1 - Burlov-Vasiljev et al.,1992; 2 - Lockwood et al.,1992; 3 - Arvesen et al.,1969; 4 - Neckel and Labs, 1984; 5 - Photometers 1976-1982; 6 - mean of five distributions; 7 - mean distribution (Makarova et al., 1991).

REFERENCES

Abbot, C.G., Fowle, F.E., Aldrich, L.S.:1932, *Ann.Smithson.Inst.Astrop. Obs.*5

Anon.,Standard Specifications for Solar Constant and Air Mass Zero Solar Spectral Irradiance,ASTM Standard ,E490-73a, 1974,Annual Book of ASTM Standards, part 41, Philadelphia, PA

Arvesen, I.C., Griffing, R.N. and Pearson, B.D.: 1969, *Appl.Opt.* 8,p.2215

Burlov-Vasiljev, K.A., Gurtovenko, E.A., Matvejev, Yu.B.:1992, Ed.Donnelly R.F., *Proceedings of the Workshop on Solar Cycle 22*, 49

Hardrop, J.: 1980, *Astron.Astrophys.* **88**, 334;

Hardrop, J.:1980, *Astron. Astrophys.* **91**, 221

Hinteregger, H.E.: 1981, *Adv. Space Res.*, 1, (12)

Johson, H.L.,Mitchell,R.I.:1975, *Revista Mexicana Astron. Astrophys.* 1, (3), 299

Kazachevskaya, T.V.,Nusinov, A.A.:1986, *Cont.of Astron.Observ.Skalnato-Plaso*, 15, 593-596

Kazachevskaya, T.V.,Bukusova, L.L. et.al.: 1991, *Adv.Space Res.*11, (1), 165-167

Lean, J.L.:1987, *J. Geophys. Res.*, **92**, 839-868

Lokwood, G.W., Tug, H., White, N.M.: 1992, *Asrophys. J.* **390**, 668

Makarova, E.A.,Kharitonov, A.V.:1972, *"Distribution of Energy in the Solar Spectrum and the Solar Constant"*, Moscow, Nauka; 1974, NASA TT-F-1803, Washington D.C.

Makarova, E.A.,Kharitonov, A.V. and Kazachevskaya, T.V.:1991," Solar Radiation Flux", Russia, Moscow, Nauka

Neckel, H., Labs, D.: 1984, *Solar Phys.* **90**, 205

Stair, R.,Johnsion, R.G.: 1956, *J. Nat. Bur. Stand.* **57**, 205.

Torr, M.R., Torr, D.G., Ong, R.A., Hinteregger, H.E.:1979, *Geophys. Res. Lett.* **6**, (10), 771

White, O.R. Ed.: 1977, *"The Solar Output and Its Variations"*, Colorado Assoc. Univ. Press, Boulder, Colorado

Zhitnik, I.A., et al.: 1989, *Trudi FIAN*, USSR **195**, 19

LYMAN-ALPHA LINE INTENSITY AS A SOLAR ACTIVITY INDEX IN THE FAR ULTRAVIOLET RANGE

A.A. Nusinov, V.V. Katyushina

Institute of Applied Geophysics, Rostokinskaja st., 9, Moscow 129226, Russia

ABSTRACT. The relations between variations of far UV (FUV) emission in 115-210 nm waveband and L_α 121.6 nm and $F_{10.7}$ are studied. The changes of FUV flux are found to lag changes of $F_{10.7}$ - as a rule for 1 day. It is shown that such a difference may be caused by two factors: 1) differences between the rates of decrease of local sources' (active regions) brightness in FUV and 10.7 cm; 2) differences between limb-darkening curves for different wavelengths. One may expect the fluxes at different wavelengths to exhibit phase shifts of one relative to another. Cross-correlation analysis reveals no time-delay between emission fluxes within the FUV waveband, in spite of different laws for limb-brightening (darkening) for different spectral intervals. The absence of a phase delay can be caused by relatively small contribution of active regions to the flux of the whole Sun at these wavelengths. Thus the Lyman-alpha line intensity variation reflects variations of Solar FUV emission more precisely than $F_{10.7}$. Therefore, using the L_α intensity for flux intensity calculations of other FUV wavelengths is preferable to using the $F_{10.7}$ index.

1. INTRODUCTION

Calculations and prediction of solar emission in the FUV are important for a number of applied problems dealing with physical processes in the upper atmosphere of the Earth and planets. Direct measurements of FUV from satellites are rather difficult and costly. Therefore models of emission variations are desirable substitutes to measured fluxes in this region (for example, 3-component model (Lean et al., 1982; Lean, 1987) or two-component model (Cook et al., 1980).

The input parameters of the models are sunspot number, radio emission flux at 10.7 cm ($F_{10.7}$) and characteristics of active regions in the optical wavelengths. It is usually supposed that $F_{10.7}$ changes simultaneously with a flux of UV-emission and that they are related by simple laws (usually linearly, see e.g. (Vidal-Madjar and Pissamay, 1980).

UV emission differs from radio emission because of the mechanisms for its generation and evolution; they are emitted by different heights in the solar atmosphere. Therefore a one-to-one correspondence between them is fulfilled only on the average and will likely differ for different time scales. Vidal-Madjar and Pissamay (1980) and Nicolet and Bossy (1985) have shown the difference between coefficients of linear regression of Lyman-alpha intensity to $F_{10.7}$, for different phases of the solar cycle. Kazachevskaja and Gonyukh (1988) obtained similar differences for rising and falling solar activity with a time scale of a few days, that imply non-synchronism of the time variations of FUV and radio emission.

Solar Physics 152: 201–206, 1994.

Hinteregger (1981) and Nusinov et al. (1989) have shown the advantage of using of "ultraviolet" activity index for UV models: especially the emission flux in the L_α line (121.6 nm). The non-synchronism of FUV fluxes and activity indices (L_α-emission or $F_{10.7}$) decreases the precision of the models and must be considered.

The aim of this work is to study the regularities of the relation between solar emission in the 130-210 nm waverange and a flux in L_α line and $F_{10.7}$ index, taking into account approprate phase shifts.

2. OBSERVATIONS

For this study we have used the data set (corresponding to L_α line and 130-210 nm range) of daily average values of solar emission fluxes measured onboard SME satellite (see e.g. (Katyushina et al, 1991) from 1 Jan. 1982 to 2 Jul. 1988 in 1 nm spectral interval and daily $F_{10.7}$ data (Solar Geophysical Data, 1982-1990). To reduce the data volume and directly compare the results with data obtained onboard AE-E satellite (Hinteregger, 1981) the fluxes were combined in sixteen 5-nm intervals: 130-135 nm,...205-210 nm. To further simplify the comparison, data were also normalized to the 1-st value in each interval and the trend corresponding to the 11-year cycle was removed. Residuals were then found (corresponding to the 27-day cycle) by extracting the smoothed values (calculated as 81-day running mean). These residuals were used to calculate cross-correlation functions for I_{L_α}, $F_{10.7}$ and emission in each of the 16 wavelength intervals using time delay limits of ±30 days.

3. RELATION BETWEEN L_α FLUXES AND 10.7 RADIOEMISSION

The correlation between I_{L_α} and $F_{10.7}$ has been studied using a cross-correlation function r(t), where t is the relative delay time. In most cases r(t) approaches maximum at $t=t_m=1$ day ($t_m=2$ days in 1987). In table 1 the values of $r(t_m)$ are given for the seven one year periods (for 1988 the volume of sample N=155). The estimates of their RMS error are about s≈0.05. One can see that the effect of the phase delay exists regularly, for all observation periods.

TABLE 1: Time of I_{L_α} delay relative $F_{10.7}$ for different years.

Year	1982	1983	1984	1985	1986	1987	1988
t_m	1	1	1	1	1	2	1
$r(t_m)$	0.74	0.76	0.83	0.48	0.64	0.72	0.73

The cross-correlation function for 1981-1983 $F_{10.7}$ and I_{L_α} data set (N=815) is shown in Figure 1 by the solid line. The values r(t)±1.96s (corresponding to 5% confidence level) are shown also by dashed lines. It is evident that changes of r(t) near t=0 are small, i.e. the value $t_m = -1$ seems to be statistically insignificant. Nevertheless, if the changes of $F_{10.7}$ and $I_{L\alpha}$ were simultaneous, the plot r(t) would be symmetrical relative to t=0. One can see that r(t) for values t<0 are greater than for corresponding positive t, and these differences are statistically significant. The systematic appearance of shift of r(t) relative to t=0 indicates that $I_{L\alpha}$ as a rule lags behind $F_{10.7}$. Some uncertainty is introduced by the fact that $I_{L\alpha}$ is given as daily mean value and $F_{10.7}$ is given for 17.00 UT.

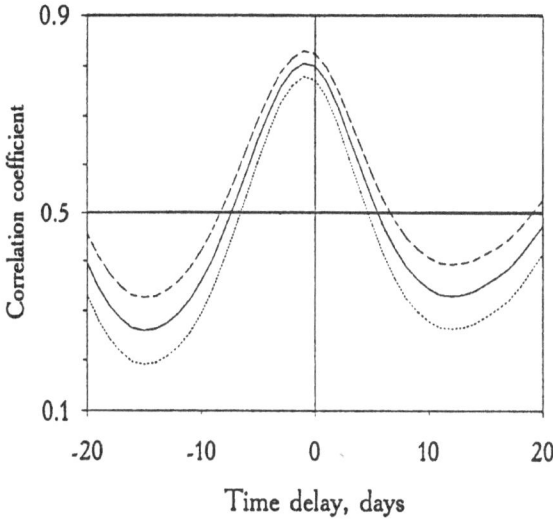

Fig.1. Cross-correlation function for Lyman-alpha and $F_{10.7}$.

Increasing or decreasing the time interval of the data does not influence significantly the value of t_m. Sometimes when the intervals are not long (~30 days) t_m may differ from 1 day. For example when the Sun's activity was low (1985 and 1986) there are periods during which r(t) approached the maximum value at t_m=0. In 1983 there where observed intervals with t_m=-1, i.e. $F_{10.7}$ lags relative to $I_{L\alpha}$. However such periods are rare.

4. CORRELATION BETWEEN ACTIVITY INDEXES AND FLUXES OF FUV

Cross-correlation functions were produced for flux variations in 130-210 nm range, using both $I_{L\alpha}$ and $F_{10.7}$ as base functions. The results are similar (differences are found only for maximum values of r) and allow us to distinguish the following characteristics: 1) variations of FUV fluxes are synchronous with respect to L_α for all wavelengths; and 2) their dominant period is about 26 days. The dependence of the maximum cross-correlation coefficient, r, is given in Figure 2 for the whole wavelength range studied using both L_α and $F_{10.7}$ as a base function. It can be seen that r approaches 0.7-0.8 for

L_α and decreases with increasing wavelength. The r value calculated for L_α always exceeds that for $F_{10.7}$, and their difference is statistically significant (for this data volume). Thus using L_α flux as an index of solar activity to model FUV provides higher accuracy than the more traditional $F_{10.7}$ index.

To determine deviations from synchronous behavior during different short epochs of the solar cycle, an additional analysis was carried out. A sliding 81 day interval of L_α (or $F_{10.7}$) was used, and the corresponding FUV data were shifted by 1-10 days to determine the maximum cross-correlation function. The resulting distribution functions N(t) were obtained and analysis shows that both positive and negative phase shifts are possible, but maxima occur at t=0 for I_{L_α} and at t=1 day for $F_{10.7}$. The significant "broadening" of functions (and hence the large number of nonzero shifts) begins from 180-185 nm.

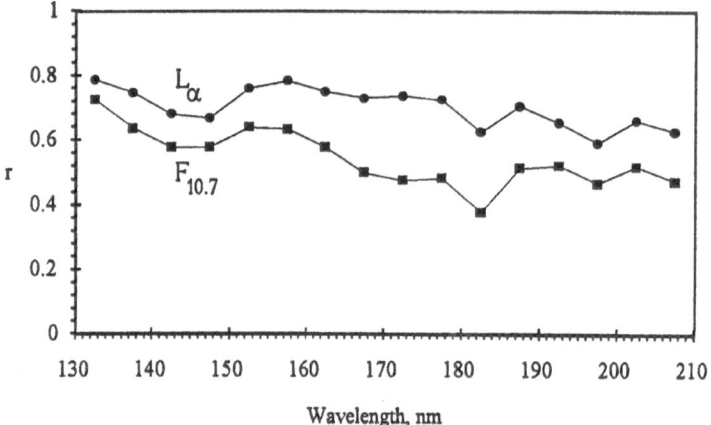

Wavelength, nm

Fig. 2. Correlation coefficients for FUV fluxes, $F_{10.7}$ and $I_{L\alpha}$.

The average time delay <t> can be estimated as a mean value for every distribution function. For different wavelengths the values of <t> are presented in Figure 3 for correlations both with I_{L_α} and $F_{10.7}$. It is seen that the conclusion is confirmed that the I_{L_α} and FUV variations. are approximately synchronous. For $F_{10.7}$ and FUV the values of <t> differ significantly from 0 and as a rule are close to 1 day.

To estimate the improvement of accuracy using L_α flux instead of $F_{10.7}$ we compare the RMS for linear regression of FUV fluxes on L_α (or $F_{10.7}$ with 1 day displacement). For L_α it is 1.2-1.7 times better than for $F_{10.7}$, and this difference is statistically significant.

5. DISCUSSION

Let us consider the possible causes of the delay in FUV relative to $F_{10.7}$. We have shown earlier (Bocharova and Nusinov, 1983) that there exists an asymmetry in the dependence of active region (AR) radio emission flux on the distance θ from the central meridian (CM) as governed by the "aging" of AR brightness (exponential decrease with time). According to Bocharova and Nusinov (1983), the 10.7 cm radio emission flux from an AR is determined by the corresponding plage area S ($F \sim S^{3/2}$) and by the limb-darkening ffunction. In particular, $F \sim \cos(0.75\theta)$ for 10.7 cm. One can suppose that for FUV radiation a dependence on AR aging also exists but in the FUV case the flux is directly proportional to S, since the flux is emitted from a rather thin layer.

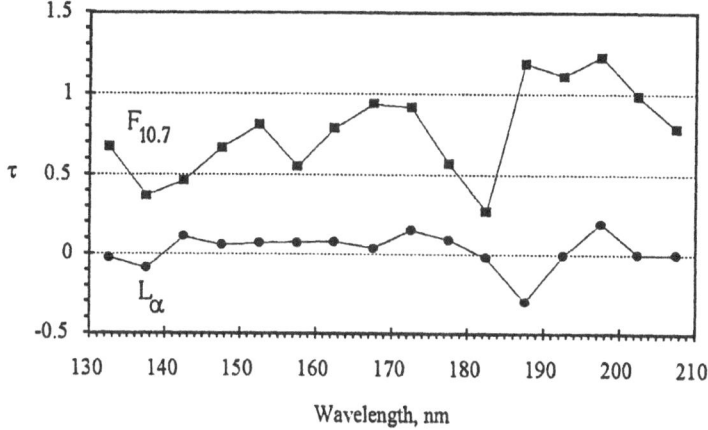

Fig. 3. Mean FUV delay time relative $F_{10.7}$ and $I_{L\alpha}$.

With these flux relations, the time difference for maximum radiation is governed by the difference in their respective limb darkening curves and in the characteristic time t_r of "aging". Furthermore, the aging time, t_r cannot differ strongly for UV and radio emission, since they are often observed simultaneously and track each other for many return passages of the same AR. Otherwise one type of radiation would still be present and the other would have disappeared. Apparent temporal shifts between radiation of different wavelengths must be entirely due to differences of limb darkening curves. The smaller the decrease in the brightness toward the limb, the sooner the radiation maximum occurs. It was shown in Krasinetz and Nusinov (1991) that a time delay of 1 day between $F_{10.7}$ and I_{La} corresponds to a limb darkening function of $\cos(0.96\theta)$.

One can evaluate the height, h, at which the L_α line is generated using the thin layer approximation and the geometric condition of the source observed behind the limb, $R \approx 6.96 \cdot 10^5$ km: $R/(R+h) = \sin(90^\circ/0.96)$. It follows that $h \approx 1.5 \cdot 10^3$ km, a value close to estimates of VAL models (Vernazza et al., 1985).

In empirical models of the Earth's atmosphere (e.g. Jacchia, 1977; Hedin, 1987) $F_{10.7}$ is used as an input parameter, with a lead time of 1-1.5 days. Evidently, this empirical value is (at least partly) due to the fact that FUV emission lags $F_{10.7}$ by ~1 day. This small lag is about the minimal time for the upper atmosphere to react to a variation in FUV.

6. THE MAIN CONCLUSIONS

1. There exists a phase shift between time series of $F_{10.7}$ and FUV fluxes. In particular, the values of the "ultraviolet" activity index I_{La} are delayed for ~1 day relative to $F_{10.7}$.

2. The apparent delay is evidently due to different aging of active regions and to limb darkening functions.

3. The use of L_α emission, rather than $F_{10.7}$, as an activity index for calculations of FUV fluxes avoids errors arising from the phase shift between FUV and $F_{10.7}$.

4. Using the L_α flux improves the accuracy of such calculations from 10-20% due to a better correlation between FUV and La radiations.

The synchronous behavior of the L_α-line and FUV radiations, especially for 27-day variations, and the higher (compared with $F_{10.7}$) correlation coefficient indicate that L_α flux should replace the traditional $F_{10.7}$ index in calculations and forecasts of FUV fluxes.

REFERENCES

Bocharova, N. Yu., Nusinov, A. A.: 1983, *Solnechnye dannye* (Rus. Solar Data) N 1, 106.

Cook, J. M., Brueckner, G. E., Van Hoosier, M. E.: 1980, *J. Geophys. Res.* **A85**, 2257.

Hinteregger, H. E.: 1981, *Adv. Space Res.* **1**, No 12, 39.

Hedin, A. E.: 1987, *J. Geophys.Res.* **A92**, 4649.

Jacchia, L. G. :1977, Spec. Report N 375 Smithsonian Inst. Astrophys. Observ.

Katyushina, V. V., Krasinetz, M. V., Nusinov, A. A., Barth, C. A., Rottman, G. J.:1991, *Geomagnetizm i Aeronomya* (Rus. Geomagnetizm and Aeronomy) **31** , 225.

Kazachevskaja, T. V., Gonyukh, D. A.: 1988, *Solnechnye dannye* (Rus. Solar Data) N 9,88.

Krasinetz, M. V., Nusinov, A. A.: 1991, *Solnechnye dannye* (Rus. Solar Data) N 7, 88.

Lean, J. L., White, O. R., Livingston ,W. S. et al.: 1982, *J. Geophys. Res.* **A87,** 10307.

Lean , J.: 1987, *J. Geophys. Res.* 1987 **D92**, 839.

Nicolet, M., Bossy, L.: 1985, *Planet Space Sci.* **33**, 507.

Nusinov, A., Kazachevskaja, T., Katyushina, V.: 1989, *Solar-Terrestrial Prediction* : Proc. of a Workshop at Leura. Australia **1**, 546.

Solar Geophysical Data. Prompt reports.: 1982-1990, NOAA USA.

Vernazza, J.E., Avrett, E.H., Loeser, R.: 1985, *Astrophys. J. Suppl. Ser.* **45**, 635.

Vidal-Madjar, A., Pissammay, B.:1980, *Solar Phys.* **66**, 259.

MODELED SOFT X-RAY SOLAR IRRADIANCES

W. KENT TOBISKA

TELOS/JPL, MS 264-765, 4800 Oak Grove Dr., Pasadena, CA 91109, U.S.A.

Abstract. Solar soft X-rays have historically been inaccurately modeled in both relative variations and absolute magnitudes by empirical solar extreme ultraviolet (EUV) irradiance models. This is a result of the use of a limited number of rocket data sets which were primarily associated with the calibration of the AE-E satellite EUV data set. In this work, the EUV91 solar EUV irradiance model has been upgraded to improve the accuracy of the 3.0 to 5.0 nm relative irradiance variations. The absolute magnitude estimate of the flux in this wavelength range has also been revised upwards. The upgrade was accomplished by first digitizing the SOLRAD 11 satellite 4.4 to 6.0 nm measured energy flux data set, then extracting and extrapolating a derived 3.0 to 5.0 nm photon flux from these data, and finally by performing a correlation between these derived data and the daily and 81-day mean 10.7 cm radio flux emission using a multiple linear regression technique. A correlation coefficient of greater than 0.9 was obtained between the dependent and independent data sets. The derived and modeled 3.0 to 5.0 nm flux varies by more than an order of magnitude over a solar cycle, ranging from a flux below 1×10^8 to a flux greater than 1×10^9 photons cm^{-2} s^{-1}. Solar rotational (27-day) variations in the flux magnitude are a factor of 2. The derived and modeled irradiance absolute values are an order of magnitude greater than previous values from rocket data sets related to the calibration of the AE-E satellite.

1. Introduction

Extreme ultraviolet (EUV) empirical models are important from an aeronomical perspective. The solar EUV irradiance is a fundamental thermospheric and ionospheric energy input though there are few measured data. The satellite data sets that exist extend through various levels of solar activity though they often contain missing days and large uncertainties. Rocket measurements have greater accuracy compared to the satellite data yet only provide measurements for very short time intervals. Given these conditions, models are important.

Some unresolved solar EUV and soft X-ray questions are posed. For example, during solar cycle minimum conditions is an EUV/soft X-ray value for the 5 to 57.5 nm total integrated flux interval higher than irradiances measured by rockets in the 1970s? Richards and Torr (1984), Ogawa and Judge (1986), Link *et al.* (1988), and Winningham *et al.* (1989) suggest that some or all of the total flux in that wavelength range is higher during low solar activity by up to a factor of 2. This implies that either the entire range of 5 to 57.5 nm has greater flux or that the shorter wavelengths have dramatically more flux and the longer wavelengths have little increased flux compared to those variations presently modeled. In the latter scenario, the total flux of 5 to 57.5 nm is only slightly higher; the longer wavelengths contribute the bulk of the measured photons while the shorter wavelengths provide substantially more secondary ionization energy into the lower thermosphere and E-region ionosphere.

Solar Physics **152**: 207–215, 1994.

A corollary question is whether or not a value for the soft X rays (0.1 to 10 nm) should be used which is substantially higher by more than an order of magnitude at all levels of solar activity? Barth *et al.* (1988) and Siskind *et al.* (1990) suggest from lower thermospheric nitric oxide data and model comparisons that these soft X rays should be scaled upwards significantly (up to 60 times).

Additionally, measured E-region electron densities are higher by 30 to 50%, depending upon date, compared to state-of-the-art ionospheric model calculations (PRIMO workshop, 1991 CEDAR meeting; M. Buonsanto, private communication, 1991). Even slight increases in solar soft X-ray irradiance values on the order of a few percent would improve the modeled electron density values.

In the past, this solar soft X-ray flux has been modeled with data from AFGL rocket data (Hinteregger *et al.*, 1981) which has been extrapolated into a time series based upon measurements by the AE-E satellite. The AE-E instruments themselves did not measure below 15 nm. The low, moderate, and high solar activity measurements in the wavelength range of 4.4–6.0 nm by the SOLRAD 11 (SR11) satellites (Kreplin, 1970; Kreplin *et al.*, 1977; Kahler and Kreplin, 1991; Kreplin and Horan, 1992) provide new material for modelers, especially for relative day-to-day solar variability. Thus, the SR11 data may provide plausible answers to the questions posed above.

2. Derivation of the SOLRAD 11 3.0 – 5.0 nm Data

In order to develop linear regression coefficients between the SR11 4.4–6.0 nm energy flux data and ground-observed proxy data which can be used to model the flux, the SR11 data were processed in several steps. The published data were first digitized and recalibrated so that the digitized data matched the published data. Then the data were transformed from 4.4–6.0 nm energy flux values into derived 3.0–5.0 nm photon flux values. This process enabled the derived SR11 data to be compared with previous rocket measurements related to the AE-E calibration as described by Hinteregger *et al.* (1981) (these data are referred to herein as HEH) and permitted linear regressions to be conducted with proxy data sets.

2.1. DIGITIZATION OF THE SOLRAD 11 44 – 60 Å DATA

The published 4.4–6.0 nm energy flux data of Kreplin and Horan (1992) were photocopied then scanned into a TIFF format digital image that is readable by a MacIntosh computer. The image was manually edited to remove all textual and axes detail while leaving the 4.4–6.0 nm energy flux time series intact. MacDrawPro® software was used to perform these tasks.

The image time series was next converted to a vector of x-y coordinates (uncalibrated in the time and flux value dimensions) using the NIH Image 1.45 software. Figure 1 demonstrates the method by which data were vectorized. The scanned time series is rotated 90° clockwise, ANDed with a mask of lines to create a scatter plot, then analyzed to read out the Y and X values of the points.

The linear x-y values were recalibrated back to the original logarithmic values of the published energy flux time series by the inverse of the relationship

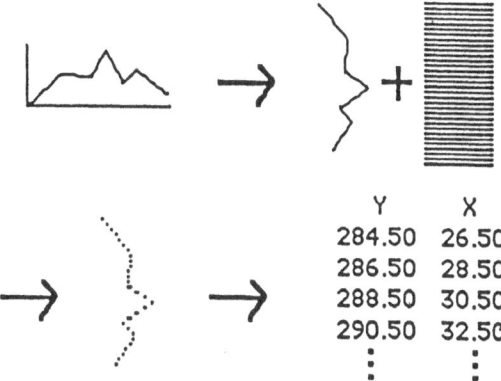

Y	X
284.50	26.50
286.50	28.50
288.50	30.50
290.50	32.50
⋮	⋮

Fig. 1. The scanned TIFF image digital time series is rotated 90° clock-wise, ANDed with a mask of lines to create a scatter plot, then analyzed to read out the linearized Y and X values of the points (NIH Image 1.45 handbook).

$$y = a + b \cdot a \log_{10} z \qquad (1)$$

where z is the recalibrated, digital, logarithmic energy flux value, y is the linear value from the scanned image, and a and b are the linear offset and slope coefficients from the scanned data. The estimated temporal uncertainty introduced into the data by this digitization method was $\pm 1\%$ and the flux magnitude uncertainty was up to 10%. F.E. Eparvier (private communication, 1993) indicated that his independent digitization of these data yielded values within 5% of the absolute numbers used in this study. Figure 2 (a) shows the digitized SR 4.4–6.0 nm energy flux beginning with SR7A and continuing through SR11.

2.2. TRANSFORMATION: SOLRAD ENERGY FLUX TO PHOTON FLUX

The derived 4.4–6.0 nm photon flux, $F_{44\text{-}60}$, was obtained by converting the calibrated 4.4–6.0 nm energy flux, $E_{44\text{-}60}$, to photon flux by

$$F_{44\text{-}60} = E_{44\text{-}60} \frac{\lambda}{12400 \cdot 1.6022 \times 10^{-12}} \qquad (2)$$

where λ is 45.5 Å and is the wavelength at which 50% of the flux has entered the photometer bandpass beginning with the filter cutoff at 43 Å (cf. figure 4, Kreplin and Horan, 1992).

The 3.0–5.0 nm photon flux was derived from the 4.4–6.0 nm photon flux by the following algorithms. The photon flux for the combined coronal and chromospheric components is defined as

$$F_{30\text{-}50} = F_{44\text{-}60} \frac{A_1}{B_1} \qquad (3)$$

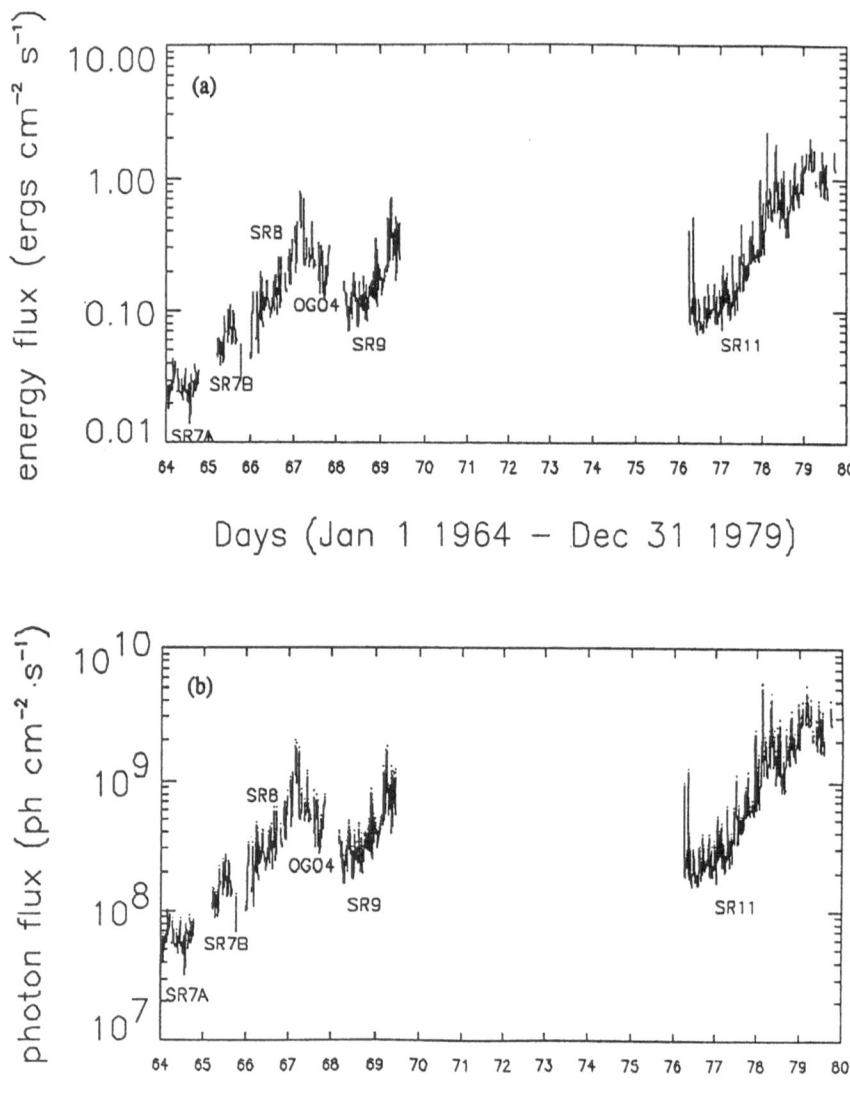

Fig. 2. (a) Daily SR7A, 7B, 8, 9, 11, and OGO4 4.4–6.0 nm energy flux, E_{44-60}, in ergs cm^{-2} s^{-1} between January 1, 1964 and December 31, 1979. (b) Daily derived 3.0–5.0 nm photon flux, F_{30-50}, in photons cm^{-2} s^{-1} $\Delta\lambda^{-1}$ for the same time period and satellite data sets.

where F_{44-60} is the derived SOLRAD 4.4–6.0 nm photon flux. The fraction of the 4.3–5.0 nm photon flux in the SOLRAD bandpass (cf. figure 4, Kreplin and Horan, 1992), A_1, is set to a value of 0.90. It is determined by integrating the digitized bandpass (photometer response) values in the 4.3–5.0 nm interval and then by finding the fractional contribution of that interval (90%) to the total integrated values under the 4.3–5.3 nm curve. Next, the fraction of the photon flux in the 4.3 – 5.0 nm interval compared to the 3.0–5.0 nm interval, B_1, varies from solar minimum to solar maximum conditions. The value used here is determined by an empirical linear relationship where

$$B_1 = p_1 - \frac{F_{10.7} - 68}{175} \, p_2 \qquad (4)$$

where $F_{10.7}$ is the 10.7 cm flux for a given date, 68 is the value of $F_{10.7}$ at solar minimum, and 175 is the range of $F_{10.7}$ values between solar minimum (68) and solar maximum (243). The fractional photon flux between 4.3–5.0 nm for solar minimum, p_1, is 0.86 while the fractional flux for solar maximum, p_2, is 0.07 as determined from the SC#21REFW and F79050N data files provided by Hinteregger et al., (1981). Figure 2 (b) shows the derived SR 3.0–5.0 nm photon flux, F_{30-50}, for the same time period and satellite data as Figure 2 (a).

3. Data – Proxy Correlation and Comparison

A multiple linear regression technique was used to obtain the data and proxy correlation. The same Interactive Data Language (IDL) code used to produce EUV91, described by Tobiska (1991), was also used in this work. The $F_{10.7}$ daily and 81-day mean (F_{81}) values were the independent data sets. These ground-based measurements are representative of the transition region and coronal solar emissions which are also the source layers for the 3.0–5.0 nm flux.

The dependent data were the SR11 measurements. Even though SR7 through SR11 were available, the SR11 data alone were selected for correlation. These data represented low to high solar activity measured by two intercalibrated instruments (SR11A and SR11B were flown at the same time). These instruments (referred to herein as SR11) were the standard to which other SR instruments were compared. The SR11 time frame also allows comparisons between the SR data and HEH data.

Most of the flux in the 3.0–5.0 nm interval comes from solar transition and coronal source regions. From the source region code in the SC#21REFW data file, the small chromospheric component of this wavelength range consists of two unspecified ion lines at 46.67 Å and 47.87 Å. These two lines in the HEH data contributed 37% of the flux in this wavelength interval at solar minimum and 12% of the flux at solar maximum. The coronal component is clearly dominant and therefore the combined source region contributions are modeled using $F_{10.7}$ and F_{81}.

Figure 3 (a) shows the newly modeled 3.0–5.0 nm flux (herein temporarily called EUV93) for the period between 1962 and 1990 while Figure 3 (b) shows the comparison between the SR11 data and the EUV93 values for 500 days during the rise of solar cycle 21. The correlation coefficient, r, between the SR11 and the $F_{10.7}$ data is 0.95

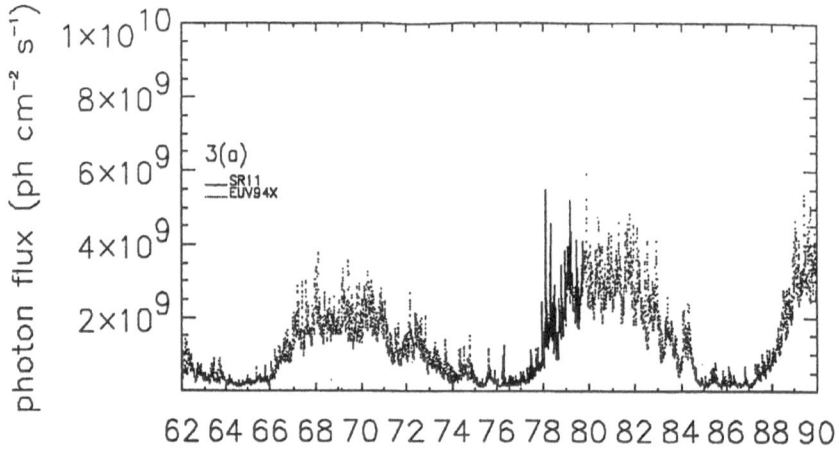

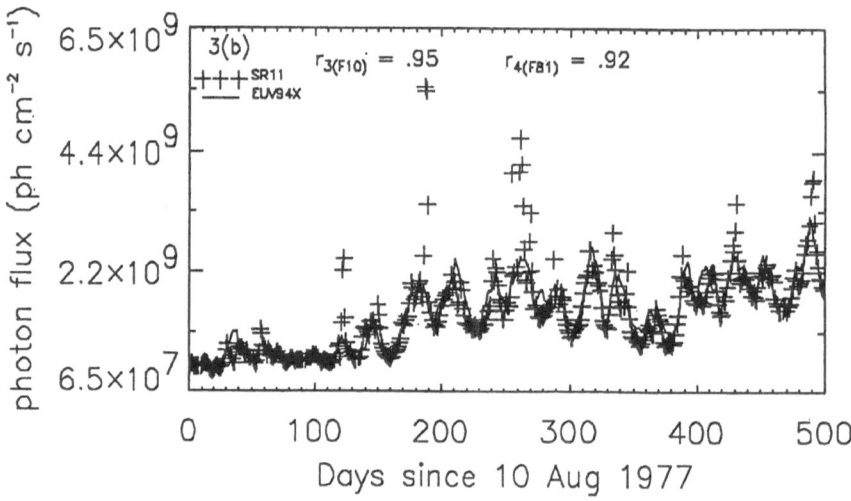

Fig. 3. (a) Daily modeled (EUV93) 3.0–5.0 nm photon flux (combined coronal and chromospheric components) between January 1, 1962 and December 31, 1989 (dots) and SR11 derived
data (solid line). (b) Daily EUV93 (line) and derived SR11 3.0–5.0 nm photon flux (plus symbol)
for 500 days beginning August 10, 1977.

while r between SR11 and F_{81} is 0.92. The absolute magnitude and the relative solar rotation peak-to-valley variation in EUV93 compares very well with the derived data. A typical EUV93 27-day peak-to-valley ratio is 2.0 while a typical derived data 27-day peak-to-valley ratio is 2.2.

Comparisons between derived data of 3.0–5.0 nm photon flux, the EUV91 model, and the HEH values are shown in Table I. The SR values for solar minimum conditions (76200) are 7.4 times larger than the HEH values and are 27.4 times larger at solar max-imum (79050). These large ratios have produced debate regarding the ratio of derived to measured flux for the two sets of instruments (SOLERS22 meeting, 1993).

TABLE I

Comparison of photon fluxes in the 3.0–5.0 nm interval

Date (YYDDD format)	Data (photons cm^{-2} s^{-1} $\Delta\lambda^{-1}$) origin		
	HEH	EUV91	SR11
76200 (SC#21REFW)	2.3×10^7	1.6×10^7	1.7×10^8
79050 (F79050N)	1.9×10^8	1.6×10^8	5.2×10^9

Figure 4 (a) shows the comparison of the SR11, EUV93, and HEH absolute flux values for the 3.0–5.0 nm interval. Between mid-1977 and mid-1979, during the rise of solar cycle 21, there consistently is an order of magnitude difference between the SR-derived/EUV93 flux and the HEH data. Also, the absolute value of the derived 3.0–5.0 nm flux during the solar cycle rise increased by at least a factor of 10. Figure 4 (b) shows the ratios of these data for the same time frame. The SR11/HEH ratio averages around 27 for most solar activity conditions outside of solar minimum.

The absolute differences between SR11 and HEH data may represent calibration differences between the instruments to be potentially resolved with new data sets such as YOHKOH SXT. The empirical model in this work does not attempt to resolve these absolute differences. Instead, it uses the SR11 data as the primary source, i.e., the dependent data set, to estimate the relative time variability of the 3.0–5.0 nm photon flux using the 10.7 cm daily and 81-day mean proxies.

4. Summary

Improved estimates of 3.0–5.0 nm soft X-ray photon flux relative variation values have been obtained. The estimates were accomplished by digitizing the SOLRAD 11 4.4 to 6.0 nm energy flux data, extracting and extrapolating a derived 3.0 to 5.0 nm photon flux, and by performing a correlation between these derived data and the daily and 81-day mean 10.7 cm radio flux emission using a multiple linear regression technique. The derived 3.0–5.0 nm photon fluxes correlate well with 10.7 cm daily and 81-day mean values with correlation coefficients of 0.95 and 0.92, respectively. The relative variability on 27-day solar rotational time scales during the rise of solar cycle 21 is a factor of 2.2 for the derived 3.0–5.0 nm data.

When comparing the derived SR data and the HEH data, the SR11 derived 3.0–5.0 nm data have ratios near 27 during the rise of solar cycle 21. In addition, the derived

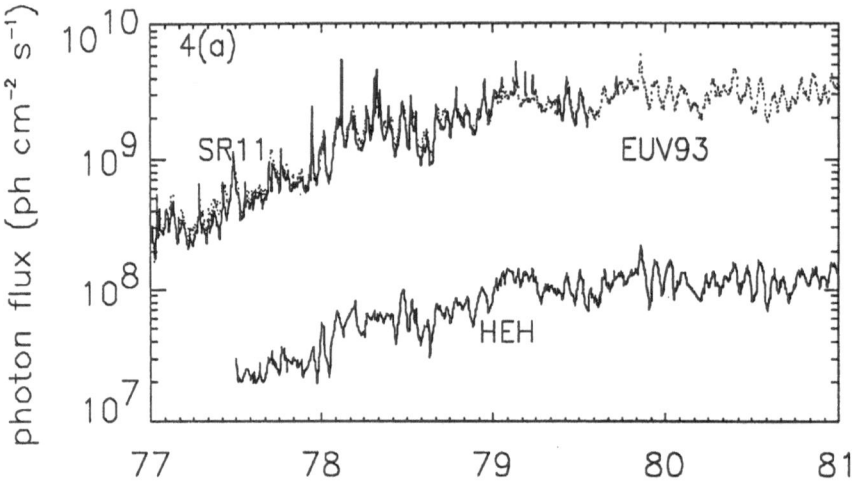

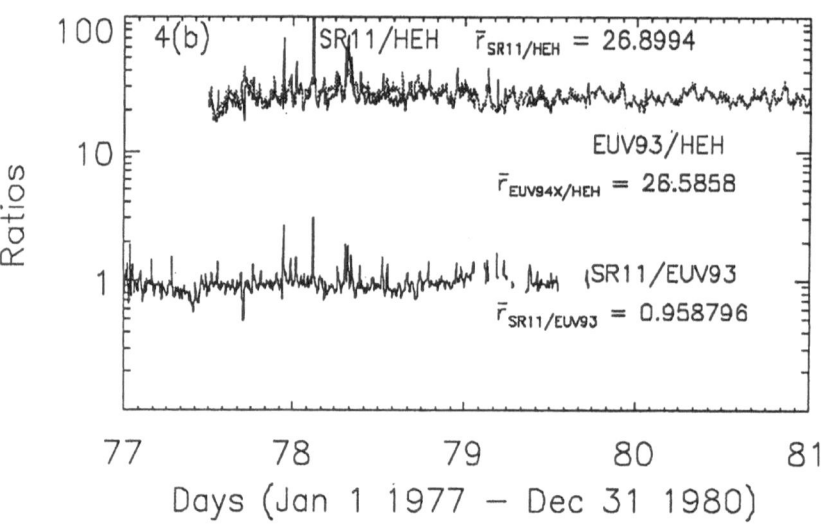

Fig. 4. (a) Daily SR11 (labeled solid line), EUV93 (dots), and HEH (labeled solid line) absolute 3.0–5.0 nm photon flux between January 1, 1977 and December 31, 1980. (b) Daily ratios of SR11 to HEH (labeled solid line), EUV93 to HEH (dots), and SR11 to EUV93 (labeled solid line) for the same time period.

and modeled 3.0–5.0 nm photon fluxes vary by more than an order of magnitude over rising solar cycle activity time scales from 1×10^8 to 1×10^9 photons cm^{-2} s^{-1}.

The revised soft X-ray photon fluxes for all levels of solar activity suggest that important questions in E-region ionosphere and lower thermosphere studies which are related to time-varying solar energy input can be investigated in more detail.

Acknowledgments

Mr. Bill Hoffman and Mr. Paul Fisher of the Jet Propulsion Laboratory Galileo Project provided much appreciated assistance in digitizing the SOLRAD data sets. The author also thanks Dr. Robert Kreplin, Dr. Anatoly Nusinov, and Dr. Frank Eparvier who provided many helpful comments on this work.

References

Barth, C. A., Tobiska, W. K., Siskind, D. E., and Cleary, D. D.: 1988, *Geophys. Res. Letters* **15**, 92.
Hinteregger, H. E., Fukui, K., and Gilson, B. R.: 1981, *Geophys. Res. Letters* **8**, 1147.
Kahler, S. W., and Kreplin, R. W.: 1991, *Solar Phys.* **133**, 371.
Kreplin, R. W.: 1970, *Ann. Geophys.* **26**, 567.
Kreplin, R. W., Dere, K. P., Horan, D. M., and Meekins, J. F.: 1977, in O. R. White (ed.), *The Solar Output and Its Variation,* Colorado Associated University Press, Boulder.
Kreplin, R. W. and Horan, D. M.: 1992, in R. F. Donnelly (ed.), 'Proceedings of the workshop on the solar electromagnetic radiation study for solar cycle 22,' p. 405.
Link, R., Gladstone, G. R., Chakrabarti, S., and McConnell, J. C.: 1988, *J. Geophys. Res.* **93**, 14,631.
Ogawa, H. S., and Judge, D. L.: 1986, *J. Geophys. Res.* **91**, 7089.
Richards, P. G., and Torr, D. G.: 1984, *J. Geophys. Res.* **89**, 5625.
Siskind, D. E., Barth, C. A., and Cleary, D. D.: 1990, *J. Geophys. Res.* **95**, 4311.
Tobiska, W. K.: 1991, *J. Atmos. Terr. Phys.,* **53**, 1005.
Winningham, J. D., Decker, D. T., Kozyra, J. U., Jasperse, J. R., and Nagy, A. F.: 1989, *J. Geophys. Res.* **94**, 15,335.

COSMIC RAYS AS AN INDICATOR OF SOLAR ACTIVITY

M. MINAROVJECH, V. RUŠIN, AND M. RYBANSKÝ

Astronomical Institute, Slovak Academy of Sciences
059 60 Tatranská Lomnica, The Slovak Republic

Abstract. The modulation of galactic cosmic rays, as derived from ground-based neutron monitors over a solar cycle appears to be a good indicator of solar activity. Especially good correlation was found between short-term variations of cosmic ray flux and solar UV emission in the region of 250 nm during one year (1982).

1. Introduction

Current efforts to study total solar irradiance variations, in our opinion, are in understanding its changes over the different temporal variations. A basic question is, how is the total solar irradiance connected with solar activity?

Correlation between variations of total solar irradiance and different indices of solar activity can show us which spectral regions should have maximum influence on its variation, e.g. Pap (1992). This information allows us to model the changes, and, on behalf of the model, to extrapolate a set of measurements in the past.

In this paper the coronal index of solar activity (CI) and solar UV flux variations are compared with variations in cosmic ray flux.

2. Results and Conclusions

As was shown in the paper of Rybanský et al. (1993), we prepared the coronal index of solar activity (CI), derived from ground-based green corona observations over the period 1939-1991. Comparing the temporal variations of different solar indices we found that the course of solar activity might be derived from the depth of galactic cosmic ray modulation (the modulation changes only by 20 percent from minimum to maximum, and is inversely correlated), which is derived from ground-based neutron monitors. Many authors have discussed solar irradiance changes but without discovering physical mechanisms that explain those changes. Recently, Nagashima et al. (1991) compared a distribution of magnetic fields on the solar surface and modulation of cosmic ray at the Earth, and received satisfactory results. Figure 1 displays running averages of 27-daily values of CI and cosmic ray data obtained at Deep River station (geomagnetic latitude 46°, cut-off rigidity: 1.02 GV) and Tokyo station (geomagnetic latitude 36°, cut-off rigidity: 11.6 GV), respectively. The course of CI for the same period is shown for comparison.

Coefficients of cross-correlation function were found to have a relatively flat course, and between the CI and Deep River amounts 0.84 for a time shift of 72 days. A similar value for CI and Tokyo is 0.82 for a time shift of 66 days. This means CI changes precede cosmic ray variations by approximately two months (on average).

It is also interesting to note a striking similarity (see Figure 2) of temporal short-term variations obtained from NIMBUS-7 measurements in the 245-250 nm channel and Deep river cosmic ray data for 1982. The 245-250 nm data were adopted from Donnelly (1988). All the data in Figure 2 represents a difference of daily values from 27-days running averages (long-term variations were removed). This similarity was not found between the CI and the solar irradiance data.

Solar Physics **152**: 217–219, 1994.

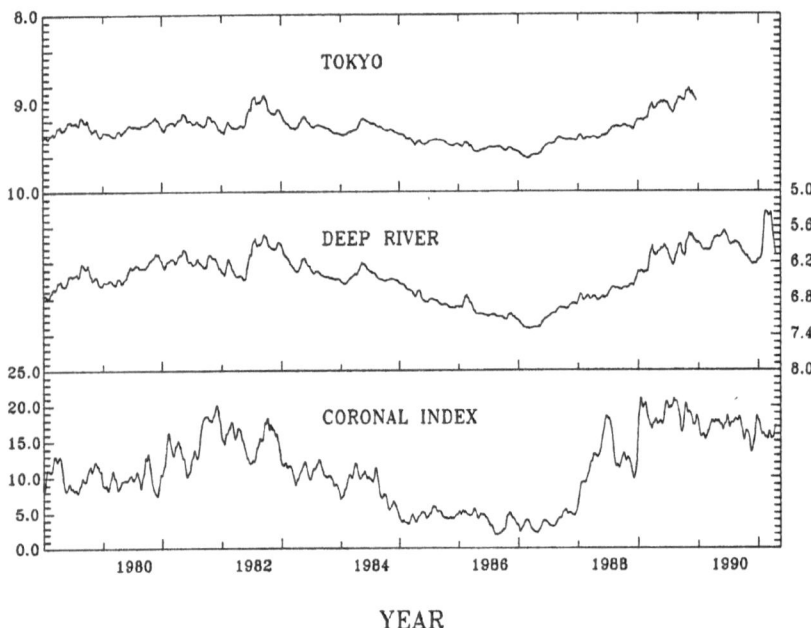

Fig. 1.   Plot of the running averages of daily values of CI and cosmic ray data from Deep River and Tokyo stations.

We lack the experience to theorize on the relationship between cosmic ray and solar activity, however, it does exist. The temporal variation in the cosmic ray flux, should be mainly induced by changing conditions in the heliosphere (different parameters of solar wind and the large-scale solar magnetic field over the solar cycle), even if the other parameters are not negligible (diffusion, convection and energy changes). We would simply like to stress that data from neutron monitors, obtained by ground-based cosmic ray stations, are easy to access and independent of weather.

Acknowledgements

The authors are grateful to the Slovak Grant Agency for Science (Grant No. 59/1993) for partial support of this work. Travel costs (V.R.) were partially defrayed by U. S. National Science Foundation Grant ATM-9224968 and SOLERS 22 officials.

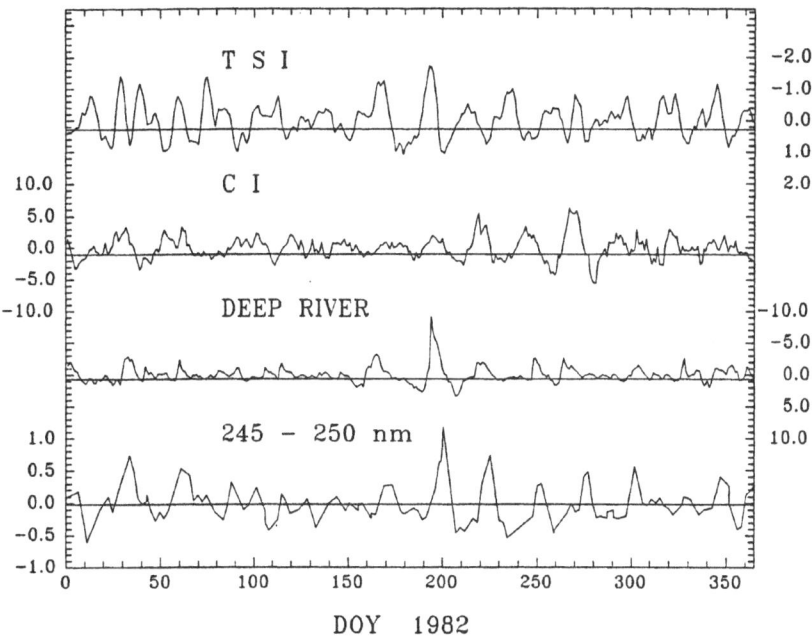

Fig. 2. Short-term temporal variations of 245-250 nm solar UV flux, Deep River neutron cosmic ray, CI, and total solar irradiance (TSI) for 1982.

References

Donnelly, R. F.: 1988, *Annales Geophys.* **6**, 417.
Nagashima, K., Fujimoto, X. and Tatsuoka, R.: 1991, *Planet. Space Sci.* **39**, 1617.
Pap, J.: 1992, *Astron. Astrophys.* **214**, 524.
Rybanský, M., Rušin, V., Minarovjech, M. and Gašpar, P.: 1993, *Solar Phys.* in press.,

MERIDIONAL MOTIONS OF SUNSPOT GROUPS
DURING ELEVEN ACTIVITY CYCLES

G. LUSTIG[1] AND H. WÖHL[2]

[1]Institut für Astronomie, Universitätsplatz 5, A-8010 Graz, Austria
[2]Kiepenheuer-Institut für Sonnenphysik, Schöneckstr.6, D-79104 Freiburg, Germany

ABSTRACT. Greenwich data (1874-1976) are used for a time-dependent analysis of meridional motions of sunspot groups. We obtain the latitude-dependence of meridional motions of sunspot groups with respect to a mean latitude determined for half-year intervals. The daily meridional motions of groups are also given separately for growing and decaying sunspot groups. The development is determined from changes of sunspot areas. Our results are compared with the reductions performed by Howard (1991b) using the Mt. Wilson sunspot data from 1917 until 1985: Although we have smaller errors, we do not find any significant drift. We also do not find different trends in the meridional motions of growing as compared to decreasing sunspots.

1. Introduction

Meridional motions of sunspots have been studied by a number of people with material from Greenwich and Kanzelhöhe (Balthasar and Wöhl 1980, Balthasar et al. 1986, Hanslmeier and Lustig 1987, Lustig and Hanslmeier 1987, Lustig and Wöhl 1991, Tuominen et al. 1983). Others have investigated meridional motions and various other phenomena for individual sunspots, using either the white-light or the magnetographic data sets from Mt. Wilson (Howard et al. 1984, Howard and Gilman 1986, Howard 1991a, Howard 1991b, Howard 1991c).

Howard (1991b) and Howard and Gilman (1986) used the Mt. Wilson white-light data from 1917 to 1985 to study the meridional motions of sunspots as a function of different parameters. In this paper - as a first step - we analyzed the entire set of data from Greenwich (Photoheliographic Results). Later we intend to investigate data from the Solar Observatory Kanzelhöhe between 1947 and 1991. Preliminary information about the research project was given by Lustig and Wöhl (1993).

2. Data and Analysis

Daily sunspot data from Greenwich, the "Photoheliographic Results", have been keyed into a computer. These data include the daily positions, the areas, the numbers of spots, the latitudes, and some other characterizations. We grouped the daily positions in pairs for two consecutive days. All meridional motions are expressed in degrees per 24 hours.

For this study, which includes data for the years 1874 through 1976 we use 100211

Solar Physics **152**: 221–226, 1994.
© 1994 *Kluwer Academic Publishers.*

pairs of positions for meridional motions. Of the total number of groups, 39191 grew from the first to the second day, and 61020 groups decreased in area. We exclude all motions larger than 2.0 deg/day in latitude. The data is divided into latitude zones of 2.5 or 5.0 degrees respectively; we determine these latitudes as the mean for two consecutive days.

We handle the data similarly to Howard (1991b) for the different spot groups. First we determine the average latitude, ξ_o, of the spots in each hemisphere, binned into six-month intervals. Figure 1 shows these averages for the northern and southern hemisphere for the time from 1874 to 1976. In two intervals near minimum activity no spots were in the northern hemisphere, so the average latitude was zero (1901 and 1912).

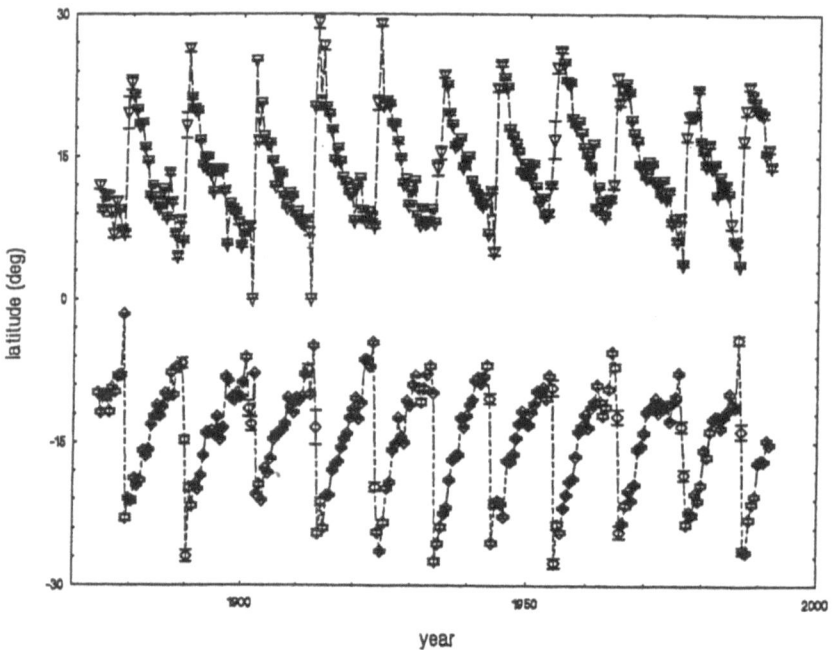

Fig. 1: The average latitudes of the sunspot groups for both hemispheres, binned into six-month intervals for the years 1874 to 1976

3. Results, Discussions and Conclusions

In Figure 2 we plot the daily latitude drifts as a function of the absolute latitude for all 100211 pairs of data for the time interval from 1874 to 1976. The northern and southern hemispheres are plotted together; averages are taken over 5 degrees in latitude. The error bars represent a standard deviation from the mean; latitude drift is positive toward the poles.

Linear fit:
drift [deg/day] = -0.00764 ($\pm$ 0.000580) + 0.00037 ($\pm$ 0.00022) · latitude [deg]

For comparison we include the curve from Howard (1991b) in our Figure 2. The drift in the two curves is similar but our daily meridional motions are smaller and not very different from zero. We cannot see a significant poleward motion for high latitude spot groups. There is an indication for an equatorward motion for low latitude groups.

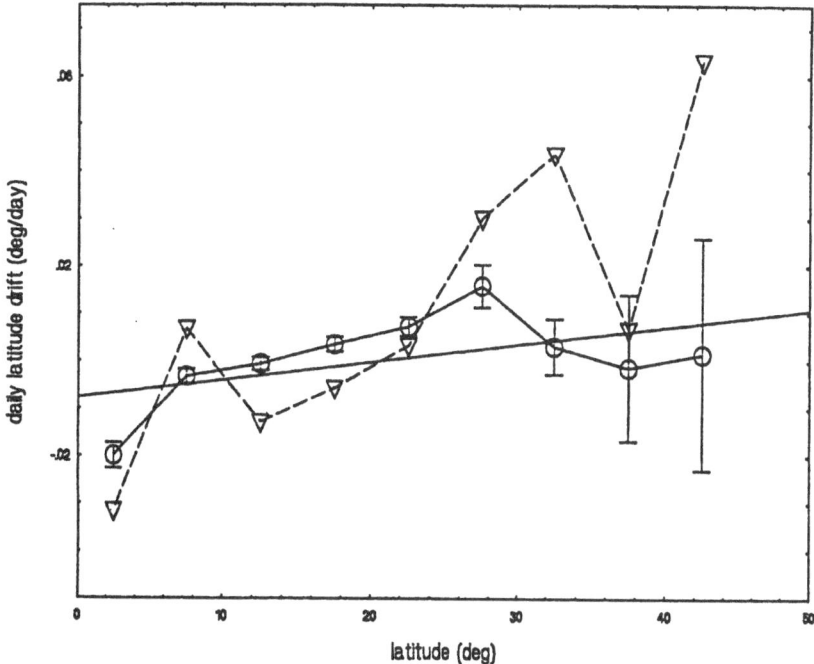

Fig. 2: Latitude drift as a function of absolute latitude for all the spot groups. Latitude drift is positive toward the poles. Greenwich data 1874 - 1976): full lines, Mt.Wilson data 1917 - 1985 (Howard, 1991b): dashed line. The straight full line is the linear least squares fit for the Greenwich data.

In Figure 3 we examine the meridional motions for growing (solid line) and decreasing (dashed line) spot areas separately.

For each case we performed a linear fit:
Growing in area:
drift [deg/day] = + 0.00533 ($\pm$ 0.01160) + 0.00056 ($\pm$0.00045) · latitude [deg]

and decreasing in area:

drift [deg/day] = - 0.01646 (± 0.01118) + 0.00031 (± 0.00043) · latitude [deg]

The slopes of the lines are not significantly different from zero . No trend can be seen and this contradicts the results for the decaying groups from the Mt. Wilson data set (Howard, 1991b).

   In Figure 4, similar to Figure 2, the latitude drifts for all spots over 5.0 degree intervals are given separately for both hemispheres. If one compares the two figures one can see very clearly that one looses details when averaging the two hemispheres. Different trends or effects can be lost.

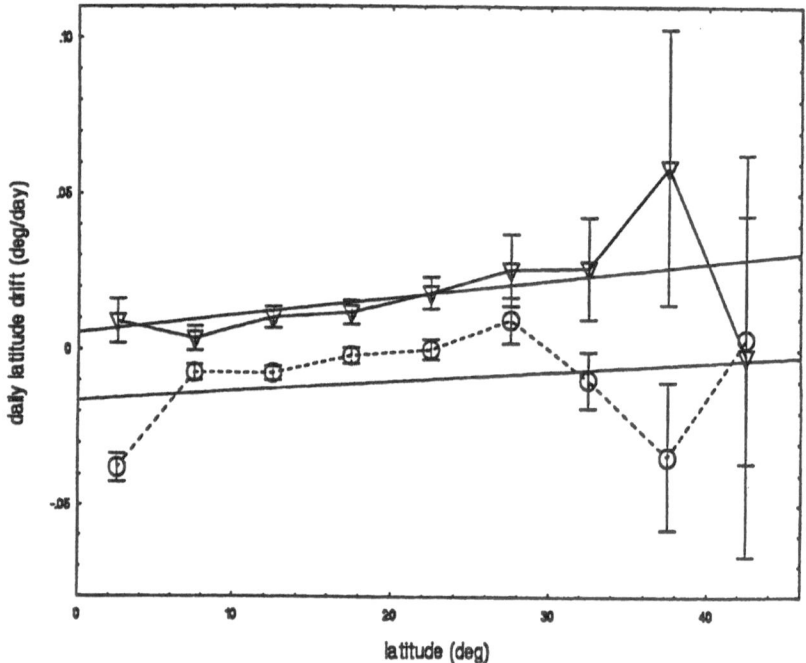

Fig. 3: The same as Fig. 2 for the Greenwich data except that spot groups growing in area (solid line) and decreasing in area (dashed line) are given separately. The straight lines are the least-squares solutions.

Finally we determine for each latitude value ξ latitude differences $(\xi - \xi_o)$ for the relevant hemisphere. In each hemisphere a positive ξ represents a position poleward of ξ_o. Near minimum activity there are spots at both high and low latitudes. At these times we handle the data in the following way: For all spots with $| \xi | > 10$ deg the values of ξ are also calculated for the two temporally adjacent values of ξ_o. If one of those gives $| \xi | < 5$ deg, that value is used instead of the original average. In Figure 5 the daily meridional drifts for spots growing in area (solid line) and decreasing in area (dashed-dotted line) are given as well as the straight lines for the linear fits for the two different types:

Growing in area:

drift [deg/day] = + 0.01244 (± 0.00565) - 0.00112 (± 0.00065) · ξ [deg]

and decreasing in area:

drift [deg/day] = + 0.01255 (± 0.00251) - 0.00072 (± 0.00029) · ξ [deg]

Only for the growing spot groups can we confirm the effect found by Howard (1991b) for the Mt. Wilson data: They also show a tendency to drift poleward on the equatorward side of ξ_0 and equatorward on the poleward side, although not significant. The decaying groups do not show the opposite direction, as in the Mt. Wilson sample. The error bars are in general smaller for the Greenwich results as compared to the results from the Mt. Wilson data; nevertheless no gradient of the linear fit is significant.

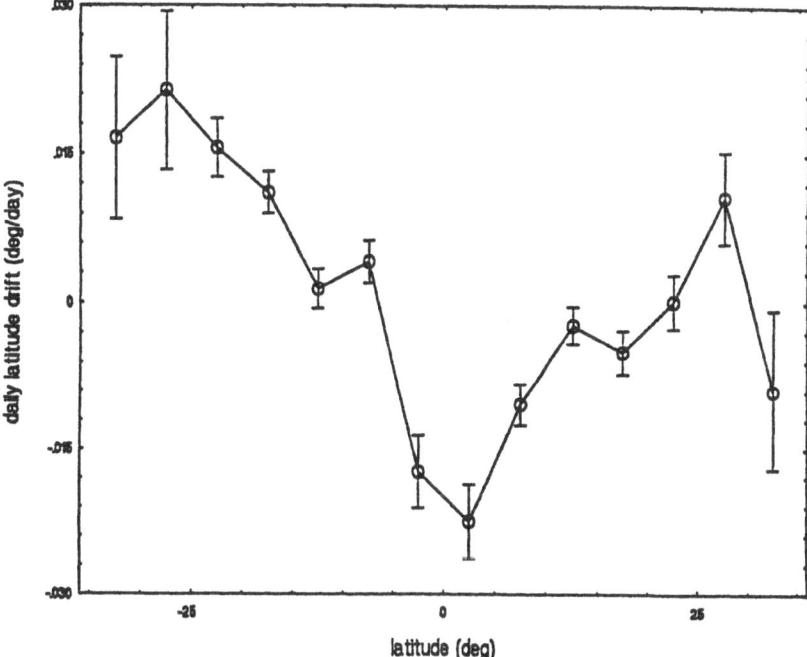

Fig. 4: Latitude drifts for all sunspot groups, separately for the northern and the southern hemispheres

The Greenwich data are weighted positions of groups of sunspots while the Mt. Wilson data are positions of individual spots. It is possible that this could account for some differences in these results (Howard and Gilman, 1986).

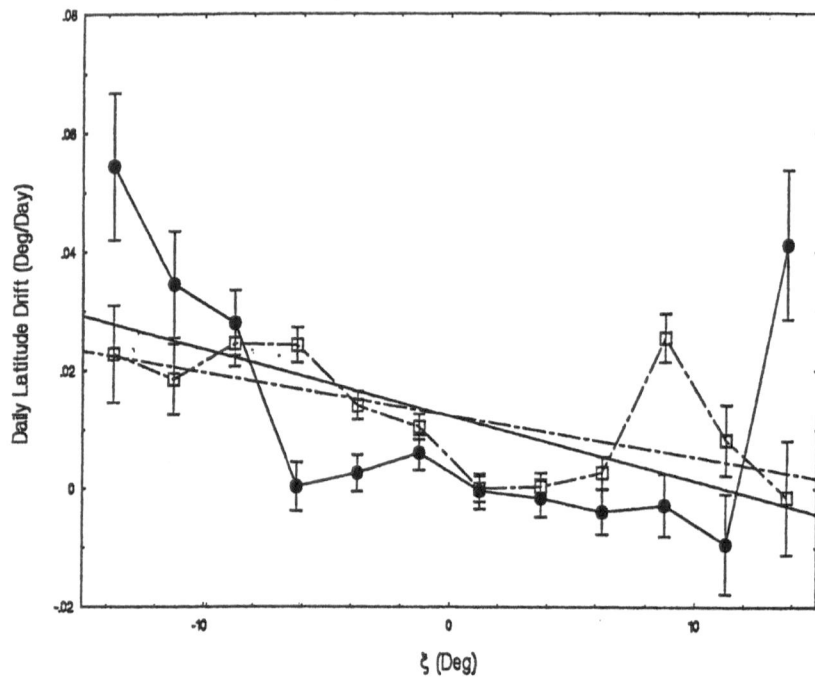

Fig. 5: Spot groups growing in area (solid line) and decreasing in area (dashed-dotted-line). The straight-lines are the least-squares solutions.

References

Balthasar, H. and Wöhl, H.: 1980, *Astron. Astrophys.* **92**, 111

Balthasar, H., Vázquez, M. and Wöhl, H.: 1986, *Astron. Astrophys.* **155**, 87

Hanslmeier, A. and Lustig, G.: 1986, *Astron.Astrophys.* **154**, 227

Howard, R.F.: 1991a, *Solar Phys.* **131**, 259

Howard, R.F.: 1991b, *Solar Phys.* **135**, 327

Howard, R.F.: 1991c, *Solar Phys.* **135**, 43

Howard, R.F. Gilman, P.A. and Gilman, P.I.: 1984, *Astrophys.J.* **283**, 373

Howard, R.F. and Gilman, P.A.: 1986, *Astrophys.J.* **307**, 389

Lustig, G. and Hanslmeier, A.: 1987, *Astron.Astrophys.* **172**, 332

Lustig, G. and Wöhl, H.: 1991, *Astron.Astrophys.* **249**, 528

Lustig, G. and Wöhl, H.: 1993, in *Alvensleben, A. von (ed.) : JOSO Annual Report 1992*, 82

Tuominen, J., Tuominen, I. and Kyröläinen, J.: 1983, *M.N.R.A.S.* **205**, 691

RADIATION-HYDRODYNAMIC WAVES AND GLOBAL SOLAR OSCILLATIONS

J. STAUDE

Astrophys. Inst. Potsdam, Solar Observatory "Einsteinturm", D-14473 Potsdam, Germany

and

N. S. DZHALILOV and Y. D. ZHUGZHDA

IZMIRAN, Troitsk city, Moscow Region, 142092 Russia

Abstract. We investigate nonadiabatic hydrodynamic waves in a nongrey, radiating, thermally conducting, homogeneous atmosphere in LTE with a finite mean free path of photons. Avoiding the Eddington approximation the remaining simplifications in the basic equations are discussed, the generalized dispersion relation is analysed, and some wave properties in a grey model are studied. The properties of waves in a stratified atmosphere are analysed as well. In connection with the predicted properties of the nonadiabatic waves we discuss observations of p–modes by measuring brightness fluctuations.

1. Introduction

Observations of global solar oscillations show the presence of the p–modes in velocity as well as in luminosity data. There are hints at deviations from an adiabatic behaviour, pointing out that waves and oscillations in stars are inevitably nonadiabatic. This is mainly due to the interaction of the acoustic waves with radiative transfer and with other processes such as thermal conduction. To simplify the complicated situation, theoretical efforts are often focused on the adiabatic case, and nonadiabatic approaches are usually restricted to a grey, homogeneous atmosphere and to further assumptions such as the Eddington approximation. An exact theoretical treatment, however, requires the self-consistent consideration of the full set of basic equations for all types of processes including the dynamical effects of radiation and for realistic models of the atmosphere as well.

For practical applications in modelling some of the approximations mentioned above are inevitable, but their accuracy should be tested by comparing their results with those from the more general equations. In the present paper our interest is focused on waves in the solar atmosphere. For this end we will discuss the terms of the basic equations of radiation hydrodynamics in the appropriate frame and fluid regime, moreover, the effects of a nongrey and a stratified atmosphere will be considered.

The most general dispersion relation for waves in a uniform, nongrey, scattering atmosphere has been derived by Prokofiev (1957), without analysing it in more detail. Golitsyn (1963) and other geophysicists considered the dispersion relation of acoustic waves in a nongrey, homogeneous medium in

Solar Physics **152**: 227–239, 1994.

LTE and analysed it for the optically thin case in order to model the atmosphere of the Earth. For astrophysical applications, however, the whole range of opacities should be considered. The basic equations of radiation hydrodynamic waves in an ideal gas with small velocities $v \ll c$ (c and v are the speed of light and of the gas, respectively) have been formulated and discussed in detail by Mihalas and Mihalas (1984) (hereafter, MM) neglecting thermal conductivity. The latter authors also derived a solution of the linearized equations for a homogeneous, infinite, grey atmosphere assuming LTE and the Eddington approximation; the resulting dispersion relation has been analysed for the study of the spatial evolution. Extensions were made by several authors: Ibáñez and Plachco (1989) included the effect of thermal conduction in the quasi-static limit, the present authors abandoned the Eddington approximation and extended the analysis of MM to both the temporal and the spatial development (Dzhalilov et al., 1992) (hereafter, Paper 1), and the nongrey case has been discussed as well (Zhugzhda et al., 1993) (hereafter, Paper 2).

Analytical solutions for a uniform medium are needed, even in numerical calculations for a stratified atmosphere, in order to allow for a correct choice of the boundary conditions and a satisfactory analysis of the numerical results. For example, our analytical solution for a uniform, grey atmosphere (Paper 1) shows, that radiation wave modes exist together with nonadiabatic acoustic modes, that both modes are coupled and that they have very different phase relations.

In the subsequent section we will discuss properties of the basic equations in order to clarify the allowed simplifications.

2. Basic Equations

The basic equations of radiation hydrodynamics for an ideal gas with $v \ll c$ have been formulated in Eqs. (1) of Paper 1 referring to the detailed discussion by MM. In order to give reasons for the assumed approximations also for our subsequent analysis, we will go in more detail through the deduction of the equations.

Our equations were given in Eulerian description (laboratory frame) with the exception of the emissivity η_ν and the opacity χ_ν. These two radiative quantities are measured in the co-moving frame, where they are isotropic, but evaluated at the inertial laboratory frame frequency ν of the radiation. Restricting to terms of order v/c the hydrodynamic equations (the equations of continuity, of gas momentum, and of gas energy) remain valid as formulated in Paper 1, however, the radiative transfer equation as well as the radiation energy and radiation momentum equations should be completed by additional Doppler terms with factors $\dot{v}/c$.

Having in view applications to the atmosphere of the Sun and other cool stars, further simplifications are possible because $v \ll c$ is valid with an accuracy of many orders. In the solar photosphere the sound speed is $c_0 \approx 7$ km/s, and the amplitudes of the p-modes are $v_0 \lesssim 1$ m/s. However, we cannot simply omit terms with an explicit factor v/c. Instead of it we have to compare the orders of magnitude of the complete terms in each equation as it has been done by MM for different regimes of fluid dynamics. Following MM (Sections 93 and 94) we find that the equations in Paper 1 remain valid, without the Doppler terms, in two regimes, namely in the streaming limit and in the static or "true" diffusion limit.

In the streaming limit we have to assume $\lambda_p/l \gtrsim 1$, where λ_p is the mean free path of photons, and l is a typical length scale of the gas (the wavelength Λ or the scale height of a stratified medium; in the first mentioned case $l/\lambda_p \approx \tau_\Lambda$, the optical thickness of one wavelength). The static diffusion limit is matched for $t_d \ll t_f$ or $\lambda_p/l \gg v/c$, where $t_d \approx l^2/c\lambda_p$ is the radiation diffusion time and $t_f \approx l/v$ is the fluid flow time. t_f has to be compared with the radiation flow time $t_R \approx l/c$; however, for optically thin regions $t_R/t_f = O(v/c) \ll 1$.

A rough evaluation for the solar photosphere (Vernazza et al., 1981) results in the following numbers: With $c_0 = 7$ km/s, $v_0 = 1$ m/s, and a wave period of 5 min we estimate a wavelength of $\Lambda = l \approx 2\,10^3$ km and $t_f \approx 5$ min (with $v = c_0$) or $t_f \approx 580$ hours (with $v = v_0$). From the inverse value of χ_ν we estimate $\lambda_p \approx 54$ km for $\tau_0 = 1$ but $3.2\,10^5$ km for $\tau_0 = 2\,10^{-4}$, that is at the temperature minimum. τ_0 is the optical depth at the radiative standard wavelength of $\lambda = 500$ nm. The values of λ_p/l are $2.6\,10^{-2}$ and $1.5\,10^2$, and $t_d \approx 0.27$ s and $4.6\,10^{-5}$ s for the two depths, respectively. This gives $t_d/t_f \approx 9\,10^{-4}$ and $1.5\,10^{-7}$ for both depths and $v = c_0$, but $1.3\,10^{-7}$ and $2.2\,10^{-11}$ for $v = v_0$. For $v = c_0$ or $v = v_0$ we have $v/c \approx 2\,10^{-5}$ or $3\,10^{-9}$, respectively. That means, in any case one of the two limits is matched and our approximations are justified. Only in the dynamic diffusion limit all terms, including the Doppler terms, should be retained, but the corresponding conditions $t_f \lesssim t_d$ and $v/c \gtrsim \lambda_p/l$ are never matched in the solar mantle.

Consequently, the basic equations are valid as formulated in Eqs. (1) of Paper 1 or Eqs. (1–6) of Paper 2, and the self-consistent system of equations of radiation hydrodynamics of an ideal gas with $v \ll c$ in a thermally conducting, uniform atmosphere in LTE in the initial state but without introducing the Eddington approximation, can now be written as follows:

$$\frac{d\rho}{dt} + \rho \nabla \cdot \mathbf{v} = 0 \tag{1}$$

$$\rho \frac{d\mathbf{v}}{dt} + \nabla p = -\frac{1}{c} \int_0^\infty \oint \chi_\nu (S_\nu - I_\nu) d\nu \, d\Omega \mathbf{n} \tag{2}$$

$$\rho T \frac{ds}{dt} = \lambda_T \Delta T + \frac{\mathbf{v}}{c} \int_0^\infty \oint \chi_\nu (S_\nu - I_\nu) d\nu\, d\Omega \mathbf{n} - \int_0^\infty \oint \chi_\nu (S_\nu - I_\nu) d\nu\, d\Omega \quad (3)$$

$$\frac{1}{c} \frac{\partial I_\nu}{\partial t} + \mathbf{n} \cdot \nabla I_\nu = \chi_\nu (S_\nu - I_\nu) \quad (4)$$

$$p = \Re \rho T \quad (5)$$

$$S_\nu = B_\nu(T) = \frac{2h\nu^3}{c^2 (\exp(h\nu/k_B T) - 1)}, \quad (6)$$

where

$$n_x = \sqrt{1 - \mu^2} \cos \phi, n_y = \sqrt{1 - \mu^2} \sin \phi, n_z = \mu = \cos \theta,$$

and

$$\frac{d}{dt} = \frac{\partial}{\partial t} + \mathbf{v} \cdot \nabla.$$

I_ν is the intensity, $S_\nu = \eta_\nu/\chi_\nu$ the source function, $\Re$ the gas constant, $B_\nu(T)$ the Planck function, k_B the Boltzmann constant, and λ_T the thermal electronic conductivity. The other symbols have their usual meanings. Let us consider linear waves of frequency ω and wavenumber $\mathbf{k} = k\mathbf{I_z}$, where $\mathbf{I_z}$ is the unit vector along the z-axis.

If the wave propagates in the positive direction, perturbations of all physical quantities are proportional to $\exp i(kz - \omega t)$, and the set of Eqs. (1)–(6) reduces to the following set of linear equations for the amplitudes of the physical quantities

$$\omega \rho' - k\rho_0 v_z = 0 \quad (7)$$

$$i\omega \rho_0 v_z - ikp' = \frac{2\pi}{c} \int_0^\infty \int_0^\pi \chi_\nu (S'_\nu - I'_\nu) \sin \theta \cos \theta\, d\nu\, d\theta = \frac{Int1}{c} \quad (8)$$

$$\frac{i\omega \Re \rho_0 T'}{\gamma - 1} - ikp_0 v_z = k^2 \lambda_T T' + 2\pi \int_0^\infty \int_0^\pi \chi_\nu (S'_\nu - I'_\nu) \sin \theta\, d\nu\, d\theta = Int2 + k^2 \lambda_T T'$$

$$\quad (9)$$

$$\frac{p'}{p_0} = \frac{T'}{T_0} + \frac{\rho'}{\rho_0} \quad (10)$$

$$\left(-\frac{i\omega}{c} + ik \cos \theta\right) I'_\nu = \chi_\nu (S'_\nu - I'_\nu) \quad (11)$$

$$S'_\nu = \frac{\partial B_\nu(T_0)}{\partial T_0} T', \quad (12)$$

where p_0, ρ_0, and T_0 are gas pressure, mass density, and temperature in the unperturbed atmosphere, respectively, and $\gamma = c_p/c_v$ is the adiabatic exponent (the ratio of specific heats). After elimination of ρ', p', and v_z from Eqs. (7)–(10) the following equation has been obtained

$$\frac{i\omega T' p_0 (\omega^2 - k^2 c_0^2)}{(\gamma - 1) T_0 (\omega^2 - k^2 c_*^2)} = \frac{\omega k c_*^2}{c(\omega^2 - k^2 c_*^2)} Int1 + Int2 + k^2 \lambda_T T' \,, \qquad (13)$$

where c_0 is the adiabatic sound speed and c_* is the isothermal sound speed. Eq. (11) of radiative transfer and Eq. (12) permit us to express the intensity perturbations in terms of temperature perturbations

$$I'_\nu = \frac{\chi_\nu T'}{\chi_\nu + i(k \cos \theta - \omega/c)} \frac{\partial B_\nu (T_0)}{\partial T_0} \,. \qquad (14)$$

This equation permits us to reduce the integral $Int2$ to

$$Int2 = \frac{2\pi(\gamma - 1) T_0}{p_0} T' \int_0^\infty \int_{-1}^1 \frac{\partial B_\nu (T_0)}{\partial T_0} \frac{\chi_\nu (\mu - \omega/kc) d\nu d\mu}{\mu - \omega/kc - i(\chi_\nu/k)} \,. \qquad (15)$$

The further analysis will be restricted to the case $c_* \ll c$. After substitution of Eq. (15) in Eq. (13) we obtain the dispersion relation for nonadiabatic hydrodynamic waves in a uniform, nongrey atmosphere

$$\frac{i\omega(\omega^2 - k^2 c_0^2)}{\omega^2 - k^2 c_*^2} = \int_0^\infty \chi_\nu \frac{\partial B_\nu (T_0)}{\partial T_0} \left[1 - \frac{\chi_\nu}{k} \operatorname{arccot} \frac{\chi_\nu}{k} \right] d\nu + \frac{k^2 \lambda_T}{c_\nu \rho_0} \,, \qquad (16)$$

where the integration over μ has been performed. If we suppose that the opacity doesn't depend on frequency the dispersion Eq. (16) reduces to

$$\frac{i\omega T_R (\omega^2 - k^2 c_0^2)}{\omega^2 - k^2 c_*^2} = 1 - \frac{\chi}{k} \operatorname{arccot} \frac{\chi}{k} + \frac{T_R}{T_T} \left(\frac{k}{\chi} \right)^2 \,, \qquad (17)$$

where

$$T_R^{-1} = \frac{16\gamma \chi c_0}{B_0} \,, \quad B_0 = \frac{\rho_0 c_p c_0}{\sigma T_0^3} \,, \quad T_T^{-1} = \frac{\chi^2 \lambda_T}{\rho_0 c_\nu} \,.$$

T_R is the characteristic time of temperature disturbances due to radiative cooling as described in Spiegel's formula (see below), B_0 is the Boltzmann number, σ the Stefan-Boltzmann constant, and T_T is the characteristic time of temperature disturbances of length χ^{-1} due to electronic conductivity.

3. Exemplary Results for a Grey Atmosphere

In a subsequent paper the basic equations of the preceding section will be analysed in detail for the grey case. Only a few preliminary results will be

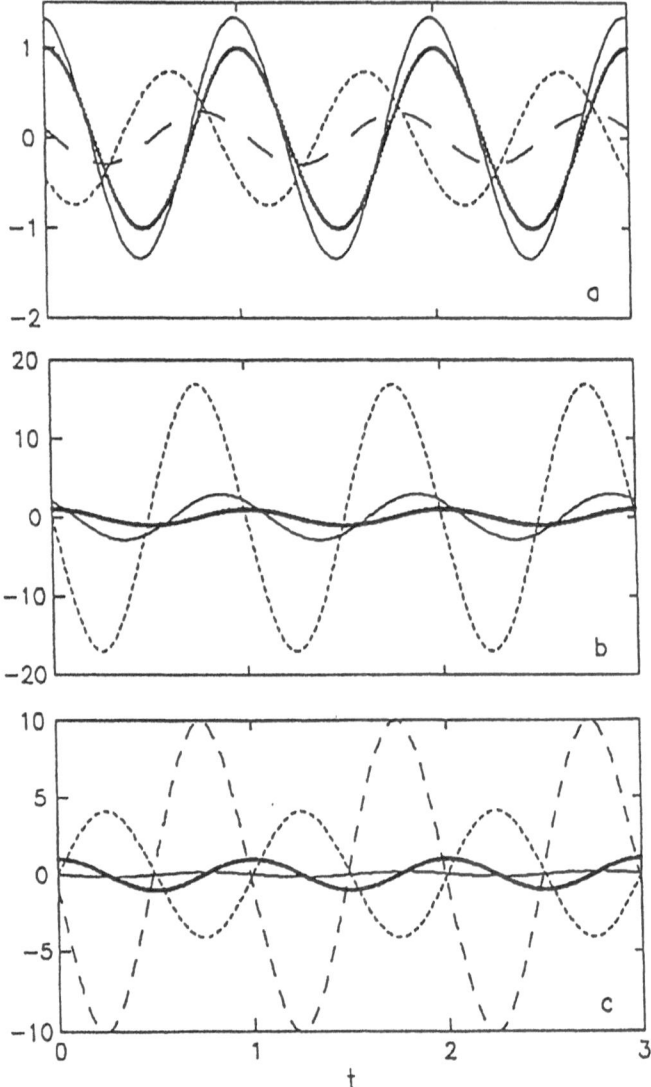

Fig. 1.   The time behaviour of plasma velocity v (thick solid line), gas pressure p' (thin solid), temperature T' (dashed), and intensity I' (dotted) at $\mu = 1$ for acoustic (a), radiation (b), and thermal (c) wave modes for $\Omega = 1$, $B_R = 0.1$, $B_T = 10^{-5}$, and $\xi = 2.7\,10^{-5}$. The velocity amplitude is equal to one for all 3 wave modes. The amplitudes and phases of p' and T' for the radiation mode practically coincide. The amplitudes for the thermal mode were changed to present them in the same scale: T' is divided by 100 and p' is multiplied by 100. The time t is given in units of the wave period.

given here. Let us introduce some dimensionless parameters for the frequency $\Omega = \omega/c_0\chi$, phase velocity $V_{ph} = v_{ph}/c_0 = \omega/c_0 \operatorname{Re} k$, conductivity $B_T = (\chi c_0 T_T)^{-1}$, radiative cooling $B_R = \chi c_0 T_R$, and sound speed $\xi = c_0/c$.

A spatial analysis (solution for real ω but complex k) of the dispersion relation Eq. (17) shows, that 3 different wave modes with very different properties are excited in our homogeneous, radiative, conducting atmosphere. For any comparison with observations it is desirable to have not only the relation $\omega = \omega(k)$, but the ratios of amplitudes and the phase relations too in order to distinguish between the different modes.

Fig. 1 represents an example of such a detailed portrait of waves for a special set of our dimensionless parameters. The time dependence of v, p', T', and I' is given for the acoustic, the thermal, and the radiation wave modes. There are strong differences in the amplitudes and in the phase shifts. For instance, the amplitudes of T' and p' for the thermal mode are so different, that we had to scale the ordinates to present them in the same graph. I' and T' are in antiphase for the thermal mode, but nearly in phase for acoustic and radiation waves. For the acoustic mode the dependence of V_{ph} on Ω and on B_T is shown in Fig. 2a.

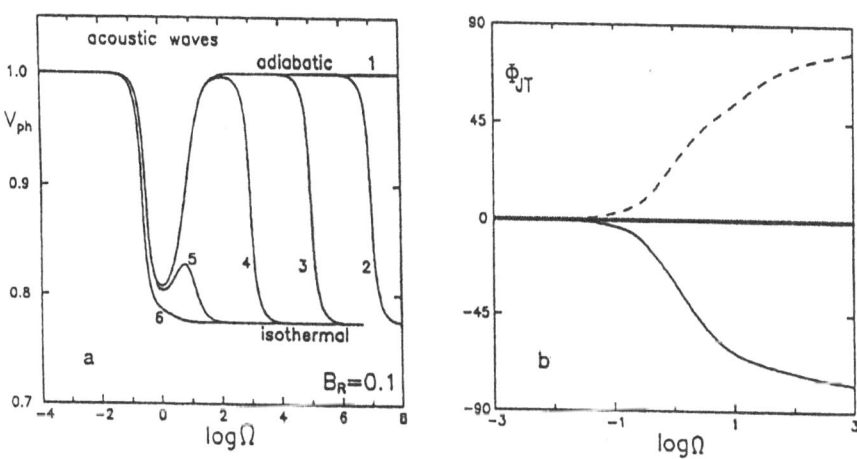

Fig. 2.   a: The dependence of the phase velocity of acoustic waves on frequency and on the conductivity parameter B_T. The numbers in increasing order correspond to $B_T = 0, 10^{-7}, 10^{-5}, 10^{-3}, 0.1$, and 1, respectively.
b: The phase shift between the fluctuations of brightness and temperature for upward running (solid curve), downward running (dashed curve), and standing (thick solid curve) acoustic waves for full-disk observations.

The phase shift between brightness (J) and temperature fluctuations versus Ω is shown in Fig. 2b. This is a crude modelling of brightness fluctuations from full-disk observations, obtained by integrating I' over μ from 1 to 0, that is from the centre to the limb of the Sun. There are clear phase shifts for running acoustic waves, and the assumption of equal phases (and equal amplitudes too) between brightness and temperature which is often made in the interpretation of observed data is wrong.

4. Waves in Grey and Nongrey Atmospheres

The question arises whether the wave problem for the nongrey atmosphere can be reduced to the simple case of an equivalent grey atmosphere. For this purpose a special procedure has been proposed in Paper 2 allowing to reduce the dispersion relation Eq. (16) for the nongrey case to that for the grey model, Eq. (17), by introducing a "wave mean opacity" $\bar{\chi}_w$ depending on the wavenumber k. $\bar{\chi}_w$ should satisfy the equation

$$\bar{\chi}_w\left(1 - \frac{\bar{\chi}_w}{k}\text{arccot}\frac{\bar{\chi}_w}{k}\right) = \frac{1}{B}\int_0^\infty \chi_\nu \frac{\partial B_\nu}{\partial T_0}\left(1 - \frac{\chi_\nu}{k}\text{arccot}\frac{\chi_\nu}{k}\right)d\nu , \qquad (18)$$

$$\text{where} \quad B = \int_0^\infty \frac{\partial B_\nu}{\partial T_0}d\nu .$$

In the stationary case there exist at least 3 different procedures for calculating a mean opacity which could reduce the radiative transfer problem of a nongrey atmosphere to that of a grey model (see, e.g., MM). Eq. (18) introduces another mean opacity for applications to wave problems. However, in Paper 2 it has been shown that Eq. (18) has two solutions for $\bar{\chi}_w$. Only in two limiting cases it is possible to derive a unique $\bar{\chi}_w$: In the opaque limit for the wave problem, if the diffusion approximation is valid, we have optically thick disturbances measured over one wavelength Λ, $\chi_\nu/k \gg 1$. In this case the first root of Eq. (18) reduces to the ordinary formula for the Rosseland mean opacity, that is $\bar{\chi}_w = \bar{\chi}_R$, while the second root approaches zero. In the opposite case of the Newton approximation for optically thin disturbances we have $\chi_\nu/k \ll 1$. In this transparent limit Eq. (18) reduces to

$$\bar{\chi}_w = \frac{1}{B}\int_0^\infty \chi_\nu \frac{\partial B_\nu}{\partial T_0}d\nu = \bar{\chi}_N , \qquad (19)$$

which has been introduced as a definition of the "Newton mean opacity", a new type of mean opacity for wave problems (arithmetic mean instead of harmonic mean as applied for $\bar{\chi}_R$).

In the general case Eq. (16) remains irreducible to Eq. (17). However, this does not exclude the use of a grey approximation in practical applications. Its accuracy depends on the difference between $\bar{\chi}_R$ and $\bar{\chi}_N$ which provides a measure for the discrepancy between the nongrey and the grey approximations. This idea can be illustrated for the radiation mode described by Eq. (16). That is a non-oscillatory smoothing of temperature disturbances which take place, when the temperature disturbances are accompanied by relatively small pressure disturbances and acoustic waves do not appear. In that case the solution for the dispersion relation Eq. (16) reduces to the formula for the inverse relaxation time t_R^{-1}

$$t_R^{-1} = \frac{16\sigma T_0^3}{\rho c_p B} \int\limits_0^\infty \chi_\nu \frac{\partial B_\nu}{\partial T_0} (1 - \frac{\chi_\nu}{k} \operatorname{arccot} \frac{\chi_\nu}{k}) d\nu \ , \qquad (20)$$

that is a generalization of the classical formula of Spiegel (1957) for the relaxation mode of the perturbed grey atmosphere:

$$t_R^{-1} = \frac{16\chi\sigma T_0^3}{\rho c_p} (1 - \frac{\chi}{k} \operatorname{arccot} \frac{\chi}{k}) \ . \qquad (21)$$

Eq. (21) differs from the original formula of Spiegel (1957) by a factor γ. This difference is due to the neglect of hydrodynamic processes by Spiegel, who considered only the energy equation. (Please note that there is a misprint in Eqs. (25–28) of our Paper 2: c_v should be replaced by c_p.) The calculations for typical parameters of the solar atmosphere in Paper 2 have shown, that the difference between the relaxation times for the grey and nongrey approximations is about 10–20%. It is due to the difference between $\bar{\chi}_R$ and $\bar{\chi}_N$, which is about a dozen percents. Nevertheless this difference can be large. The value of $\bar{\chi}_N$ depends strongly on the behaviour of χ_ν in the infrared part of the spectrum. Consequently, the difference in the mean opacities might be large for stars which are cooler than the Sun.

We compared the numerical solutions of Eqs. (16) and (17) also for the acoustic mode in the solar atmosphere. There are some small differences in frequencies and relaxation times of nonadiabatic waves between the grey and nongrey approximations. Our conclusion coincides with the results of numerical calculations by Christensen-Dalsgaard and Frandsen (1984). The differences between the waves and relaxation modes in grey and in nongrey models is connected not only with different relations between frequencies and wavelengths, but also with the dependence of the amplitudes and phases of the intensity perturbations on the ange θ and the optical frequency ν, see Eq. (14). The ratio τ_ν/k varies in a wide range of values due to the strong dependence of χ_ν on ν for a real astrophysical plasma. The angular dependence of I' for a grey atmosphere will be analysed in a subsequent paper.

5. Intensity Fluctuations Due to Waves in a Nongrey Stratified Atmosphere

The solution of the radiative transfer equation and its analysis can be performed for a stratified atmosphere without solving the complete set of equations of radiation hydrodynamics. The linearized equation of radiative transfer in a stratified atmosphere can be written as follows

$$\left(\frac{1}{c}\frac{\partial}{\partial t} + \mathbf{n}\cdot\nabla\right)I'_\nu = \chi_{0\nu}\left[S'_\nu - I'_\nu + \frac{\chi'_\nu}{\chi_{0\nu}}(S_\nu - I_\nu)\right]. \tag{22}$$

The difference of this equation from Eq. (11) for a uniform atmosphere results not only from the height dependence of all quantities, but also from the influence of the opacity fluctuations on the fluctuations of intensity. We can write the solution of Eq. (22) for the LTE case, when $S_\nu = B_\nu$ and the time derivative can be omitted exactly as in the case of a uniform atmosphere. The fluctuation of the emergent intensity is equal to

$$I'_\nu = \int_0^\infty e^{-\tau_\nu/\mu}\left[\frac{dB_\nu(\tau_\nu)}{dT_0}T' + \frac{\chi'_\nu}{\chi_{0\nu}}(B_\nu(\tau_\nu) - I_\nu)\right]\frac{d\tau_\nu}{\mu}, \tag{23}$$

where τ_ν is the optical depth and the unperturbed outgoing ray intensity is

$$I_\nu(\tau_\nu,\mu) = \int_{\tau_\nu}^\infty e^{-(\tau'_\nu-\tau_\nu)/\mu}B_\nu(\tau'_\nu)\frac{d\tau'_\nu}{\mu}. \tag{24}$$

The relative fluctuations of intensity are defined by

$$\frac{I'_\nu}{I_{0\nu}} = \frac{\int_0^\infty e^{-\tau_\nu/\mu}B_\nu(\tau_\nu)Q'_\nu(\tau_\nu,\mu)\frac{d\tau_\nu}{\mu}}{\int_0^\infty e^{-\tau_\nu/\mu}B_\nu(\tau_\nu)\frac{d\tau_\nu}{\mu}}, \tag{25}$$

where

$$Q'_\nu(\tau_\nu,\mu) = \frac{1}{B_\nu(\tau_\nu)}\left[\frac{dB_\nu(\tau_\nu)}{d\ln T_0}\frac{T'}{T_0} - \frac{\chi'_\nu}{\chi_{0\nu}}\int_{\tau_\nu}^\infty e^{-(\tau'_\nu-\tau_\nu)/\mu}\frac{dB_\nu(\tau'_\nu)}{d\tau'_\nu}d\tau'_\nu\right] \tag{26}$$

The fluctuations of opacity depend on the disturbances of thermodynamic quantities, that can be defined only by solving the general set of radiation hydrodynamics equations,

$$\frac{\chi'_\nu}{\chi_{0\nu}} = \frac{d\ln\chi_{0\nu}}{d\ln T_0}\frac{T'}{T_0} + \frac{d\ln\chi_{0\nu}}{d\ln\rho_0}\frac{\rho'}{\rho_0}. \tag{27}$$

The intensity fluctuations were considered by many authors in calculations of the so-called visibility of p–modes and g–modes of global solar oscillations. Toutain and Gouttebroze (1988, 1993) obtained the most general formula, that takes into account nongrey effects and opacity fluctuations. Eq. (23) and the formula of Toutain and Goutebroze can be reduced to each other by differentiation by parts. In the case of a uniform atmosphere Eq. (23) gives the correct result in the absence of an influence of opacity fluctuations on the intensity fluctuations, that is connected with equality $B_\nu = I_\nu$ for a uniform atmosphere. This is an essential difference between stratified and uniform atmospheres, that does not permit to consider all important effects in the simple model of a uniform atmosphere. It is often assumed, that intensity fluctuations are proportional to temperature fluctuations and that the phase shift between velocity and intensity equals the phase shift between temperature and velocity. The derived formula shows, however, that this is not correct, because intensity fluctuations are proportional to density fluctuations too. Of course, all details depend on the relative values of the amplitudes of temperature and density fluctuations as well as on the relative influence of density and temperature fluctuations on the opacity fluctuations. For qualitative analysis of Eq. (25) a linear approximation of B_ν can be introduced

$$B_\nu = a_\nu + b_\nu \tau_\nu. \tag{28}$$

After substituting this approximation and Eq. (27) into Eq. (26), Eq. (25) for the relative fluctuations of the emergent intensity reduces to

$$\frac{I'_\nu}{I_{0\nu}} = -\frac{\int_0^\infty e^{-\tau_\nu/\mu} \left[\left(\frac{d\tau_\nu}{d\ln T_0} + \mu \frac{d\ln \chi_{0\nu}}{d\ln T_0} \right) \frac{T'}{T_0} + \mu \frac{d\ln \chi_{0\nu}}{d\ln \rho_0} \frac{\rho'}{\rho_0} \right] \frac{d\tau_\nu}{\mu}}{a_\nu/b_\nu + \mu} \tag{29}$$

This formula shows clearly the centre-to-limb variation of the relative intensity fluctuations. The influence of opacity fluctuations decreases towards the limb, and the intensity fluctuations near the limb depend practically only on the fluctuations of the source function (compare Eqs. (29) and (23)).

The different terms in the integrand of Eq. (29) have been estimated for an example and compared with the results of Toutain and Gouttebroze (1988). The VAL3C model of the average solar atmosphere (Vernazza et al., 1981) has been used in a recalculated form (Staude, 1981) (Obridko and Staude, 1988), including opacity calculations from the UV (Lyman continuum) up to the near IR (1.65 μm). We assumed a value of $\Omega = 1$ corresponding to a period of 5 min at $\tau_0 = 0.2$, moreover, $B_T = 10^{-5}$ is applied. The quantity Q in the relation

$$T'/T_0 = Q\rho'/\rho_0 \tag{30}$$

can only be determined from the full set of the basic equations. Toutain and Gouttebroze (1988) assumed adiabatic oscillations. In this case $Q = \Gamma_3 - 1$ (Γ_3 is the third adiabatic exponent) which approaches 2/3 for a neutral or fully ionized hydrogen gas but $\lesssim 0.1$ in the partially ionized lower chromosphere. From our basic equations in Section 2 we obtain

$$Q = (\gamma - 1)\frac{1 - \eta^2(\gamma - i\xi)/\Omega}{1 - \gamma + \xi(B_{\mathrm{T}} - i\eta^2/\Omega)} \qquad (31)$$

where $\eta = \omega/kc_0$. This is quite another expression as compared with the adiabatic case, including a phase shift. For our exemplary set of parameters we obtain completely different values of Q for the different wave modes: $Q = 0.464 \exp(i\,217°)$ for the acoustic mode, $Q = 20.1 \exp(i\,170°)$ for the radiation mode, but $Q = -1$ for the thermal mode. Let us introduce the notations $X = d\tau_\nu/d\ln T_0$, $Y = d\ln\chi_{0\nu}/d\ln T_0$, and $Z = (1/Q)d\ln\chi_{0\nu}/d\ln\rho_0$ for the 3 terms in the integrand of Eq. (29). The estimates for our exemplary depth $\tau_0 = 0.2$ show that $|Y/Z| \gg 1$ in most cases (one order of magnitude for acoustic waves, much more for radiation and thermal waves), so that Z can usually be neglected. $Y/X \ll 1$ in the UV, -0.3 in the visible (500 nm), but -30 in the IR. That is, in the centre of the solar disk we have to take into account the opacity fluctuations in the IR and to some extent also in the visible spectral region, but the numbers strongly differ from the results of Toutain and Gouttebroze (1988).

6. Conclusions

We analysed the basic equations of radiation hydrodynamics in a homogeneous, nongrey, radiating, thermally conducting atmosphere. Having in mind applications to waves in the solar atmosphere it seems sufficient to restrict the study either to the streaming limit or to the true diffusion limit. That means, our basic equations in the laboratory frame are correct, without the Doppler terms, in a form similar to that in our Papers 1 and 2.

Avoiding the Eddington approximation a generalized dispersion relation has been derived. For the grey case this relation has been investigated by calculating a detailed wave portrait for an exemplary set of parameters in the solar atmosphere; a more systematic analysis will be the subject of a subsequent paper. It becomes evident that there exist three different wave modes, the acoustic, the radiation, and the thermal waves, which have very different amplitudes and phase shifts. There are intersections of the three branches, and close to the frequencies of these junctions wave coupling is possible. On the Sun these frequencies are close to those of the observed p–modes. Therefore the influence of radiation and thermal waves on the acoustic waves should be considered in any analysis, contrary to the usual practice of neglecting it. For instance, the customary assumption of equal phases and amplitudes of the fluctuations of brightness and of temperature is wrong.

The introduction of a 'wave mean opacity' provides a method to estimate the accuracy of a grey model as compared to the nongrey case. For a nongrey atmosphere the angular dependence of I' is different for different ν. This property of waves in a nongrey model could be important for helioseismological investigations, if observations of I' in in different parts of the continuous spectrum are available. Preliminary results from ground-based and space-borne observations did not yet show measurable phase shifts between the different channels of multi-channel photometers (Jiménez et al., 1990)(Schrijver et al., 1991). We hope luminosity oscillations planned to be measured aboard the satellites CORONAS and SOHO and simultaneously obtained ground-based velocity data will provide a better basis for facing the modelling with observations.

Our theoretical approach for a homogeneous model takes into account the nonadiabatic effects in great detail and more general than most earlier papers, but future investigations should consider the basic effect of the gravitational stratification. Our preliminary estimate of I' in a stratified atmosphere shows that such an analysis should always take into account the nonadiabatic effects, otherwise the interpretation of observations could be completely misleading.

Acknowledgements

The authors gratefully acknowledge support of the present work by the German Space Agency (DARA) under grant No. 50 QL 9207 2.

References

Christensen-Dalsgaard, J., and Frandsen, S.: 1984, *Mem. Soc. Astron. Ital.* **55**, 285.
Dzhalilov, N. S., Zhugzhda, Y. D., and Staude, J.: 1992, *Astron. Astrophys.* **257**, 359.
Mihalas, D. and Mihalas, B. W.: 1984, *Foundations of Radiation Hydrodynamics*, Oxford University Press, Oxford.
Golitsyn, G. S.: 1963, *Izvestia Akad. Nauk SSSR, Ser. Geofiz.* **6**, 960.
Ibáñez S., M. H. and Plachco M., F. P.: 1989, *Astrophys. J.* **336**, 875.
Jiménez, A., Álvarez, M., Andersen, N. B., Domingo, V., Jones, A., Pallé, P. L., and Roca Cortés, T.: 1990, *Solar Phys.* **126**, 1.
Obridko, V. N., and Staude, J.: 1988, *Astron.Astrophys.* **189**, 232.
Prokofiev, V. A.: 1957, *Prikladnaya Matematika i Mekhanika* **21**, 775.
Schrijver, C. J., Jiménez, A., and Däppen, W.: 1991, *Astron. Astrophys.* **251**, 655.
Spiegel, E. A.: 1957, *Astrophys. J.* **126**, 202.
Staude, J.: 1981, *Astron.Astrophys.* **100**, 284.
Toutain, T., and Gouttebroze, P.: 1988, *Proc. Symp. 'Seismology of the Sun & Sun-like Stars'*, ESA SP-286, 241.
Toutain, T., and Gouttebroze, P.: 1993, *Astron.Astrophys.* **268**, 309.
Nowak, T., and Ulmschneider, P.: 1977, *Astron.Astrophys.* **60**, 413.
Vernazza, J. E., Avrett, E. H., and Loeser, R.: 1981, *Astrophys. J. Suppl.* **45**, 635.
Zhugzhda, Y. D., Dzhalilov, N. S., and Staude, J.: 1993, *Astron.Astrophys.* **278**, L9.

EXCITATION OF SOLAR GRAVITY WAVES

B.N. ANDERSEN

Norwegian Space Centre, PO Box 85, Smestad, N-0309 Oslo, Norway

Abstract. The interaction between convection and gravity waves are simulated numerically in a model closely corresponding to the physical conditions in the solar interior. The penetration of convective elements into the stably stratified interior is shown to generate gravity waves. The energy efficiency of this generation is less than 0.1 %. The simulations also show that the convective overshoot region is very shallow, 0.02-0.06 pressure scaleheights.

1. Introduction

The idea that the penetration of convective matter from the solar convection zone into the radiatively stable solar interior may generate internal gravity waves has been discussed by several authors (Press 1981, Hurlburt et al 1986, Zahn 1992, Montalban 1993). Aside from the possible generation of gravity waves the study of convective overshoot is interesting because it contributes to determine the physical and dynamical conditions of the convection zone boundary layer. The depth of the convective overshoot layer varies in different studies from fractions of a scaleheight to more than one (Hurlburt et al 1986, Narasimha and Roxburgh 1993, Roxburgh and Vorontsov 1993, Berthomieu et al 1993).

This work is a continuation of the work described in Andreassen et al. (1992) and Andersen et al. (1993), hereafter designated Papers 1 and 2.

2. Method

We have used an efficient numerical code for solving the appropriate hydrodynamical equations with physical parameters close to those describing the solar interior. The code is two-dimensional and uses pseudospectral methods on the fully compressible hydrodynamic problem. The strength of the code is that it treats the wave mechanics consistently. The upper and lower boundaries are open, thus transmitting wave energy created in the model. The model is repeated periodically in the horizontal direction. The details of the model and code are given in Papers 1 and 2.

For this study some developments have been made in the code. Besides making it more efficient, the radiative flux has been treated in a more consistent manner. This explicitly means that the thermal driving of the convection zone is maintained by the background heat flux, and thus the simulations may be carried out for very long time series with a quasi-stationary situation developing both in and below the convection zone.

Solar Physics **152**: 241–246, 1994.
© 1994 *Kluwer Academic Publishers.*

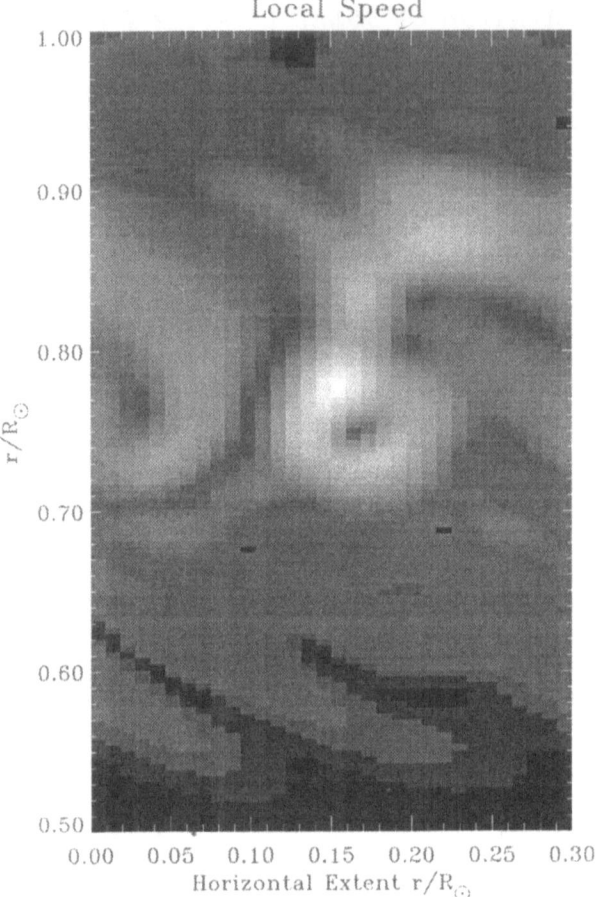

Fig. 1. The amplitude of the speed as function of depth and horizontal scale for timestep 175. The convective pattern is clearly seen as well as the diagonally shaped signatures of gravity waves below the convection zone. The gray scale has has been enhanced for the low amplitudes and is thus only indicative for the relative amplitudes.

The equations of mass, momentum and energy may be expressed as:

$$\rho_t + \rho(u_x + v_z) + u\rho_x + v\rho_z = 0$$

$$u_t + uu_x + vu_z(1/\rho)p_x = \Pi_{xx} + \Pi_{xz}$$

$$v_t + uv_x + vv_z(1/\rho)p_z + \rho g = \Pi_{zx} + \Pi_{zz}$$

$$p_t + up_x + vp_z + \gamma p(u_x + v_z) = \Psi$$

Here the t, x and z indices indicate derivation in time and the horizontal and vertical directions, respectively. The horizontal and vertical velocity components are given by u and v. The right hand terms are diffusion terms and are described in detail in Papers 1 and 2. In addition to the description there a term describing the background heat flux is included. The functional forms of the diffusion terms used in the current calculations are

$$\Pi_{ij} = \nu_0 \varpi_i(z) \frac{\partial}{\partial x_j} \left(\nu(x_j) * \frac{\partial v_i}{\partial x_j} \right)$$

for the momentum equations and

$$\Psi = \epsilon_0 \varpi_i(z) \frac{\partial}{\partial x_j} \left(\nu(x_j) * \frac{\partial T}{\partial x_j} \right) + \kappa \frac{\partial^2 T_0}{\partial z^2}$$

for the heat equation, where the indices i and j indicate the x and z directions and

$$\varpi_z(\xi) = \sqrt{1 - \xi^2}, \varpi_x(\xi) = 1$$

and

$$\epsilon(k), \nu(k) = \begin{cases} 0, & k < \alpha\sqrt{N} \\ 1 - \sqrt{\alpha}N/k^2, & k \geq \alpha\sqrt{N}. \end{cases}$$

in spectral space; here $*$ denote convolution, $\alpha \approx 1$ is a constant, and ϖ is the Chebyshev norm which gives uniform truncations across the nonuniform vertical mesh. The parameter κ gives the total thermal conductivity as function of depth. Since $\varpi = 0$ at the boundaries, the equations become locally hyperbolic along these.

The solar model which used in these calculations is adapted from the "Model 6" (Christensen-Dalsgaard 1991). A plane parallel approximation, extending from 0.5 to 1 solar radii with a width equivalent to about 0.3 solar radii was used. The upper part of the model (from 0.92 to 1.0) has been modified to be subadiabatic in order to study the transmission of gravity waves through the convection zone. By doing this, a description of the H and He ionization zones is also avoided.

3. Results

The initial static solar model is randomly perturbed in density (≤ 0.1 %) within in the region corresponding to the solar convection zone. This perturbed model is then developed in time. The simulation is then carried out for 1200 timesteps, each timestep being the sound travel time of one scaleheight at the bottom of the model. Initially large velocity amplitudes of about 5 % of the sound speed are generated in the convection zone. With these pertubations a clear convective pattern is generated. The initial velocities are damped and a quasi- stationary convective pattern is generated with typical velocities of about 0.5 % of the sound speed. This convective pattern is maintained from about 175 timesteps to the end of the

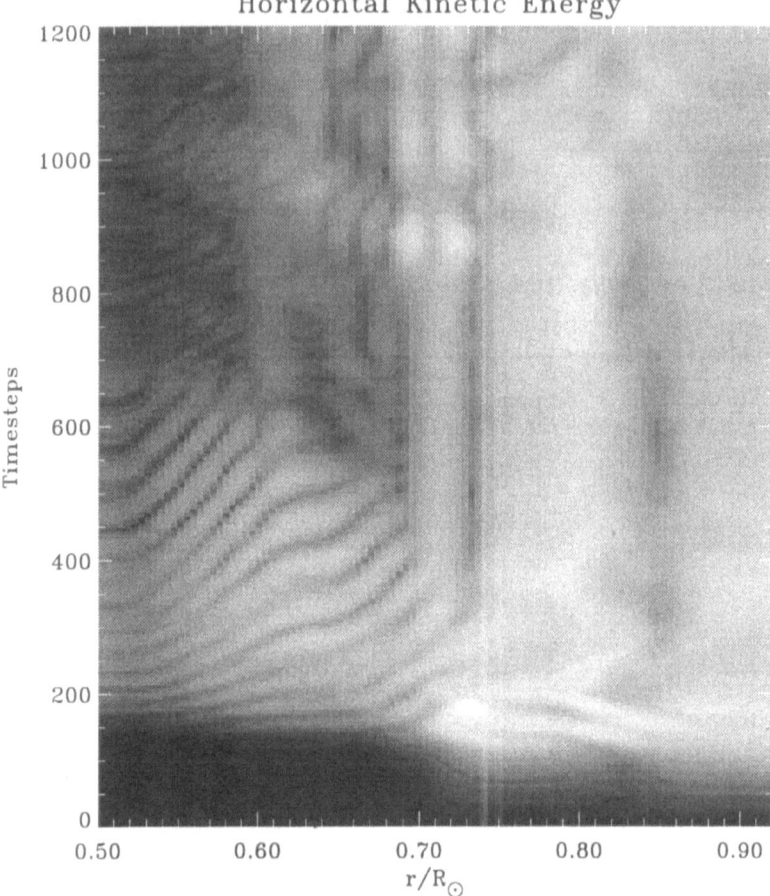

Fig. 2. The horizontal kinetic energy is averaged horizontally for all depth points. The time dependence of the logarithm of this value is indicated as gray levels. The quasi-stationary structure within the convetion zone is seen clearly after timestep 240. The convectively generated gravity wave pattern below the convection zone is also visible.

simulation. A typical snapshot of the velocity pattern is given in Fig. 1. Indications of a gravity wave velocity pattern can clearly be seen below the convection zone. It is also possible to identify the individual convective elements impacting on the lower boundary of the convetive zone and generating a spectrum of gravity waves.

Since a major fraction of the kinetic energy in a gravity wave is in the horizontal direction, the horizontally averaged kinetic energy is used as tracer for gravity waves. In Fig. 2 this averaged kinetic energy is shown as a function of depth and time. The wave pattern being generated as function of depth and time is clearly seen.

In order to study the waves and their frequency properties in more detail the power spectra of the horizontally averaged kinetic energy was calculated as a function of depth. The result is shown in Fig. 3. A relatively chaotic behavior within the convection zone is seen, with no clear ridges or peaks in the power spectrum. This is consistent with the convective nature of this region. However, below the convection zone there is a clear ridge at a frequency of about 0.02 per timesteps. In addition a weaker ridge at twice of the frequency of the main ridge is also seen. In the same units the Brunt-Väisälä frequency varies nearly linearly from a value of 0.055 per timestep at 0.5 solar radii to 0.04 per timestep at 0.65 solar radii. Both the frequency rigdes are below the Brunt-Väisälä values locally, thus confirming their nature as gravity waves.

In Fig. 2 there seems to be a doubling of the typical wave frequency in the last half of the simulation; this causes the higher frequency ridge. The lower frequency ridge seems to originate from the initial larger amplitude pertubations in the convection zone while the higher frequency rigde comes from the quasi-stationary convective structures. The properties of these structures are more descriptive of the physical conditions in the solar convection zone than the initial high amplitude pertubations. These higher frequency waves persist to the end of the simulation with approximately constant amplitude. The energy density in the waves at the end of the simulation is stable at about 0.1 % of the typical kinetic energy density in the convection zone.

In all the figures a very sharp boundary at the bottom of the convection zone is seen, closer investigation of the data show that the vertical convective overshoot is effectively damped within 0.02-0.06 pressure scaleheights.

4. Conclusions

The numerical simulations presented here indicate that convective overshoot at the bottom of the convection zone continously generates internal gravity waves. The kinetic energy density of the waves in these calculations indicate an exitation efficiency of the convection of about 0.1 % persisting through the calculations and relatively independant of the amplitude of the exciting convective elements. However, the wave frequencies seem to depend on the amplitude of the excitation, the larger amplitudes resulting in waves of lower frequencies.

The convective overshoot layer has a thickness of typically about 0.03 pressure scaleheights at the bottom of the convection zone. This small value seems to be consistent with the latest interpretations of observational data and theoretical predictions (Gough and Sekii 1993, Monteiro et al 1993, Roxburgh and Vorontsov 1993).

5. Acknowledgements

The numerical calculations in this paper have been made with financial support from "Tungregneutvalget" in the Research Council of Norway.

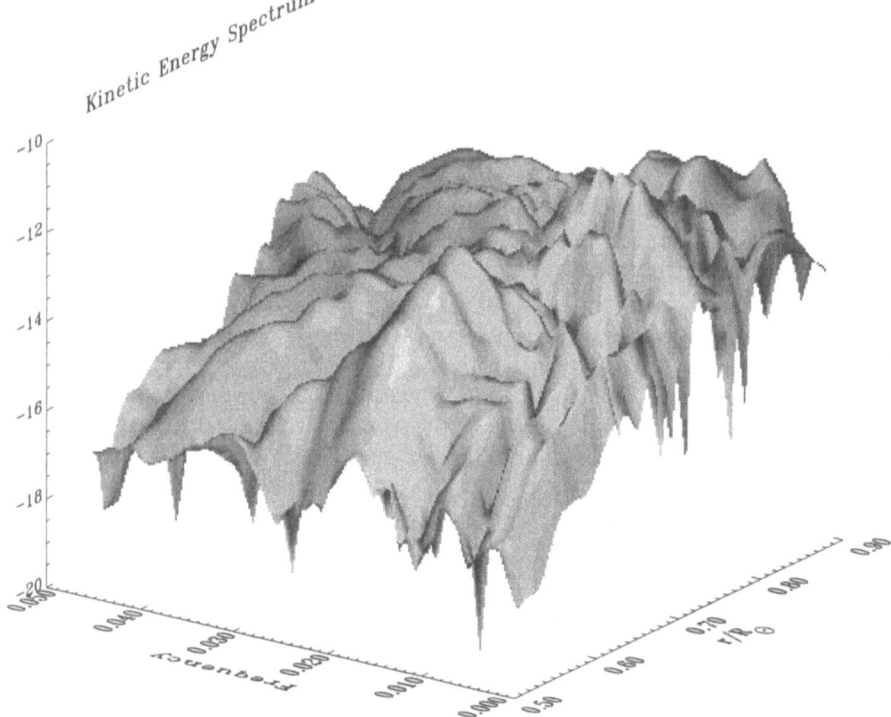

Fig. 3. The depth dependence of the power spectra of the horizontally averaged kinetic energy is shown. A clear rigde at a frequency of 0.02 per timestep is seen below the convection zone. A lower amplitude rigde is seen at twice this frequency. Within the convection zone several depth dependent peaks are seen, but no clear depth independant pattern.

References

Andersen, B.N., Andreassen, Ø., Leifsen, T. and Wassberg C.E.: 1993, *ASP Conference Series* **42** ,49.

Andreassen, Ø., Andersen, B.N. and Wassberg C.E: 1992, *Astron.Astrophys.* **109**, 301.

Berthomieu, G., Morel, P., Provost, J. and Zahn, J.-P.: 1993, *ASP Conference Series* **40**, 60.

Christensen-Dalsgaard, J.: 1991, *Lecture Notes in Physics* **38**, 11.

Gough, D.O. and Sekii, T.: 1993, *ASP Conference Series* **42**, 177.

Hurlburt, N.E., Toomre, J., Massaguer, J.M.: 1986, *Astrophys.J.* **311**, 563.

Narasimha D. and Roxburgh, I.W.: 1993, *ASP Conference Series* **42**, 73.

Montalban, J.: 1993 *ASP Conference Series* **40**, 278.

Monteiro, M.J.P.F.G, Christensen-Dalsgaard, J. and Thompson, M.J.: 1993, *ASP Conference Series* **40**, 557.

Press, W.H. :1981 *Astrophys. J.*, **245**, 286.

Roxburgh, I.W. and Vorontsov, S.V.:1993, *ASP Conference Series* **42**, 169.

Zahn, J.-P.: 1992, *Astron. Astrophys.* **252**, 179.

SOLAR NOISE SIMULATIONS IN IRRADIANCE

B. N. ANDERSEN

Norwegian Space Centre, PO Box 85, Smestad, N-0309 Oslo, Norway

T. LEIFSEN

Institute of Theoretical Astrophysics, University of Oslo, PO Box 1029, Blindern, N-0315 Oslo, Norway

and

T. TOUTAIN

Space Science Department, PO Box 299, ESTEC SC, NL-2200 AG Noordwijk

Abstract. The global signature of granulation, meso- and supergranulation is calculated using values for intensities and lifetimes from spatially resolved observations. These simulations are compared with observations from ACRIM, IPHIR and the SOVA-1 photometers. The results indicate that the overall shape of the background signal in the simulations reproduce the observations at low frequency. However when the granulation lifetimes are about 500 seconds the simulated data do not correspond to the observations between 1 and 2 mHz.

1. Introduction

Several attempts have been made to search for the observational signatures of solar global gravity modes (Hill et al 1991, and references therein). However, no unambigious results have been obtained. These negative results are caused by a combination of a low amplitude solar oscillations signal, instrumental noise, sometimes bad window functions and the influence from the non-oscillatory solar surface structures. In the best space based observations it seems as if this solar noise signal may be the limiting factor in the search for solar g-modes. It is necessary to understand how the effects on low resolution radiance and irradiance from the time development of solar surface structures like granulation, mesogranulation and supergranulation enter into the observed time series. Without this understanding it will be difficult to discern these observational signatures from low frequency solar g-modes.

In addition the calibration given by comparing high and low resolution solar measurements may make it possible to deduce information about stellar surface structures from the observations of stellar irradiance time series.

2. Solar Noise Model

Harvey (1985) used a simple model for the continuum background power spectrum. This parameterization has been used relatively successfully in comparison with powerspectra of long timestrings of ground-based velocity measurements. This model was extended to irradiance measurements in the VIRGO proposal. Harvey et al. (1993) have used this model to study the solar noise spectrum for chromospheric radiance oscillations. The main problem with this type of power spectrum parameterization is that no time series is produced, and thus a realistic test of reduction procedures cannot be made.

In this study we have made a simple model of the contribution to the irradiance observables from solar granulation, mesogranulation and supergranulation.

Solar Physics **152**: 247–252, 1994.
© 1994 *Kluwer Academic Publishers.*

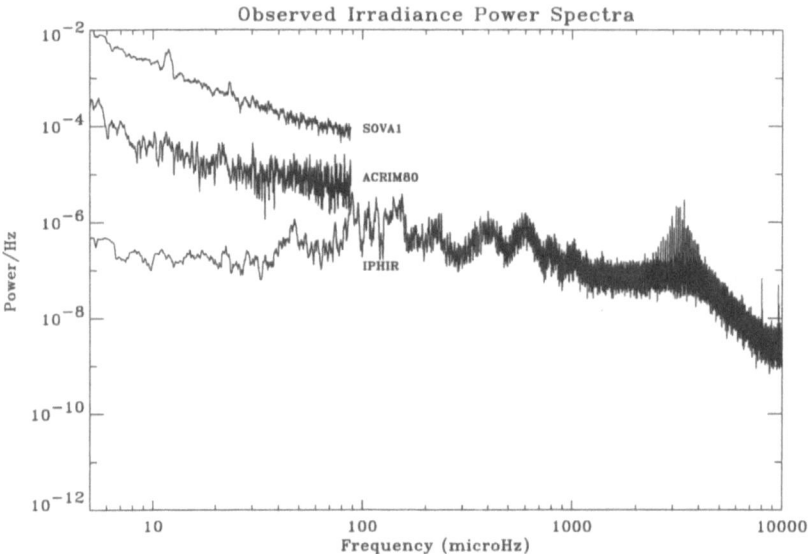

Fig. 1. The power spectra of the timestrings from the IPHIR, ACRIM 1980, and the SOVA-1 data are shown. The data have been treated with a 21 point running mean in frequency. The first two daily harmonics can be seen in the SOVA-1 data.

The input to this model are the observed physical parameters (intensity, spatial distribution and lifetimes) taken from high resolution measurements and theoretical modelling. The simulations produce "noise" time series at any given time resolution. The model is described in more detail in Andersen (1991, 1992)

The solar surface was divided into elements of 8*8 arcsec. This number was chosen as a compromise in spatial resolution and computing time. In addition, this pixel size encompasses approximately one mesogranular cell. The time evolution of the granules and mesogranules were followed for each pixel. Within each pixel granules and mesogranules are created at a random point in their evolution. The effect of rotation is taken into account for supergranulation, but not for granulation and mesogranulation. It is assumed that a constant number of supergranules exist on the solar surface. When one supergranule dies another is created at a random position and with random age. The intensity distribution of supergranules is assumed to be a bright ring with diameter 32 arcsec.

The current calculations consist of three month timestrings with a timestep of 60 seconds. The typical time constants and amplitudes of the super-, meso- and normal granulation are $7 * 10^4$ seconds, $0.008, 8000$ seconds, $0.01, 500$ seconds and 0.04, respectively. Except for the value of the granular lifetime these data are typical values deduced from high resolution measurements. The shorter granular lifetime was chosen as some observers (Harvey et al. 1993, Isaak, private communication)

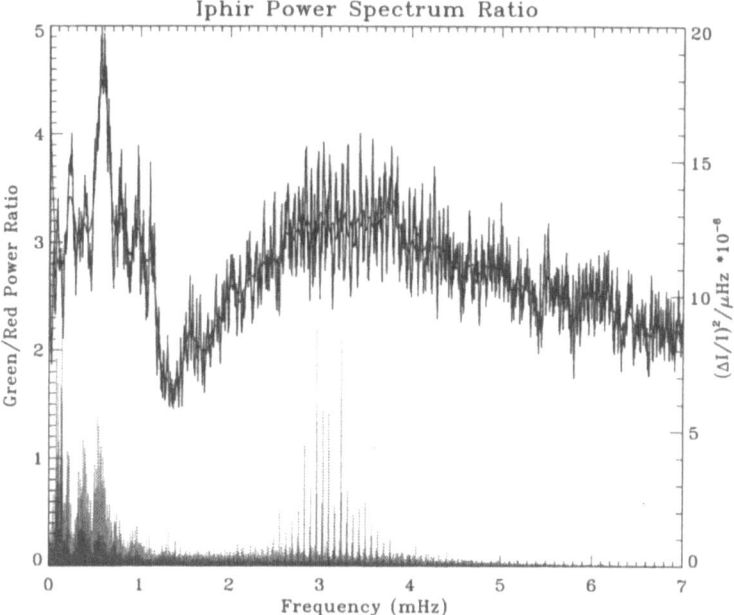

Fig. 2. The power spectra of the green and red channels of the IPHIR instrument are shown in different shades of grey at the bottom of the figure. The green data have the largest amplitude. In addition the ratio of the smoothed power spectra is shown. This ratio should be smooth for the regions where the solar signal or the pure instrumental noise is dominant.

have indicated that this may be deduced from their observations. The simulations presented earlier (Andersen 1991,1992) have used granular lifetime values in the region 700 to 1100 seconds.

3. Results and Comparison with Observations

Several long time series have been observed of both total and spectral solar irradiance. The data from the ACRIM instrument on SMM and the IPHIR instrument on the Phobos mission have been available for some time. Recently observations were carried out with the SOVA-1 and 2 instruments on the EURECA platform. Ongoing observations of total irradiance are being performed by the ACRIM-II instrument on UARS.

In this study we have used data from ACRIM, IPHIR and SOVA-1 photometers. The power spectra of these observational data are shown in Fig. 1. The latter two experiments provide information about the wavelength dependence of the signal. In this study we have concentrated on data from around 500 nm or, as for the ACRIM

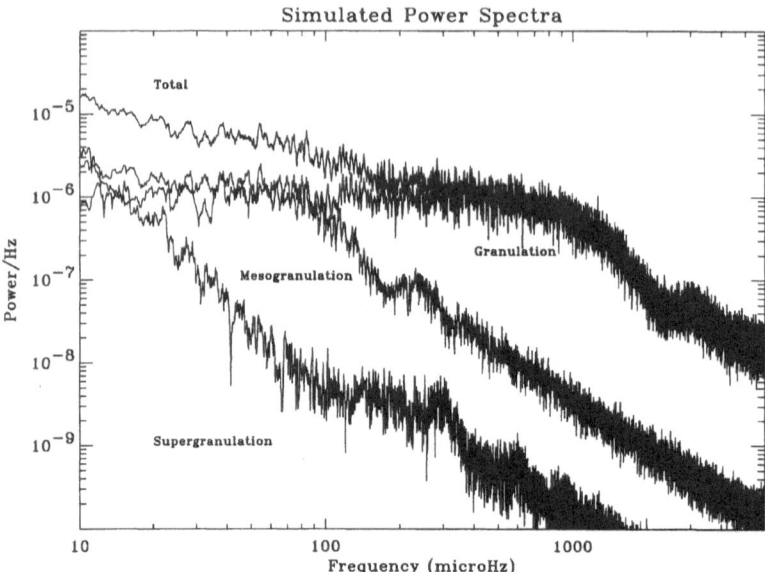

Fig. 3. The power spectra of the total noise simulations as well as the contributions from granulation, meso- and supergranulation are shown. The power spectra have been smoothed with a running mean of 21 frequency points.

data, total irradiance measurements. The IPHIR data are strongly influenced by the excessive pointing errors of the spacecraft. By studying the power spectra ratio of the 500 nm and 865 nm channels, shown in Fig. 2, we see that an abrupt change in the ratio occurs at about 1.3 mHz. This change clearly has a non-solar origin and is probably caused by the guiding error correction. These data are therefore not used below 1.3 mHz in the analysis.

The photometer observations from the SOVA-1 experiment have not yet been fully analyzed, but problems exist with the influence of instrumental effects in the 5 minutes band. In addition the low inclination of the orbit causes orbital occultations. For this study we have used the orbital averages for the ACRIM and SOVA-1 data. From Fig. 1 we see that the SOVA-1 data are significantly higher than the ACRIM data by a factor of 10-50 depending on frequency. This is partly due to the higher sensitivity to solar variations in the spectral irradiance measurments as compared to the total irradiance, but this effects should only give a factor of about two. Closer scrutiny of the SOVA-1 photometer data show unexplained peaks at the daily harmonics. This implies that further data reduction has to done with the SOVA-1 data before they can be compared directly with the simulations and other observations.

In Fig. 3 we show the contribution to the solar noise signal from the three different components in the simulation. We see that above 150 μHz the granulation is

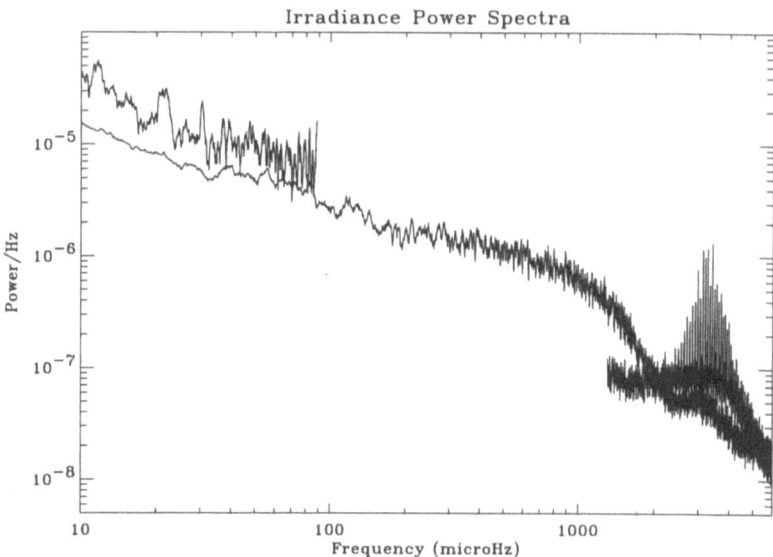

Fig. 4. The power spectra of the low frequency ACRIM and IPHIR high frequency data are shown in comparison with the total simulated data. The power spectra have been smoothed with a running mean of 41 frequency points.

the dominant signal. In Figure 4 we have compared the total power of the simulations with the observed data. We see that there are no good data available in the frequency range 100 mHz to 1 mHz. This is very unfortunate as it is in this range the differences between the different time constants of meso and normal granulation would be the largest. The current simulation gives a factor of about 2 lower power spectrum than the ACRIM data at low frequency. This is a reasonable answer taking into account that there may be a significant instrumental noise in ACRIM at these frequencies. Between 1.3 and 2.1 mHz the fit is poor. The short granular lifetime in the simulation causes the signal to lie above the observations below 2 mHz. Between 2 and 5 mHz the simulation is a factor of about three below the observations and the shape of the signal agrees well. Above 5 mHz the observations drop of significantly faster than the simulations.

These results imply that a 500 s granular lifetime (or shorter) is incompatible with the IPHIR observations. It will require good observational data in the 100 mHz to 1 mHz to determine the lifetime of the global granulation signature.

4. Acknowledgements

The authors are indepted to the Principal Investigator of SOVA, Dr. D. Crommelynck for making the preliminary data from the SOVA-1 photometers available.

References

Andersen, B.N.: 1991, *Adv. Space Res.* 11, No. 4, 93.

Andersen, B.N.: 1992, *Proceedings of Mini-Workshop on Diagnostics of Solar Ocillations Observations*, University of Oslo, 15.

Harvey, J.W.: 1985, *ESA SP-235*, 199.

Harvey, J.W., Duvall, T.L., Jefferies, S.M. and Pomerantz, M.A.: 1993, *ASP Conference Series*, **42**, 111.

Hill, H., Fröhlich, C., Gabriel, M. and Kotov, V.A.: 1991, *"Solar Interior and Atmosphere"*, University of Arizona Press, Eds. A.N. Cox, W.C. Livingston, M.S. Matthews, 562.

SECULAR VARIATIONS IN THE SPECTRUM OF SOLAR P-MODES

A. Jiménez, P.L. Pallé, C. Régulo and T. Roca Cortés
Instituto de Astrofísica de Canarias. E-38205 La Laguna . Tenerife. Spain

Abstract

The solar p-mode spectrum of very low l is measured with high accuracy for a long enough period of time so as to allow the search for solar cycle variations. In this paper solar cycle variations of the frequency and energy of the modes are confirmed. Moreover, a slight variation,within errors, of its rotational splitting with the solar cycle, is suggested.

1. Introduction

The discovery of the solar acoustic spectrum of solar modes of very low l (<4), using resonant scattering spectrophotometers to measure the solar radial velocity of the whole Sun (Claverie et al., 1979; Grec et al., 1980) provided an invaluable tool to probe the solar interior structure and dynamics near the solar core (Ulrich and Rhodes, 1983; Kosovichev, 1993). The possible variation of this spectrum with the solar activity cycle was already pointed out in Pallé et al. (1986) and Jiménez et al. (1988), where the energy contained in the 5-min range showed a slight variation with solar cycle, being minimum at solar maximum. Pallé et al. (1990a, 1990b) later on confirmed this result using data all along a cycle of solar activity.

The change in frequencies of the p-modes with solar cycle was already suggested by Woodard and Noyes (1985) using ACRIM data on board SMM. Further evidence has been found (see Woodard ,1987, Fossat et al. ,1987, Pallé et al. 1989, 1990a, 1990b, Anguera Gubau et al, 1992, and Elsworth et al.,1990a and 1990b) which confirmed the suggestion.

Finally rotational splitting of such modes has proved very difficult indeed to measure. The first measurements came from Claverie et al (1982) which already showed the difficulty in interpreting such measurements because the mode lifetime has similar time scales, confusing the measurements. However, several groups have provided such measurements for the l=1 mode very recently (Toutain and Fröhlich,1992; Toutain, 1993; Loudagh et al, 1993).

In this work, using the longest data already collected with the same instrument from the same site (Observatorio del Teide, Tenerife), an analysis is presented where parameters like frequency, energy and splittings are measured, and their variations with solar activity cycle are studied.

Solar Physics **152**: 253–260, 1994.

2. Observations and analysis

Radial velocity data have been obtained at Observatorio del Teide (Izaña, Tenerife) since 1975 using a resonance scattering solar spectrophotometer based on potassium vapour. The instrument is extensively described in Brookes et al. (1978) and subsequent small modifications and updatings are mentioned in Pallé et al. (1989). However, it was not until 1984 that regular, daily non-stop measurements were taken. The raw data obtained every day is detrended from Earth´s daily motion in a standard way already described in earlier papers (see Pallé et al.,1986; van der Raay et al.,1986) to obtain the daily residuals. Afterwards they are grouped in series with data of ≅30 contiguous days. A total of 76 series have been used for analysis in this work, spanning from 1980 to 1992. These series are then analyzed to obtain their power spectra. The monthly power spectra are computed using an interactive sine wave fitting procedure with a sampling interval of 0.1μHz. The frequencies and energies per unit mass, contained in an interval of 2 times the linewidth around each peak, are obtained as the centroids and the area under the peaks, respectively, as already explained in Anguera Gubau et al. (1992) and Pallé et al. (1990b). From this analysis, the frequency shifts (see Figure 1) as a function of time and energies (see Figure 2) as a function of frequency, have been obtained.

A subset of the obtained data series has been used to look for the rotational splitting of the modes. The aim of this analysis is to measure the splitting in the two extremes of the activity cycle. If some change in the sun's rotation occurs correlated with magnetic activity, it will be maximum when looking at exact minimum and maximum of activity. For this reason, we have chosen those series taken at solar minimum, during which the RZ number was less than 15 and, at solar maximum, when it was bigger than 150. In this selection, we have also taken into account the quality of the data and the duty cycle, in order to choose the best series which fulfil the conditions mentioned above; a total of 9 series for minimum and 9 series for maximum were finally used. Table I summarizes the data used .

Table I . Data used for the splitting measurements presented in this work.

SOLAR MINIMUM			SOLAR MAXIMUM		
MONTHLY SERIE	RZ	D. C.	MONTHLY SERIE	RZ	D. C.
1984: 09/02 - 10/01	14.2	29%	1989: 06/01 - 06/30	196.0	42%
1985: 03/29 - 04/23	14.7	31%	1989: 08/01 - 08/30	170.7	27%
1985: 08/1 - 08/30	11.0	34%	1989: 09/01 - 09/30	176.7	26%
1985: 08/31 - 09/27	4.0	26%	1990: 07/01 - 07/31	149.4	40%
1986: 05/19 - 06/17	10.5	34%	1990: 08/01 - 08/31	200.3	33%
1986: 07/16 - 08/14	10.8	43%	1991: 02/01 - 02/28	167.5	24%
1986: 08/15 - 09/13	7.8	37%	1991: 06/01 - 06/30	170.7	42%
1987: 01/12 - 02/10	8.0	24%	1991: 07/01 - 07/31	174.1	42%
1987: 02/12 - 03/12	8.8	32%	1991: 08/01 - 08/31	180.0	40%

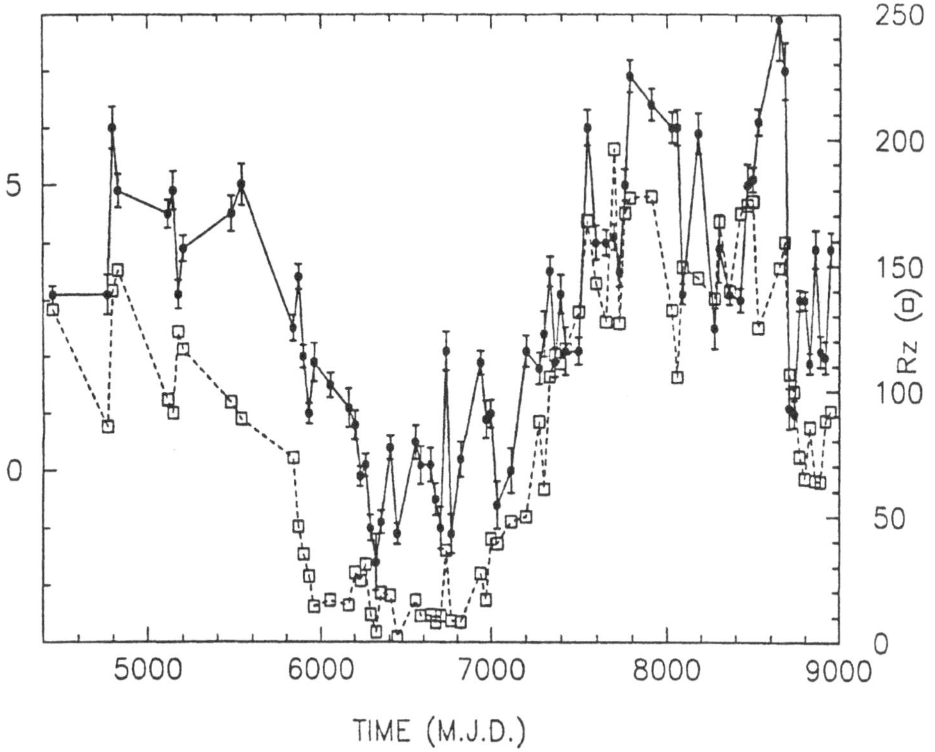

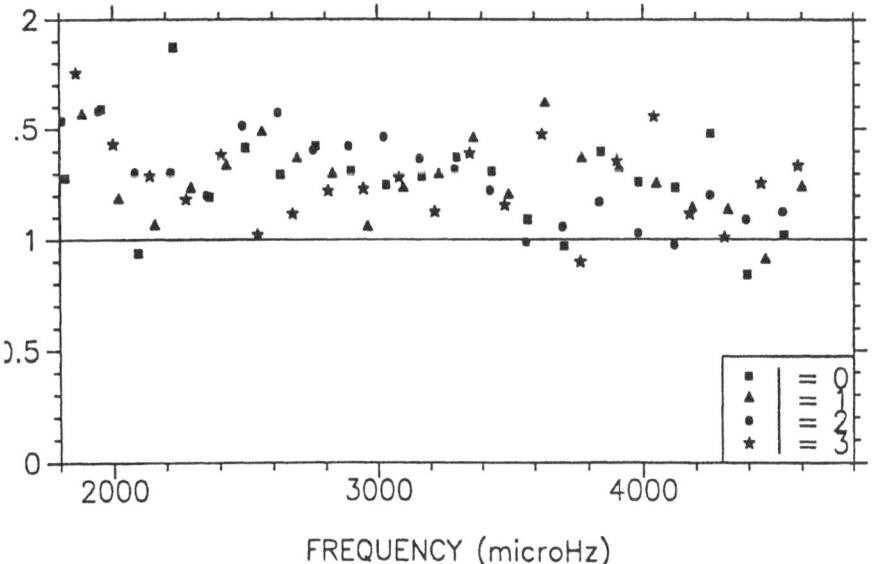

Fig 1. The frequency shifts (obtained as crosscorrelation of each monthly spectra with an average of 6 at solar minimum) along time is presented starting August 1980 up to the end of 1992. Also plotted is the value of RZ .

Fig 2. The ratio of the energies (per unit mass) of each measured mode, at minimum and maximum, is presented as a function of frequency. The mean value found is 1.30 ± 0.02.

To compute the splittings we have used the following data products : 1) the monthly power spectra, 2) the average spectra at both, maximum and minimum, and 3) the "folded" spectra of the averaged ones at maximum and minimum. The folded spectra are obtained in bits of 30 µHz each centred around the derived frequencies of each mode. The method used for measuring the splittings consists of fitting appropiate functions to the features present in the frequency spectrum of the solar oscillations. If a stochastic nature for the solar p-modes is assumed, their profile will be similar to the one corresponding to a freely decaying damped oscillator, that is a Lorentz profile :

$$P[\upsilon] = \frac{I_0 \cdot \frac{\gamma^2}{4}}{[\upsilon - \upsilon_0]^2 + \frac{\gamma^2}{4}} + noise$$

where : I_0 stands for the height of the line, γ for the half width at half maximum and υ_0 for the frequency at maximum.

An interval of ~40 µHz around the l=1 modes is taken and a sum of 6 Lorentzian profiles, plus locally measured white noise in that interval, is fitted. In this way the two components of the l=1 (m=±1) modes together with their sidebands are fitted. The fitting method is a standard, non-linear, least-squares method (the Marquart method), which requires an initial guess for the parameters and minimizes a merit function. Analyses of modes with frequencies between 1.8 and 3.2 mHz have been considered because lifetimes grow shorter very fast at higher frequencies. Figures 3a and 4a show an example of these determinations (at the monthly spectra), e.g. the l=1, n=18 observed at solar minimum and maximum respectively. Figures 3b and 4b show the same modes (at the averaged spectra) and figures 3c and 4c, at the folded spectra as described before. The thin lines are the best Lorentzian fitted profiles. It sould be said that in the spectra, modes with l=0 are unsplit, while those with l=2 show a confused structure mixed with l=0 sidebands. Modes with l=3 have too low an amplitude to be unambigously identified. These are the reasons why only results for l=1 are presented at this stage of the work.

Once all the selected l=1 modes have been fitted in the three kinds of spectra, the splitting is then computed. Figure 5 shows the sidereal splitting measured using the best unambiguous determinations in the average spectra. The folded spectra only yield one value at both, minimum and maximum. The mean splittings have been computed using statistical standard weights (1/error^2). Table II shows all the splitting values obtained in this work.

SOLAR MINIMUM

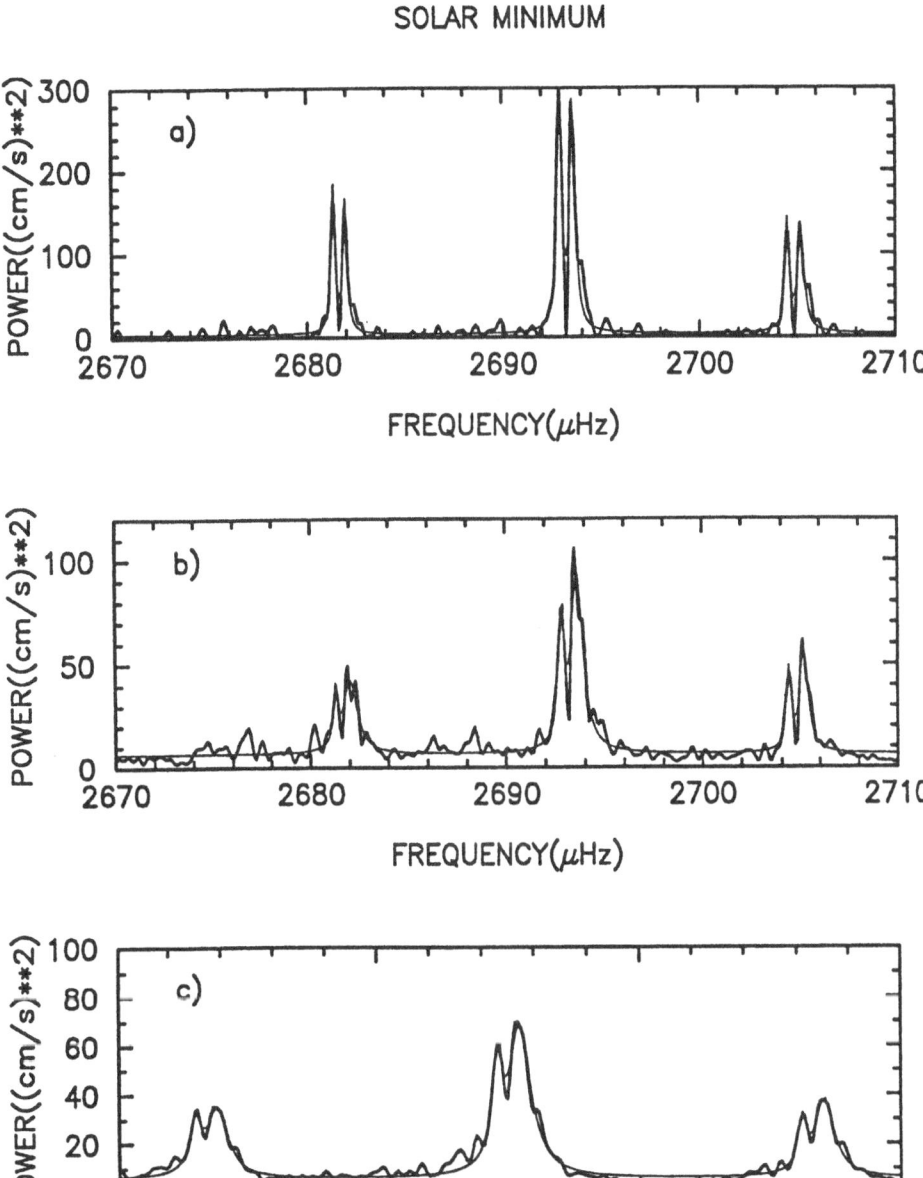

Fig 3. Examples of the power spectra obtained showing a mode, the one with l=1and n=18, (a): for one of the monthly series (86/7/16 - 86/8/14) , (b): for the average spectrum , (c): the folded one . These examples correspond to the solar minimum. The thin lines are the best fitted Lorentzian profiles.

A. JIMÉNEZ ET AL.

SOLAR MAXIMUM

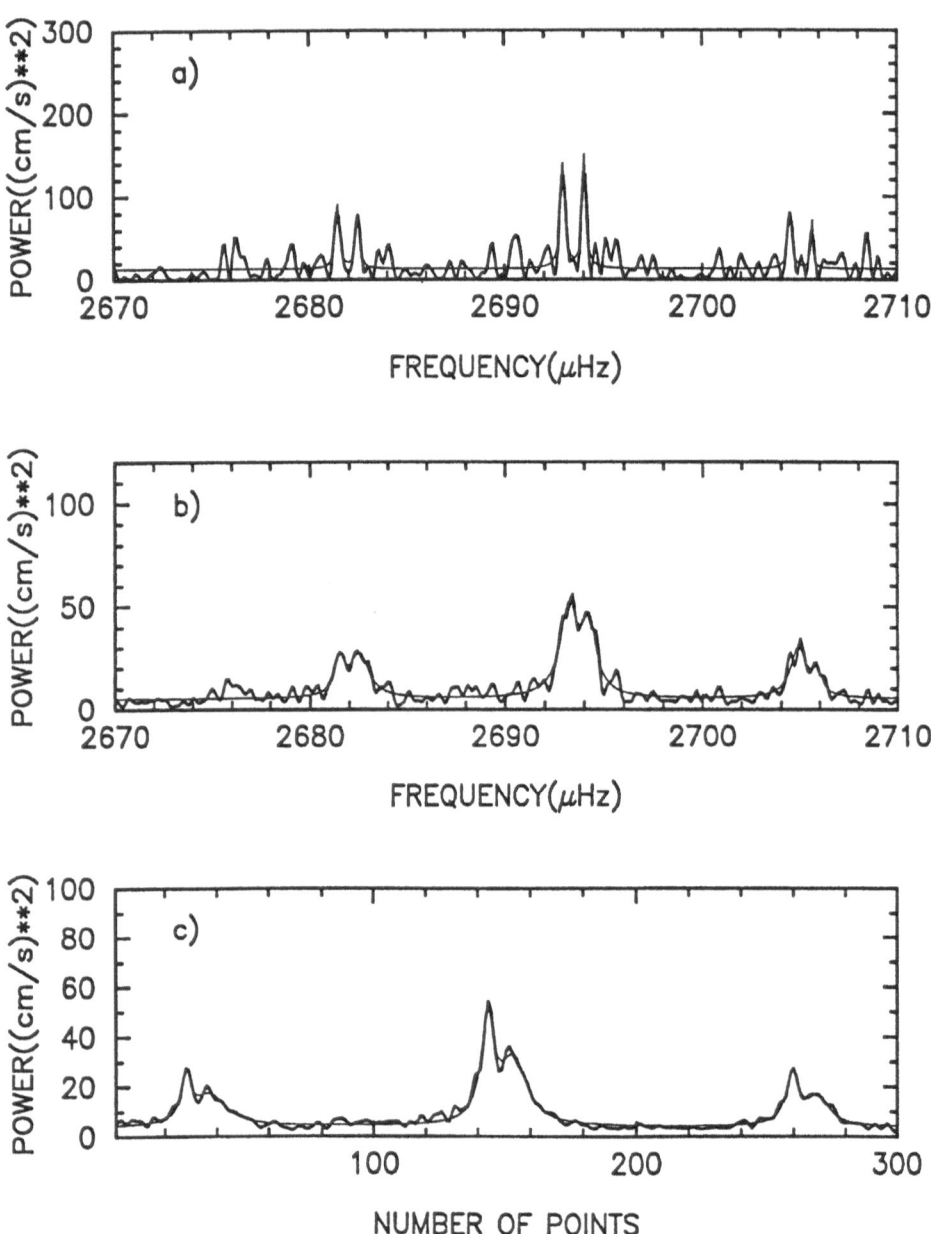

Fig 4. Examples of the power spectra obtained showing a mode, the one with l=1and n=18, (a): for one of the monthly series (91/2/1 - 91/2/28) ,(b): for the average spectrum , (c): the folded one . These examples correspond to the solar maximum. The thin lines are the best fitted Lorentzian profiles.

Table II. Average sidereal splitting obtained using the data products mentioned.

	SIDEREAL L=1 SPLITTING (nHz)	
	Solar minimum	Solar maximum
Monthly spectra	479.1 ± 23.1	529.0 ± 12.0
Average spectra	497.7 ± 29.8	510.2 ± 22.0
Folded spectra	476.8 ± 9.7	525.5 ± 12.7

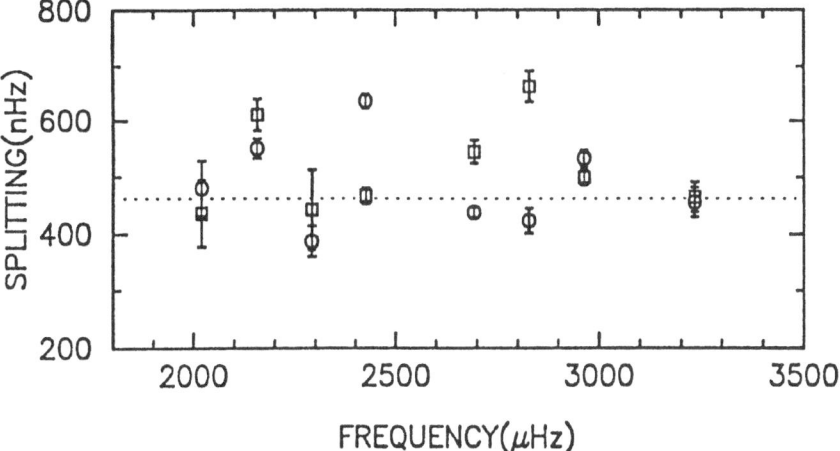

Fig 5. Sidereal splitting measured for l=1 modes in the average spectra. Circles corresponds to solar minimum and squares to solar maximum. The dotted line corresponds to the rotational splitting value at solar surface (462 nHz).

3. Discussion and conclusions

The frequency variations show a very good correlation with the cycle, confirming past work on this phenomenon; however, even shorter time correlations are visible between the two series. The energy ratio clearly confirms the effect found by Pallé et al (1990b), showing no definite change along frequency. Looking at the individual determinations of the l=1, splitting seems not to have a clear dependence on the solar activity cycle, although the mean values yield to a slight increase of the splitting at maximum, within errors. Taking the extreme values of our determinations, the splitting at minimum can be situated between 476.8 and 497.7 nHz and at maximum between 510.2 and 529.0 nHz (see Table II). However, the folded spectra offer the best determinations which yield a clear difference between minimum (476.8 nHz) and maximum (525.5 nHz).The

mean value recently reported by Toutain is 474.0 ± 23 nHz   was obtained with the
IPHIR instrument in 1988 (in the ascending phase of the cycle). The one reported by
Loudagh et al (1993), 495 ± 30 nHz, was obtained with IRIS network in summer 1991
at solar maximum. These values agree quite well with the values reported here at the
appropiate phase of the solar cycle, also suggesting a slight increase when the cycle
approaches its maximum. Although all measurements are higher than the splitting rate
at the surface, it should not be forgotten that they are compatible with it within 3 times
the errors

Acknowledgements

This work is one of the end products of a longstanding research project which has been made
possible with the effort of many individuals. We wish to thank here all the members, past and
present, of the helioseismology team at the IAC. We also are grateful to the past Birmingham
University helioseismology group which actively collaborated with us. Finally, thanks for the Solar
Data Reports from Boulder.

References

Anguera Gubau, M., Pallé, P.L., Pérez Hernandez, F., Régulo, C., and Roca Cortés, T.: 1992,
    Astron. Astrophys. **255**, 363.
Brookes, J.R., Isaak, G.R., and van der Raay, H.B.: 1978, *Monthly Notices Roy. Astron. Soc.*
    185, 1.
Claverie, A., Isaak, G.R., Mc Leod, C.P., van der Raay, H.B., and Roca Cortés ,T.: 1979, *Nature*
    282, 591.
Claverie, A., Isaak, G.R., Mc Leod, C.P., van der Raay, H.B., Pallé, P.L., and Roca Cortés ,T.:
    1982, *Nature* **299**, 704.
Elsworth, Y., Howe, R., Isaak, G.R., McLeod, C.P., and New, R.: 1990a, *Nature* **345**, 322.
Elsworth, Y., Howe, R., Isaak, G.R., McLeod, C.P., and New, R.: 1990b, *Nature* **347**, 536.
Fossat, E., Gelly, B., Grec, G., and Pomerantz, M.: 1987, *Astron. Astrophys.* **177**, L47.
Grec, G., Fossat, E., and Pomerantz, M.: 1980, *Nature*, **288**, 541.
Jiménez, A., Pallé, P.L., Perez, J.C., Régulo, C., Roca Cortés, T., Isaak, G.R., McLeod, C. P.,
    van der Raay, H.B.: 1988, in *Advances in Helio-and-Asteroseismology, IAU Symp.* **123**, 205.
Kosovichev, A.G.: 1993, in *Proceedings Vth IRIS workshop & GOLF 93 annual meeting*, 27.
Loudagh, S.,Provost, J., Berthomieu, G., Ehgamberdiev, S., Fossat, E., Gelly, B., Khalikov, S.,
    Lazrek, M., Pallé, P., Régulo, C., Sanchez, L., and Schmider F.X.: 1993, *Astron. Astrophys.*
    275, L25.
Pallé, P.L., Perez, J.C., Régulo, C., Roca Cortés, T., Isaak, G.R., McLeod, C. P., and van der
    Raay, H.B.: 1986, *Astron. Astrophys.* **169**, 313.
Pallé, P.L., Régulo, C., and Roca Cortés, T.: 1989, *Astron. Astrophys.* **224**,253.
Pallé, P.L., Régulo, C., and Roca Cortés, T.: 1990a, in *Inside the Sun, IAU Symp.* **121**, 349.
Pallé, P.L., Régulo, C., and Roca Cortés, T.: 1990b, in *Progress of seismology of the Sun and
    stars. Lectures Notes in Physics* , **367**, 129 .
Van der Raay, H.B., Pallé, P.L., and Roca Cortés, T.: 1986, in *Seismology of the Sun and the
    distant Stars*, Cambridge University Press, Cambridge, 333.
Toutain, T., and Frolich, K.: 1992, *Astron. Astrophys.* **257**, 287.
Toutain, T.:1993, in *Proceedings Vth IRIS workshop & GOLF'93 annual meeting*, 23.
Ulrich, R.K., and Rhodes, Jr.E.J.: 1983, *Astrophys. J.*, **265**, 551.
Woodard, M.F.: 1987, *Solar Phys.*, **114**, 21.
Woodard, M.F., and Noyes, R.W.: 1985, *Nature* **318**, 449.

Solar Cycle Variations in P-modes and Chromospheric Magnetism

Rekha Jain and B. Roberts
Department of Mathematical and Computational Sciences,
University of St. Andrews, St. Andrews, Fife KY16 9SS, Scotland

ABSTRACT: The effect on p-mode frequencies of a horizontal chromospheric canopy field is studied theoretically and the results compared with Libbrecht and Woodard's observations of frequency changes. Combined changes in field strength and chromospheric temperature cause frequency shifts that are similar in form to those observed. Frequency shifts in p-modes offer the possibility of signatures of solar activity cycles distinct from sunspot numbers and butterfly diagrams.

1 INTRODUCTION

Considerable effort has been made to measure very accurately the frequencies of p-mode oscillations and to determine how frequencies vary with the solar cycle. The observations by Elsworth et $al.$ (1990), Libbrecht and Woodard (1990), Woodard and Libbrecht (1991) and Anguera Gubau et $al.$ (1992) have shown conclusively that solar p-mode frequencies change with time and the measured frequency shifts $\Delta\nu \equiv \nu(1988) - \nu(1986)$ and $\Delta\nu \equiv \nu(1989) - \nu(1986)$ depend strongly on frequency ν and almost linearly on spherical harmonic degree l. Woodard and Libbrecht have compared p-mode frequencies from 1986 to those from 1988 and 1989. The year 1986 was a period of low solar activity, while 1988 and 1989 were periods of rising activity.

The perturbations responsible for the frequency shifts may be located in the upper atmosphere where the p-modes are evanescent. Since the decay rate of eigenmodes with height is determined by this upper atmosphere, its detailed structure is important. The magnetic field is a significant part of the chromosphere and so its effect on the frequencies of p-modes is of importance.

The influence of a chromospheric canopy magnetic field on frequency shifts in p- and f-modes was first pointed out by Roberts and Campbell (1988), Campbell and Roberts (1989) and Evans and Roberts (1990). Solar cycle variability was investigated in detail in Evans and Roberts (1992) and Jain and Roberts (1993b), following on from earlier suggestions in Campbell and Roberts (1989).

Chromospheric 'canopy' fields here refer to the magnetic fibril fields that fan out in the higher levels of the solar atmosphere to form an overlying magnetic canopy. In Campbell and Roberts (1989) this higher level was modelled by a horizontal magnetic field whose strength decreased with height in such a manner as to maintain a constant Alfvén speed. This type of atmospheric magnetic field causes the p-mode frequencies to decrease as a result of an increase in magnetic field strength. In Evans and Roberts (1990,1991,1992) and Jain and Roberts (1993a,b) the magnetic field is again taken to be horizontal, but now of constant magnitude. This magnetic field configuration produces a positive frequency shift. Goldreich et $al.$ (1991) have affirmed the importance of variations in the photospheric magnetic field strengths. They argued that these variations are responsible for positive frequency shifts at low p-mode frequencies and that the observed precipitious drop in the shifts at high frequencies is caused by the presence of a chromospheric resonance and an increase

Solar Physics **152**: 261–266, 1994.
© 1994 *Kluwer Academic Publishers.*

in the chromospheric temperature. There is at present no observational evidence for such a chromospheric resonance. Fernandes *et al.* (1992) have analysed doppler shift measurements of the Na D_1 absorption line, revealing oscillation frequencies up to 9.5 mHz but no evidence of chromospheric modes with 3 minute period. Wright and Thompson (1992) used a perturbation approach to show that frequency changes are influenced by the sub-photospheric layers (assumed to be field-free) only through the mode inertia of the oscillation, being inversely proportional to mode inertia.

These studies suggest that chromospheric magnetism and its influence on the *p*-modes is potentially important for solar cycle variability. The question arises: can changes in chromospheric magnetic activity over the solar cycle bring about the frequency shifts as observed?

To address this question a simple model is discussed, continuing the investigation begun by Evans and Roberts (1990, 1992) and Jain and Roberts (1993b). Our objective is to determine the thermal as well as magnetic effects of the chromosphere on solar *p*-mode frequencies. The convection zone is modelled as a fluid having a temperature profile that increases linearly with depth. The presence of vertical magnetic flux tubes in the convection zone and of a toroidal magnetic field at the base of the convection zone are neglected. The atmosphere above the temperature minimum is assumed to be isothermal and permeated by a uniform horizontal magnetic field.

2   THE MODEL AND DISPERSION RELATION

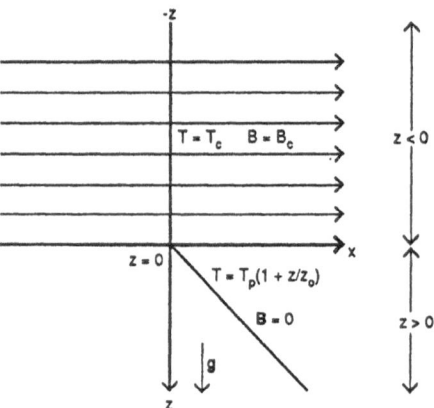

Figure 1: The equilibrium structure of the model, consisting of a field-free thermally stratified 'convection' zone in $z > 0$ and an isothermal magnetic atmosphere in $z < 0$.

The model shown in Fig. 1 is studied in an attempt to find out how the *p*-mode frequencies are influenced by a horizontal magnetic field. The model introduces two distinct layers, an upper layer representing the magnetic chromosphere and a lower one representing the convection zone. The reference level $z = 0$ corresponds to the base of the chromosphere (i.e. the temperature minimum). The chromosphere and the convection zone are represented by the regions $z < 0$ and $z > 0$, respectively, with z increasing downwards. The medium is stratified by gravity, $g\hat{z}$. The chromosphere is taken to be isothermal, with temperature T_c, within which is embedded a uniform horizontal magnetic field of strength B_c. This region models the chromospheric canopy. The region below the magnetic chromosphere, modelling the convection zone, is assumed to be field-free and to have a temperature $T_p(1 + z/z_o)$ varying linearly with depth z. Here z_o is the temperature scale-height and T_p is the temperature at the top $(z = 0)$ of this region.

The dispersion relation relating angular frequency ω and horizontal wavenumber k is (Evans and Roberts, 1990; Jain, 1992; Jain and Roberts, 1993a,b)

$$2a\omega^2 c_{sp}^2 k \frac{U\left(-a + 1, m + 3, 2kz_o\right)}{U\left(-a, m + 2, 2kz_o\right)} + \gamma g\omega^2 - kc_{sp}^2(\omega^2 + gk)$$

$$= \frac{(c_{sc}^2 + \frac{\gamma}{2}v_{Ac}^2)(g^2k^2 - \omega^4)(\omega^2 - k^2c_{sc}^2)}{gk^2c_{sc}^2 + (c_{sc}^2 + v_{Ac}^2)(\omega^2 - k^2c_{Tc}^2)\left\{k - \frac{pg}{r}\frac{\mathcal{A}_1\mathcal{A}_3}{\mathcal{A}_2}\frac{F\left(p+1,q+1;r+1;\frac{-\mathcal{A}_1}{\mathcal{A}_2}\right)}{F\left(p,q;r;\frac{-\mathcal{A}_1}{\mathcal{A}_2}\right)}\right\}} . \tag{1}$$

Here c_{sc}, v_{Ac} and $c_{Tc}\left(= c_{sc}v_{Ac}/(c_{sc}^2 + v_{Ac}^2)^{1/2}\right)$ are the sound speed, Alfvén speed, and cusp speed at the base of the chromosphere. The sound speed at the top of the convection zone is c_{sp}. The adiabatic index in the convection zone and chromosphere is γ; m is the polytropic index in the convection zone. The parameter a, which arises in the confluent hypergeometric function U, depends upon ω and k as well as the structure of the convection zone; for a neutrally stratified field-free interior, $m = .1/(\gamma - 1)$ and $2a = m(\omega^2/gk) - (m + 2)$. The parameters p, q, r, $\mathcal{A}_1$, $\mathcal{A}_2$ and $\mathcal{A}_3$, which arise in the hypergeometric function F, depend upon conditions in the magnetic atmosphere as well as ω and k; they are given in detail in Jain and Roberts (1993b).

Evans and Roberts (1990, 1992) and Jain and Roberts (1993b) have determined analytically the asymptotic behaviour of $\Delta\nu \equiv \nu(B_c', T_c') - \nu(B_c, T_c)$, the difference between cyclic frequency $\nu(\equiv \omega/2\pi)$ in a chromosphere with field strength B_c' and temperature T_c' and one with strength B_c and temperature T_c. They determined $\Delta\nu$ as a function of frequency ν and degree l by examining the behaviour of the dispersion relation (1) in the limit of small $kz_o (\sim lz_o/R_{sun})$. This limit corresponds to $lz_o \ll R_{sun}$, where $R_{sun} = 6.96 \times 10^8$m is the solar radius and $z_o = 293$ km (see Jain, 1992), yielding $l \ll 2375$; all cycle variation data sets satisfy this property. A further simplification in the form of $\Delta\nu$ is possible if attention is focussed on the case of large mode number (radial order) n (Evans and Roberts 1992). Coefficients arising in the description of $\Delta\nu$ may be substantially simplified when $n \gg 1$. In fact, the qualitative behaviour of $\Delta\nu$ is much the same at large n as at moderate or even low n (Jain 1992) so little of significance is lost by simplifying for large n but there is a considerable gain in analytical insight and simplicity.

In the absence of any temperature changes $(T_c' = T_c)$, it follows from Equation (1) that for large mode number n the frequency shift $\Delta\nu$ satisfies (Evans and Roberts, 1992; Jain and Roberts, 1993b)

$$\frac{1}{\nu}\Delta\nu = A_m\left(\frac{B_c'^2 - B_c^2}{2\mu p_p}\right)\left(\frac{\nu}{\nu_o}\right)^{2m}\left(\frac{z_o}{R_{sun}}\right)l , \tag{2}$$

where $2\pi\nu_o = (g/mz_o)^{1/2}$ and $A_m = [(m+1)\Gamma^2(m+1)]^{-1}$; Γ denotes the gamma function. Equation (2) gives a positive shift for $B'_c > B_c$. Notice the linear proportionality of $\Delta\nu$ with both degree l and the change in magnetic pressure (from $B_c^2/2\mu$ at one time to $B'^2_c/2\mu$ at a more active time), relative to the pressure p_p at the top of the photosphere. Observe, too, the flat increase in $\Delta\nu$ at low ν given by the power ν^{2m+1}.

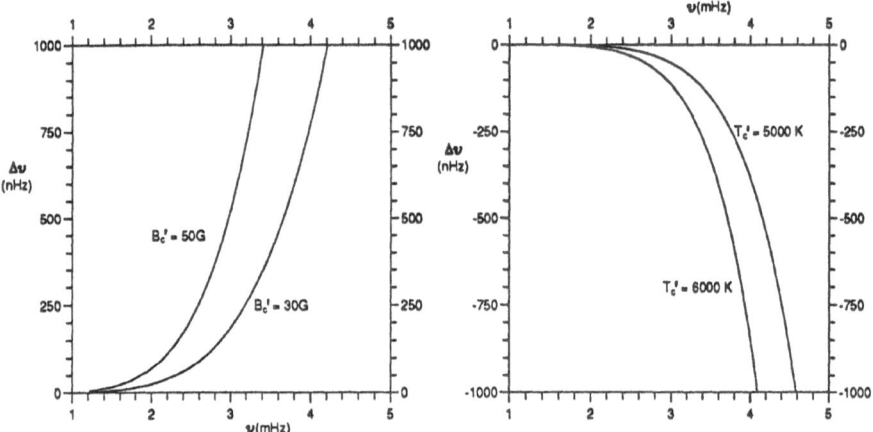

Figure 2: The calculated frequency shift $\Delta\nu$, for modes of degree $l = 75$, due solely to (a) an increase in magnetic field strength from $B_c = 5G$ to $B'_c = 30G$ or $50G$, according to equation (2); (b) an increase in the chromospheric temperature from $T_c = 4170K$ to $T'_c = 5000K$ or $6000K$; see equation (3). The adiabatic index is $\gamma = 1.5$ and the photospheric pressure is $p_p = 86.72$ N m^{-2}.

Figure 2a shows the frequency shift $\Delta\nu$ determined by equation (2) as a function of base frequency ν for two values of the final magnetic field strength, namely $B'_c = 30G$ and $B'_c = 50G$, taking an initial magnetic field strength $B_c = 5G$; the mode has a degree $l = 75$. The larger the increase in the magnetic field, the greater the frequency shift.

In a similar fashion, the frequency shift $\Delta\nu$ due to a change in chromospheric temperature can also be considered. In the absence of any magnetic field, Equation (1) leads to (Campbell and Roberts, 1989; Jain and Roberts, 1993b)

$$\frac{1}{\nu}\Delta\nu = -C_m \left(\frac{T'_c - T_c}{T_p}\right)\left(\frac{\nu}{\nu_o}\right)^{2m+2}\left(\frac{z_o}{R_{sun}}\right)l, \tag{3}$$

where $2\gamma\Gamma^2(m+2)C_m = (\gamma m/2)^{2m+1}$. It is clear that $\Delta\nu < 0$ if $T'_c > T_c$ and $\Delta\nu > 0$ if $T'_c < T_c$. Thus, $\Delta\nu$ becomes more negative with increasing chromospheric temperature $T'_c > T_c$ and the decline in frequency shift is directly proportional to l.

In Equation (2), $\Delta\nu$ is proportional to the frequency raised to the power $2m+1$. In Equation (3), the power increases to $2m+3$. Hence the frequency increase caused by a magnetic field is expected to dominate at low frequencies, whereas the decrease caused by an increase in chromospheric temperature is expected to dominate at higher frequencies.

In Figure 2b, the frequency shift from equation (3) caused solely by a chromospheric temperature rise is plotted as a function of frequency for a temperature change from $T_c = 4170K$ to $T_c' = 5000K$ or $T_c' = 6000K$. The magnitude of the frequency shift increases with frequency and is proportional to the size of the temperature change.

Finally, a combination of results (2) and (3) *suggests* a $\Delta\nu$ of the form

$$\frac{\Delta\nu}{\nu} = A_m \left(\frac{\nu}{\nu_o}\right)^{2m} \left(\frac{z_o}{R_{sun}}\right) \left\{ \left(\frac{B_c'^2 - B_c^2}{2\mu p_p}\right) - D_m \left(\frac{T_c' - T_c}{T_p}\right) \left(\frac{\nu^2}{\nu_o^2}\right) \right\} l , \quad (4)$$

where D_m is dependent both upon changes in magnetic pressure and on thermal effects; D_m is non-zero even in the absence of a magnetic field. A numerical solution of the full dispersion relation (1) shows frequency shifts of the general form given by Equation (4); see Jain (1992). It should be added, however, that while equation (4) is made plausible by the special cases that led to Equations (2) and (3), the complexity of the full dispersion relation (1) has prevented a derivation of (4).

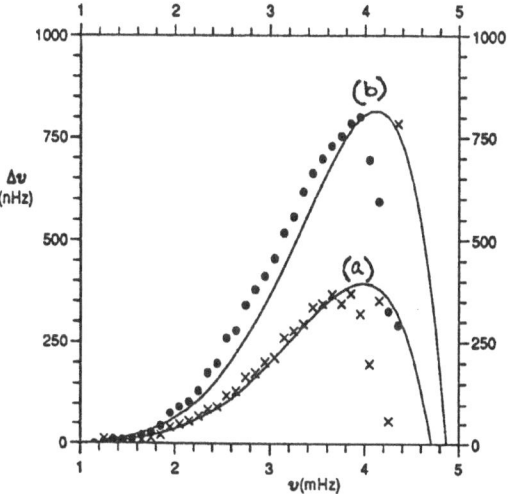

Figure 3: The calculated frequency shift equation (4) for p modes of degree $l = 75$ due to simultaneous increases in magnetic field strength and chromospheric temperature from base values $B_c = 30G$ and $T_c = 4170K$. The crosses × represent Libbrecht and Woodard's (1990) observations for frequency shifts between 1986 and 1988, while filled circles • correspond to observed frequency shifts between 1986 and 1989. (a) $T_c' = 6370K$, $B_c' = 50G$; (b) $T_c' = 7770K$, $B_c' = 61G$.

In Figure 3, the frequency shift determined by equation (4), resulting from combined increases in chromospheric magnetic field strength and temperature, is plotted as a function of frequency for: (a) temperature change from $T_c = 4170K$ to $T_c' = 6370K$, magnetic field strength change from $B_c = 30G$ to $B_c' = 50G$; (b) for $T_c = 4170K$ to $T_c' = 7770K$, magnetic field strength change from $B_c = 30G$ to $B_c' = 61G$. We have set $D_m = C_m/A_m$.

3 CONCLUSIONS

The model of the solar atmosphere discussed here allows us to examine analytically and numerically the frequency shifts in p-modes that possibly occur as a result of changes in the chromosphere. In the absence of a magnetic field ($B_c \equiv 0$), an increase in chromospheric temperature leads to a *negative* frequency shift (Campbell and Roberts, 1989; Evans and Roberts, 1990; Goldreich *et al.*, 1991; Johnston *et al.*, 1992). This is precisely opposite to the observed frequency shifts at low ν but suggestive of the observed sharp fall-off at high frequencies. In the presence of a magnetic chromosphere, however, an increase in the magnetic field strength, at fixed T_c, leads to a *positive* frequency shift, in accordance with the observations at low ν (Evans and Roberts, 1990, 1992; Jain and Roberts, 1993b).

These diametrically opposing results suggest that a combination of the two effects may be important in explaining the observed frequency shifts. We have shown here that the observed p-mode frequency shifts may be reproduced qualitatively by a model involving simultaneous increases in magnetic field strength and chromospheric temperature. However, large chromospheric temperature changes are needed to produce a sharp drop in the frequency shifts. The need for such large changes suggests that there may be other physical effects that need to be included before a full understanding of the observed frequency shifts is obtained.

This work is supported by the UK Science and Engineering Research Council.

REFERENCES

Anguera Gubau, M., Pallé, P., Peréz Hernandez, F., Régulo, C. and Roca Cortés, T.: 1992, *Astron. Astrophys.* **255**, 363.

Campbell, W. R. and Roberts, B.: 1989, *Astrophys. J.* **338**, 538.

Elsworth, Y., Howe, R., Isaak, G., McLeod, C. and New, R.: 1990, *Nature* **345**, 322.

Evans, D. J. and Roberts, B.: 1990, *Astrophys. J.* **356**, 704.

Evans, D. J. and Roberts, B.: 1991, *Astrophys. J.* **371**, 387.

Evans, D. J. and Roberts, B.: 1992, *Nature* **355**, 230.

Fernandes, D .N., Scherrer, P. H., Tarbell, T. D. and Title, A. M. 1992, *Astrophys. J.*, **392**, 736.

Goldreich, P., Murray, N., Willette, G. and Kumar, P.: 1991, *Astrophys. J.* **370**, 752.

Jain, R.: 1992, Ph D thesis, University of St. Andrews, Scotland

Jain, R. and Roberts, B.: 1993a, in T. Brown ed. *GONG 1992: Seismic Investigation of the Sun and Stars*, ASPC Series Vol. **42**, p. 53.

Jain, R. and Roberts, B.: 1993b, *Astrophys. J.* **414**, 898.

Johnston, A., Wright A.N. and Roberts, B.: 1992, in T. Brown ed. *GONG 1992: Seismic Investigation of the Sun and Stars*, ASPC Series Vol. **42**, p. 181.

Libbrecht, K. G. and Woodard, M. F.: 1990, *Nature* **345**, 779.

Roberts, B. and Campbell, W. R.: 1988, in E. J. Rolfe ed. *Seismology of the Sun and Sun-like Stars*, ESA SP-286, p.311.

Woodard, M.F. and Libbrecht, K.G.: 1991, *Astrophys. J.* **374**, L61.

Wright, A. N. and Thompson, M. J.: 1992, *Astron. Astrophys.*, **264**, 701.

SUNSPOT NUMBERS UNCERTAINTIES AND PARAMETRIC REPRESENTATIONS OF SOLAR ACTIVITY VARIATIONS

A. VIGOUROUX and PH. DELACHE

Laboratoire Cassini, associé au C.N.R.S. (U.R.A. 1362), Observatoire de la Côte d'Azur, B.P. 229, F-06304, NICE CEDEX 04, FRANCE

Abstract. Investigating the real origins of "error bars" prior to data processing can be highly rewarding. We have already shown it for solar radius determinations where uncertainties could be deduced from the dispersion of elementary measurements. In the present work, we extend our analysis to the historical monthly sunspot numbers where the "uncertainty" problem arises quite differently. This leads to a substantial revision of our initial method. Like in the radius case, we shall stress the interest of analysing procedures which have the capability of taking care of unequal error bars, such as the wavelet transform.

1. Introduction

Whereas the variability of the solar activity is well documented, for example through the sunspot number measurements, it is more difficult to be convinced that solar radius variations have been really detected. In a recent paper (Vigouroux & Delache, 1993), we have studied radius data which are obtained at Observatoire de la Côte d'Azur for more than fifteen years (Laclare, 1983). Since measurements are affected by noise sources lying in the terrestrial atmosphere such as seeing, scattering, refraction and transparency variations, a careful analysis has to take proper account of the actual sources of uncertainty. In our quoted work, we have proposed a new analysis of the solar radius through wavelet transform which takes care of the probable error distributions. As we have shown, the reality of solar radius measurements variability can be quantitatively assessed.

We propose here to apply similar techniques to historical solar activity as measured by sunspot numbers. The specific sunspot data which we analyze are the monthly mean sunspot numbers, January 1749 to December 1992, available from the National Geophysical Data Center in Boulder, Colorado (N.G.D.C.). The actual monthly values are obtained from all available observations recorded during the desired month. Of course there are periods when these observations are scarce. It may even happen that no sunspot record can been found at all during an entire month. For example, the N.G.D.C. points out that "no observations are available during February 1824. The value was interpolated from the January and March monthly means of that year". Thus, we deal with a time series which has no gap.

This monthly time series, or the yearly version of it, has already been

Solar Physics **152**: 267–274, 1994.

studied, mostly in attempting to detect periodicities of the solar cycle engine through spectral analysis, or more recently in searching for variations in the period of the sunspot cycle (Ochadlick et al., 1993) through wavelet analysis. The aim of the present paper is not to produce any new result concerning period determination, or period variations. We want to assess the degree of significance of the elements of Fourier or wavelet transforms of the data, taking care of their statistical properties. Inverting the transforms will yield to models for the variation of initial data which retain only these significant elements.

A priori, there is no uncertainty in the sunspot number; they do not depend on observing conditions as the radius measurements (Delache et al., 1993). However, one may imagine that our Sun could have shown, within a given month, a slightly different degree of daily activity. In other words, some of the day-to-day, or month-to-month, variations may not be completely deterministic. In a sense – that we are going to precise – we may consider that each of the monthly sunspot number has its own "uncertainty". We shall then proceed to the sunspot number analysis in a way somewhat similar to our previous work on the solar radius.

2. Dispersion depending on the actual value of the observed quantity

We propose to associate to every such *monthly* sunspot number the dispersion of the *daily* sunspot numbers* that it includes, considering this dispersion as the relevant "uncertainty". For that purpose, we first compute a 30-day running mean time series Z_m from the daily measurements, Z_d, which are available since 1874. We then define several classes of activity C_i for the Z_m in such a way as to obtain roughly the same number of elements in each class. Those classes have to obey two conditions, possibly in conflict:

- they should imply enough members of the Z_d series so that further statistical description is valid,
- the increment $\delta A_i = A_{i+1} - A_i$ (where A_i is the mean activity of all the Z_d values contributing to class i) from one class to the next should preferably be kept smaller than the dispersion D_i of the daily Z_d measurements contributing either to C_{i+1} or to C_i ; otherwise the observed dispersion would reflect too heavily this δA_i.

The result of the necessary compromise is given in table I which describes our choice of classes. One sees that the above conditions are correctly respected up to the last class (highest activity), where they are only marginally fulfilled.

---

* N.G.D.C. also provides this (Greenwich) daily sunspot numbers, available only from January 1874 to December 1982.

TABLE I

Activity (ie sunspot numbers) classes C_i

class number i	minimum Z_m value in the class i	maximum Z_m value in the class i	nb of Z_m elements in the class i	mean of the Z_d ("A_i") in the class i	Z_d dispersion ("D_i") in the class i
1	0.00	2.63	2659	0.96	2.98
2	2.63	6.50	2641	4.53	6.48
3	6.50	10.53	2663	8.53	9.18
4	10.53	15.43	2628	12.74	11.33
5	15.43	21.76	2669	18.57	13.36
6	21.76	28.86	2663	25.12	16.10
7	28.86	37.40	2633	32.96	19.20
8	37.40	45.97	2670	41.90	21.33
9	45.97	54.23	2645	50.01	24.00
10	54.23	62.93	2655	58.47	25.98
11	62.93	73.83	2643	68.14	28.40
12	73.83	87.23	2654	80.15	28.34
13	87.23	106.70	2651	96.18	31.60
14	106.70	137.66	2652	120.77	37.02
15	137.66	262.56	2654	169.92	47.92

Then, we consider the probability density of the daily data Z_d which contribute to all monthly averages belonging to each activity class C_i. Let us denote this probability density $P_i, (i = 1, .., 15)$. It appears that each P_i can be fitted with a Poisson distribution, provided that this distribution is not computed for the P_i, but for a quantity which is proportional to it, $b \times P_i$, *where the factor b is the same for all classes.* Figure 1 shows four examples of histograms of those distributions together with the corresponding Poisson fit that we find. Table I gives some information concerning the graph: it includes the number of monthly measurements Z_m present in the class C_i, the mean activity value A_i and the dispersion D_i of probability densities P_i of daily measurements Z_d contributing to those Z_m monthly averages.

It appears that the dispersion D_i of the Z_d data which has been used to calculate the Z_m averages depends strongly upon the mean activity A_i. For confirmation, we have plotted D_i for each P_i as a function of the average A_i of the fifteen activity classes and found that a parabolic fit adequately represents the cloud of points:

$$D_i = \sqrt{b} * \sqrt{A_i}, \tag{1}$$

where $b = 11.44$ is the parameter which we have already been using to fit the P_i. The parameter b is the *same* for all classes C_i. The quality of this parabolic fit can be evaluated through the coefficient of correlation between

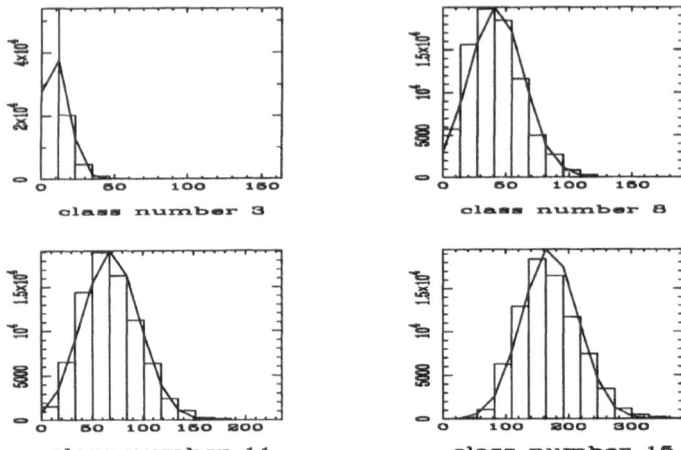

Fig. 1.   Four of the fifteen P_i with their fit, see table I.
abscissa : the Z_d value of the daily data ; *ordinate* : probability density of all Z_d which
have contributed to all monthly means Z_m belonging to the corresponding class number.

the original data and their fit: we have found it to be equal to 0.998. The
largest deviations from the fit occur in classes 12 and 15.

Discussion on the physical meaning of this fit lies outside the scope of
the present paper. In brief, let us say that it confirms the model of sunspot
occurrence which has been proposed by G.E. Morfill *et al.* (1991). Indeed,
on observational and physical grounds, those authors have suggested that a
Poisson distribution can mimic the behaviour of the solar activity over times
of the order of the solar rotation, which corresponds reasonably well to the
month over which are averaged the daily sunspot numbers.

We are thus entitled to use this model in order to assess a likely "uncertainty" to any actual value of the monthly sunspot number according to the
fitting relation.

3. Uncertainties and significance of extracted parameters

We refer the reader to our previous analysis of the radius data (Vigouroux
& Delache, 1993). Let us summarize it briefly before turning to a study of
the sunspot series.

3.1. THE METHOD IN BRIEF, A RECALL OF ITS APPLICATION TO SOLAR RADIUS DATA

Our aim was to construct a "model" of the data in which only the "significant
parts" are retained. By "parts" we mean, for example, spectral estimates ;
this would be the obvious choice if one thinks that the data is best repre-

sented by pure sine waves. One may also consider as "parts" the coefficients, in the time-frequency space, of a wavelet transform. The wavelet transform is convenient for detecting structures of different characteristic periods, appearing at different localizations in the time series. For further details on wavelets, see Daubechies (1992), Grossmann *et al* (1987) and Meyer (1992).

We proceed in two steps: first we have simulated the noise induced in the Fourier spectrum (or in wavelet time-frequency plane). Then, the second step has consisted in retaining, in the transform of the original data, only those amplitudes which are larger than a threshold level deduced form the previous transform of simulated noise.

We have then compared the results obtained through both transforms, and we have found that *for this specific radius time series, and with the errors bars which go with it*, the reconstructed signal that we have obtained through wavelet is to be preferred.

3.2. OUR METHOD AS APPLIED TO THE SUNSPOT TIME SERIES

The solar signal is made from monthly numbers from year 1749 until 1992. This corresponds to 2928 data points, but in order to get frequency independency and to keep benefit of the simple fast FFT and wavelet algorithms, we have, in this preliminary work, restricted ourselves to the more recent 2^{11} = 2048 months: so the time period covered is from May 1821 to December 1992.

In order to see the signature of these variable "uncertainties" in Fourier and wavelet transforms, we first build several fake time series containing as many data values as the original one. Each value is randomly taken according to a Poisson distribution whose dispersion corresponds to the actual "uncertainty" prevailing at this time, *but the distribution is centered around a constant*, chosen as zero for convenience. Then, from a number of transforms of our noisy series carrying *no signal*, we are able to assess the noise level at every frequency (or time-frequency) point. At any point in the transform, we are able to infer from this assessment the threshold above which an amplitude may be considered as significant at any prescribed σ-level. Then, we retain, in the transform of the original data, only those amplitudes which are larger than the threshold.

The rest of the analysis follows then exactly along identical trails. For example, we have used the same algorithm as in our previous analysis on the solar radius (see Vigouroux & Delache, 1993): it is derived from Mallat (1989) and makes use of the quadrative mirrors filters. We turn now to results of both analysis (Fourier and wavelet) and comparison of quality of reconstructed models.

In table II we give several quantities obtained by selecting significant parts of the transforms at levels 3σ and 4σ. Like in the radius case, for a same σ-level chosen as the threshold for selecting significant parts of the transforms, we find that wavelet needs less parameters than Fourier transform. However, the phenomenon is less pronounced here, probably due to the quasi periodical nature of the solar 11-year cycle. In order to determine optimum representation, as was done in the radius case, we would like to bring residues as close as possible to $\mathcal{F}$, the number of degrees of freedom ($\mathcal{F} = 2048-$ number of retained parameters). The residue ζ between original and reconstructed data is defined as:

$$\zeta = \sum_{k=1}^{2048} (\mathcal{R}_k - \tilde{\mathcal{R}}_k)^2 * W_k, \tag{2}$$

where the $\mathcal{R}_k$ are the original data, the $\tilde{\mathcal{R}}_k$ are the corresponding reconstructed points and the W_k are reciprocal of the squares of the dispersion which provide adequate weighting to each point.

Table II shows the ratio $\zeta/\mathcal{F}$. It approaches unity, the optimum value, around 4σ for the Fourier transform and at around 3σ for the wavelet transform. We have already observed such a trend in the radius case. Here too, we might interpret it as the need to apply a higher severity to the selecting process in the Fourier spectrum. The reason is that, at any single frequency, individual sinusoidal fits, while tight in some places, may be inadequate in other parts of the time series.

In order to assess the quality of the reconstructed model, we have also calculated the coefficient of correlation ρ between the original data and the "model" obtained in using the same thresholds:

$$\rho = \frac{\sum_{k=1}^{2048} \mathcal{R}_k * \tilde{\mathcal{R}}_k * W_k}{\sqrt{\sum_{k=1}^{2048} (\mathcal{R}_k^2 * W_k) * \sum_{k=1}^{2048} (\tilde{\mathcal{R}}_k^2 * W_k)}} \tag{3}$$

Table II gives the ρ values. They are quite comparable. One sees, as the previous discussion was anticipating, that the parameters which we omit when going from a 3σ to a 4σ threshold do not contribute much to the correlation, but clearly, this omission improves the ratio $\zeta/\mathcal{F}$.

At this point, we come to the conclusion that, judging from the $\zeta/\mathcal{F}$ criterion, from the correlation coefficients and from the number of parameters to be retained for modelisation (Occam razor criterion), the 4σ Fourier analysis and the 3σ wavelet analysis are quite similar. Figure 2 permits direct comparison between the original data and what is retained as significant respectively by Fourier and wavelet reconstructions. The time extent includes solar cycles # 8 to 23. The offset in the sunspot number which appears in the ordinate is such that the time average of the series is set equal to zero for convenience in the numerical algorithms. It is interesting to notice that wavelet analysis retains as significant some short time scale "accidents" in the original data whereas Fourier analysis looses it (cycles # 9, 11, 14 & 23).

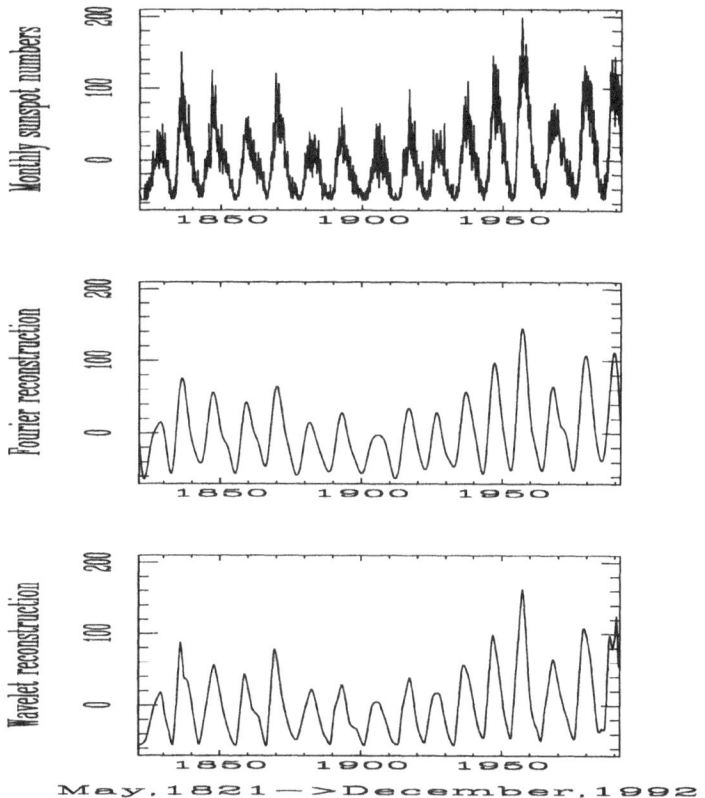

Fig. 2.   Monthly sunspot numbers, original and reconstructed time series ; *up*, original data *middle*, Fourier reconstruction (4σ threshold); *down*, wavelet reconstruction (3σ threshold).

TABLE II
Comparison between results of the two processes

transform	σ-level	number of retained parameters	$\zeta/\mathcal{F}$	coeff. of correlation
Wavelet	3σ	56	1.037	0.982
Fourier	3σ	66	0.856	0.983
Fourier	4σ	52	1.009	0.980

4. Conclusion

The well known quasi periodicity around eleven years appears clearly on Fig. 2. It is usually studied in examining its spectrum. We have compared spectra of the two reconstructed time series and found that, in both cases, *all spectral estimates corresponding to frequencies below* $(0.018 month)^{-1}$ *(peri-*

ods longer than five years) are preserved as significant. Indeed, they appear
with quasi identical amplitudes and phases. The spectrum of the wavelet
reconstructed signal shows also significant spectral estimates corresponding
to time scales down to around three years, which is obviously related to the
"accidents" already mentioned.

So, both wavelet and Fourier analysis tell us that *a large frequency band*
convey significant information (including, of course, also phase information).
Thus, a purely modal description of the data as interferences between a few
standing oscillations is a dangerous biased selection of the actual information
contained in the data. Our analysis show that data do not allow reducing
the information that they convey to such mathematical objects. For the 11-
year solar cycle, one should abandon representation in terms of independent
modes and shift either to a large band spectral description or to a time and
frequency description. But at this point, we must say that available data do
not tell us clearly which choice is the best. One should rather say that they
are equivalent *provided that all significant parts are retained.*

Physical discussion of our results, particularly in comparison to theoretical models predictions of the solar dynamo, is outside the scope of the
present paper. This will be the subject of future work, based on the new
grounds which we have described here.

Acknowledgements

We thank especially F. Laclare for having made his precious solar radius
measurements available to us, and J.A. McKinnon of the National Geophysical Data Center (Boulder) for having kindly provided us with the sunspot
series. We are also grateful to the referee for his very helpful advice.

References

Daubechies, I.: 1992, *Ten Lessons on Wavelets*, SIAM Eds., New-York

Delache, Ph., Laclare, F., Gavryusev, V., Gavryuseva, E.: 1993, *Memorie della società
astronomica italiana* **64**, 237.

Grossmann, A., Kronland-Martinet, R., Morlet, J.: 1987, in J.M. Combes *et al* (eds),
Wavelet, Time-Frequency Methods and Phase Space, Springer-Verlag, New-York, p. 2.

Mallat, S.G.: 1989, *IEEE Transf. on Pattern Anal. and Machine Intell.* **vol. 11 no. 7**,
674

Meyer, Y.: 1992, *Ondelettes et Applications*, Armand Collin Eds., Paris

Morfill, G.E., Scheingraber, H., Voges, W., Sonett, C.P.: 1991, in C.P. Sonett *et al.* (eds),
The sun in time, University of Arizona Press, p. 30.

Laclare, F.: 1983, *Astron. Astrophys.* **125**, 200.

Ochadlick, A.R., Kritikos, Jr. & H., Giegengack, R.: 1993, *Geophys. Res. Letters* **vol. 20
no. 14**, 1471.

Vigouroux, A. and Delache, Ph.: 1993, *Astron. Astrophys.* **278**, 607.

"STAR AS A SUN" OBSERVATIONS IN SEISMOLOGY OF DISTANT STARS

D. E. MKRTICHIAN

Astronomical Observatory of Odessa State University, Shevchenko Park, Odessa, Ukraine, 270014.

Abstract. The paper gives a brief description of a new procedure for seismology of pulsating CP2-stars. In analogy with the solar observations, this sensitive procedure permits one to carry out the spatial filtration of low, intermediate and high degree nonradial modes and uses periodic spatial filters associated with surface inhomogeneities in the distribution of chemical elements. The possibilities of this procedure and its observational peculiarities are discussed.

1. Introduction

For the case of "Sun as a star" observations, the photometric and spectral observations of optically unresolved discs of distant stars are sensitive only to low degree ($l \leq 3$) nonradial modes. This fact essentially decreases the diagnostic possibilities of stellar seismology. The problem of detection and identification of high degree nonradial modes is partially solved from line-profile analysis for rapidly rotating stars with high amplitudes of excited modes. However, amplitudes of nonradial pulsations (NRP) excited in stars on and near the main sequence are small, so the spectral detections of them require precise measurements of radial velocities. Such detection is possible for slowly rotating stars with sharp lines. Existing methods of detection and identification of NRP modes of stars are based on information about the contributions from different parts of the stellar disc obtained somehow from the light, line profile or radial velocity variations averaged over the visible surface. The practical power of each method is determined by how precisely it can select these contributions.

2. New Procedure for Detection and Identification of NRP in Seismology of CP2-Stars

2.1. PERIODIC SPATIAL FILTERS

In helioseismology it is possible to perform the spatial filtration of modes by getting information about the Doppler shifts or about variations of surface brightness from different local parts of the solar disc (Christensen-Dalsgaard, Gough, 1982; Christensen-Dalsgaard, 1984; Kosovichev 1986; Christensen-Dalsgaard, 1989; Christensen-Dalsgaard, Gough, 1989). These methods are very sensitive and informative; however, their direct application to stars seems impossible due to the difficulties of the optical resolution of stellar discs. The idea of the method is based on well-known observational facts and permits one to increase the sensitivity of the detection of high-degree modes in magnetic, chemically-peculiar (CP2) stars proposed by Mkrtichian (1988). The method is based on the supposition that in CP2-stars with inhomogeneous surface distribution of chemical elements (ISDCE) it is possible to select pulsational motions of different parts of the stellar surface. The selection of these motions is possible in the plane of the stellar spectrum image in the spectrograph by means of the analysis of Doppler shifts of the lines of different chemical elements which have different surface distributions.

Actually, it is well known from observations that some chemical elements on the surfaces are concentrated in the patches with high overabundances, up to +3 dex and more, and the surface distributions of different elements can not coincide.

For a rotating, pulsating star with ISDCE, the line profile averaged over the visible stellar disc can be written formally as

$$R(\lambda, \Omega t, \sigma t) = \frac{\int \int I_c(M, \gamma, \sigma t) R(M, \lambda_0 + \Delta \lambda_D(M, \Omega t, \sigma t)) dM}{\int \int I_c(M, \gamma, \sigma t) dM} \qquad (1)$$

where γ is the angle between line of sight and the point $M(\theta, \varphi)$ on the visible surface in the spherical coordinates (r, θ, ϕ); Ω, σ are respectively the angular rotational and pulsational frequencies; I_c is the intensity of continuum at the point M; $\Delta \lambda_D(M, \Omega t, \sigma t)$ is the Doppler shift of the local spectral line caused by the rotational and pulsational velocity field; $R(M, \lambda_0 + \Delta \lambda_D(M, \Omega t, \sigma t))$ is the local line profile at the point M.

Contributions of overabundant regions in the line profile averaged over the stellar disc, or its equivalent width, provide most of the contributions of the regions with normal or depleted abundances, and so they predominate in Doppler shifts of lines. This fact can essentially decrease the surface amplitude averaging effect for high-degree modes and allows one to increase the detection sensitivity for these modes (Mkrtichian, 1988).

By analogy with solar observations, we can assert that the choice of Doppler shift measurements of the spectral lines of inhomogeneously distributed chemical elements is equivalent to the observation of an optically resolved stellar disc through masks with inhomogeneous surface transparencies. The transparency of such a pseudo-mask for any fixed phase of the rotational period is first of all proportional to the projection of the visible picture of surface abundance distribution of chemical elements on the plane perpendiculiar to the observer's direction, and other local physical conditions determining the local line formation.

Hence, in analogy with the solar observations, for CP2-stars we can introduce spatial filters (SF) fixed to the stellar surface and having different sensitivities to pulsational modes of different spatial structure (i.e. different l,m,n). The long lifetimes of the abundance patches for CP2-stars exceeding

100-1000 years (Dolginov, 1988) permit us to assert that SF are effectively fixed to stellar surfaces and can be used for stellar seismology over long time intervals.

According to the definition (Christensen-Dalsgaard and Gough, 1982) the spatial filter $S_{l,m,n}$ for any mode can be expressed as

$$v_{l,m,n} = S_{l,m,n} V_{l,m,n} \tag{2}$$

where $v_{l,m,n}$ is the line-of-sight amplitude of pure pulsational line shifts which are averaged over the stellar disc and expressed in velocity units; $V_{l,m,n}$ is the amplitude of the eigenvelocity of a mode.

However, the procedure of calculating the sensitivity functions of SF for CP-stars can not be so specific and definitive as has been described earlier (for example, Dziembowski, 1977; Christensen-Dalsgaard, 1984), where the integration of the NRP velocity field over the visible surface for the constant local line-profiles has been carried out with the limb darkening weighting function.

The complicated surface distribution of physical conditions (abundance, temperature, magnetic field, etc.) , makes it necessary to do the numerical calculations of SF for every particular star by using the data about the distribution of parameters which determine the local spectral line profiles or the local equivalent widths at every point of the visible stellar surface.

The methods of mapping the spotted star surfaces worked out in the last decade (Goncharskij et al., 1982; Vogt et al., 1987, Piskunov et al., 1990) make it possible to obtain, for a given CP2-star, the necessary information for calculations of spatial filters.

It is known from observation that line profiles of the chemical elements of CP2-stars with ISDCE may be complicated, assymmetric and/or multi-component and, as a rule, change with the phase of the rotational period which causes rotational modulation of the mean line velocities.

We ensure certainity of the determination of PSF for CP2-stars for any line-profile shape if we determine the line centroid as the position of the line for the Doppler shift measurements.

In this case the line shift can be expressed as

$$V_r(l, m, n, \Omega t, \sigma t) = \left(\frac{c}{\lambda_0}\right)\left(\frac{\int R(\lambda, \Omega t, \sigma t)\lambda d\lambda}{\int R(\lambda, \Omega t, \sigma t)d\lambda} - \lambda_0\right) = $$
$$V_{rot}(\Omega t) + V_{pul}(l, m, n, \sigma t) \tag{3}$$

and consists of the two components: the rotational component, $V_{rot}(\Omega t)$, slowly changing within the $\pm v\sin i$ range, and the rapidly oscillating low-amplitude pulsational component $V_{pul}(l, m, n, \sigma t)$. The rotational component can be determined from observations or calculated from the known equatorial velocity, the inclination angle of the rotation axis and the known maps of the surface distributions of chemical elements.

Actually, from the observations, due to $\Omega \ll \sigma$, we can determine the pure pulsational line shift as the difference between the observed and the mean values over the given rotational phase, ψ,

$$v_{pul}(l, m, n, \sigma t) = V_r(l, m, n, \Omega t, \sigma t) - \int_\psi V_r(l, m, n, \Omega t, \sigma t)dt. \qquad (4)$$

Unlike the SF previously introduced for the Sun and stars (Hill, 1978; Dziembowski, 1977), spatial filters, $S_{l,m,n}(\psi)$, introduced in the same way for the CP2-stars, are periodic and change their values with the phase of the rotational period according to changes of the projection of visible surface inhomogeneties. Choosing spectral lines of different chemical elements we introduce other periodic spatial filters (PSF) with other sensitivity functions for NRP modes.

So, from the observations of Doppler shifts of lines of different chemical elements (different PSF) in pulsating CP2-stars, one can get the sets of data about the frequencies, amplitudes, phases and mean values of radial velocities determined with different PSF which, after the comparison with the expected calculated model data, can give (in analogy with the spatially-resolved solar disc observations) sufficient information for identification of modes.

2.2. MODEL CALCULATIONS OF PERIODIC SPATIAL FILTERS

The formalism of the calculations of PSF, oriented to surface maps obtained by means of the Doppler imaging method by Khokhlova and coworkers   (Goncharsky et al., 1982) is given by Mkrtichian (1993). A model calculation of pulsational line-profile variations and PSF for artificial stars with simple configurations (spots, rings) of ISDCE (Mkrtichian, 1992; 1993) shows the essential differences in observational manifestations of NRP in contrast to NRP of stars with homogeneous surface distributions of chemical elements.

The differences appear as:

1. The increase of detection sensitivities to high degree modes from the Doppler shifts of lines of ISDCE.

Figure 1 presents the results of model calculations of PSF functions for the sectoral modes and disc-integrated Doppler shift measurements of CrI 4535.14 Å spectral line. The calculations are performed for the equator-on case of a visible star with an equatorial circular spot and for the set of different radii. Cr overabundance selected equal to $[Cr/H] = +1.88 \; dex$ inside the spot and equal to $[Cr/H] = 0.0 \; dex$ outside the spot. Local profiles of the CrI 4535.14 Å line have been calculated for the Kurucz (1979) model atmosphere with $T_e = 7000°$ K, log g=4.0 and $V_t = 3.0$ km/s by means of the code ABSTAR (Ryabchikova and Piskunov, 1988).

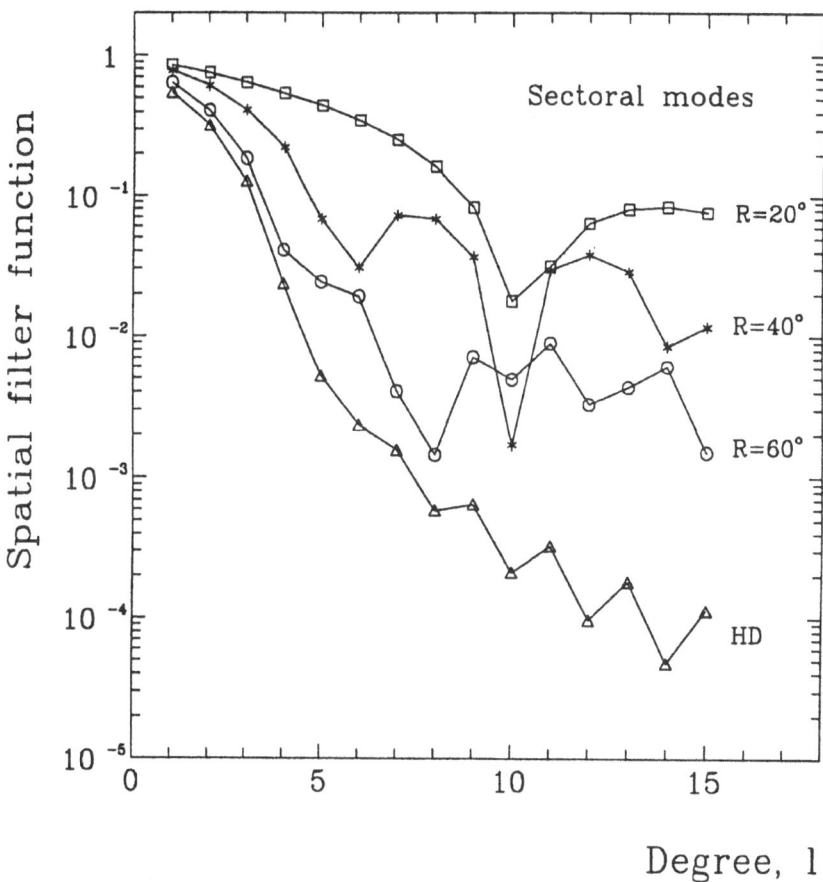

Fig. 1. Model calculations of PSF functions for an artificial star with a circular equatorial spot with Cr overabundance +1.88 dex and for the set of radii $R = 20°$(squares), $R = 40°$(asterisks), $R = 60°$(open circles) and for homogeneous distribution of Cr (triangles). Calculations are carried out for the star visible from the equator and a fixed phase of rotational period ψ corresponding to a spot passing across the disc center.

As one can see from Figure 1 the decrease of the disc-integrated amplitudes of the sectoral modes with increasing of l occurs much slower in the case of spotty distributions. For surface spot with radius R=20° the decrease of the observed amplitudes lower the 0.1 level occurs for l=|m|≥9, whereas in case of a homogeneous distribution, a similar decrease takes place for l>3.

The gain factors, G, can be calculated as the ratio of maximum of PSF functions for the spotted ($S_{l,m,n}^{spot}(\psi)$) and homogeneous ($S_{l,m,n}^{HD}$) distributions

$$G = \frac{S_{l,m,n}^{spot}(\psi)}{S_{l,m,n}^{HD}}. \tag{5}$$

As one can see from Figure 1 the maximum values of this ratio for the spot radii $R = 20°$, may increase up to the order of $10 - 10^3$ for the degrees l=5-15.

2. The essential differences of the observed amplitudes and the phases of pulsational modes determined from lines of different chemical elements having different surface distributions. The phase shifts for the zonal modes are close to 0 or π.

3. Possible modulation of amplitudes with the star's rotational period caused by changing the visible picture of surface inhomogeneties. For the "oblique pulsator" model of rapid oscillating CP2-stars (Kurtz, 1990) ISDCE can produce additional modulation of the observed amplitudes.

4. The line-profile differences appear in the false decrease of observed l,m parameters of the modes determined from the line-profile analysis.

2.3. PHENOMENOLOGICAL DESCRIPTION OF THE PROCEDURE

Below, step by step, is given the phenomenological description of the procedure of the setting up of observations, and the detection and spatial filtration of NRP modes. The procedure consists of:

1. A preliminary mapping of the stellar surface in the lines of different chemical elements.

2. Calculations of PSF functions for the mapped lines, and selection from the calculated values of the most sensitive (for desired modes) spectral lines and corresponding phases of rotational period to search for radial velocity variations.

3. Observation of radial velocities simultaneously in different spectral lines (different PSF).

4. Analysis of frequency spectra, amplitudes, phase lags, mean values of V_r for different spectral lines (or groups of lines of the same chemical elements). Comparison with the calculated values of PSF functions for different modes. Determination of the spatial structure (l,m,n) of modes.

2.4. INSTRUMENTATION

One of the optimal spectral instruments to use for the proposed procedure is an echelle-spectrograph with a matrix detector. However, because of possible phase shifts between pulsational radial velocities determined from the lines of different elements (Mkrtichian 1992, 1993), the reduction of the data must exclude the averaging of estimates of pure pulsational radial velocities determined from lines of different chemical elements. Averaging for increasing of Vr determination accuracy, must be carried out by means of an optimal combination of estimations according to the calculated PSF.

Here there is a direct analogy with the Griffin-type radial velocity spectrometer - another instrument optimal for use in the proposed procedure. Because of the reasons described above, the use of such a Griffin-type spectrometer requires the use of cross-correlation masks, rejecting only the lines of one element or lines of elements having the same surface distribution.

By choosing the cross-correlation masks, the choice of different PSF is carried out, and hence other instrumental sensitivities to NRP modes are chosen.

3. Conclusion

Periodic spatial filters introduced for CP2-stars and associated with the surface inhomogeneties give the possibility of "Star as a Sun" observations in seismology of distant CP2-stars. We have reviewed the results of model calculations of PSF functions for artificial surface distributions and the expected observational peculiarities of pulsational manifestations in pulsating CP2 stars. We have also proposed a phenomenological description of a new procedure of detection and spatial filtration of nonradial modes for asteroseismology of pulsating CP2-stars.

Acknowledgements

I am grateful to V.L. Khokhlova, A.G. Kosovichev and T.A. Ryabchikova for useful discussions and assistance. I especially thank N.S. Polosykhina for the active support of my work and G. Papikian for financing part of this work out of the funds of the enterprise "Pioneer". I thank Ukrainian "Renaissance" Foundation (Soros Foundations) for the financial support, which made it possible for me to attend this Colloquium by covering travel and accomodation expenses.

References

Christensen-Dalsgaard, J.: 1984, in Ulrich, R.K., Harvey, J., Rhodes, E.J. and Toomre, J. (eds.), *Proceedings of Conference on Solar Seismology From Space*, NASA, JPL Publ. 84-84, p. 219.
Christensen-Dalsgaard, J.: 1989, *Monthly Notices Royal Astron. Soc.* **239**, 977.
Christensen-Dalsgaard, J. and Gough, D.O.: 1982, *Monthly Notices Royal Astron. Soc.* **198**, 141.
Christensen-Dalsgaard, J. and Gough, D.O.: 1989, *Solar Phys.* **119**, 5.
Dolginov, A.Z.: 1988, in Yu.V. Glagolevskij and I.M. Kopylov (eds.), *Magnetic Stars*, Nauka, Leningrad, p. 220.
Goncharskij, A.V., Stepanov, V.V., Khokhlova, V.L. and Yagola, A.G., 1982, *Astronomicheskii Zhurnal* **59**, 1146.
Hill, H.A.: 1978, in J.A. Eddy (ed.), *The New Solar Physics*, Col. Westview Press, p. 135.

Kosovichev, A.G.: 1986, *Izv. Krym. Astrofis. Obs.* **75**, 22.

Kurtz, D.W.: 1990, *Annual Rev. Astron. Astrophys.* **28**, 607.

Kurucz, R.L.: 1979, *Astrophys. J. Suppl.* **40**, 1.

Mkrtichian, D.E.: 1988, in Yu.V. Glagolevskij and I.M. Kopylov (eds.), *Magnetic Stars*, Nauka, Leningrad, p. 195.

Mkrtichian, D.E.: 1992, in Yu.V. Glagolevskij and I.I. Romanyuk (eds.), *Stellar Magnetism*, Nauka, Sankt-Petersburg, p. 260.

Mkrtichian, D.E.: 1993, *Ph. Dissertation*, Odessa State University, Odessa.

Ryabchikova, T.A. and Piskunov, N.E.: 1988, in S.J. Adelman and T. Lanz (eds.), *Elemental Abundance Analysis*, Institut d'Astronomie de l'Univ. de Lausanne, p. 93.

Vogt, S.S., Penrod, G.D. and Hatzes, A.P.: 1987, *Astrophys. J.* **246**, 496.

ROTATING CONVECTION AND THE SOLAR DIFFERENTIAL ROTATION

Kwing L. Chan

Applied Research Corporation, Landover, Maryland 20785

and

Hans G. Mayr

NASA Goddard Space Flight Center, Greenbelt, Maryland 20771

Abstract. We discuss the implication of a numerical experiment on rotating convection and its relevance to the construction of a model for the solar differential rotation.

1. A Brief Overview

Space borne experiments have now established that the total solar irradiance varies over a full solar cycle (Willson and Hudson, 1991, Hickey et al., 1988). It is therefore clear that the solar output is linked to the dynamo process. The solar dynamo, differential rotation, and convection are closely related (Parker, 1955). Since the dynamo arises from the differential rotation and convection (through the ω and α processes), explaining the differential rotation and sorting out its relation to convection is then a prerequisite for understanding the magnetic cycle and the associated variations in solar irradiance.

Global numerical models of solar differential rotation (Gilman, 1977, Glatzmaier, 1984, Gilman and Miller, 1986) are successful in obtaining the solar angular velocity distribution at the surface. In the interior, the distributions more or less obey the well-known Taylor-Proudman theorem, so that the isorotation surfaces are cylindrical and parallel to the rotation axis. However, results of helioseismology, now confirmed by many groups (Duvall et al., 1986, Brown and Morrow, 1987, Rhodes et al., 1987, Libbrecht, 1989), show that inside the convection zone, the isorotation surfaces tend to align radially, and that beneath a thin shear layer the stable radiative region rotates more or less uniformly.

This contradiction between the observational and numerical results cannot be easily reconciled and raises the question: What is missing from the numerical models? In an earlier paper, Chan and Serizawa (1991) argued that the Taylor-Proudman theorem would not hold if the buoyance force is as important as the Coriolis force. The buoyance force depends on the distribution of the entropy [or equivalently, the superadiabatic gradient $\delta\nabla$ = $(\partial \ln T / \partial \ln p)$ - $(\partial \ln T / \partial \ln p)_{adiabatic}$], and if the entropy distribution is wrong, the buoyance force would be in error. To obtain the correct $\delta\nabla$ profile, it is necessary to perform the simulation for a period comparable to the thermal relaxation time of the layer, about 10^5 years for the solar convection zone. But the numerical calculations were usually run only for a period on the order of ten years. Thus there is a problem with direct numerical simulations. To make progress, semi-analytical approaches are therefore more practical. The idea here is to bypass the thermal relaxation calculation by making

Solar Physics **152**: 283–290, 1994.

an assumption about the distribution of δV. Since the computational load for each model is light, it is possible to construct many models and search for conditions under which the correct angular velocity distribution develops.

Semi-analytical models of solar differential rotation appeared well before the development of numerical models. One class of these models, the 'anisotropic viscosity' model, is based on the argument that the turbulent exchange of momentum along the vertical direction (radial) should be quantitatively different from that in the horizontal direction (Wasiutynski 1946). The anisotropic Reynolds stresses in the momentum equation then generate a differential rotation in the large scale. This concept was first applied in calculations by Kippenhahn (1963) and Kohler (1970). Elaborations and extensions of this approach have later been made, for example, by Durney and Spruit (1979), Rudiger (1980, 1989), and Schmidt (1982). Another class of models, the 'latitude-dependent heat transport' model, argues that the interaction of rotation with convection leads to a differential heat transport between the equator and the poles. The resulting source term in the energy equation thus sets up a meridional flow which in turn generates a differential rotation. The idea was originally proposed by Weiss (1965) and developments have been made, for example, by Durney and Roxburgh (1971), Belvedere and Paterno (1977), and Pidatella et al. (1986). Both classes of models aimed at satisfying the solar surface constraints by adjusting the heat transport parameter, the viscosity, and the Prandtl number. With the exception of a few recent specialized versions (e.g. Tuominen and Rudiger, 1989, Brandenburg et al., 1992), most of the above described semi-analytical models cannot reproduce the solar internal angular velocity distribution inferred from helioseismology.

In these semi-analytical models, the formulation is based on perturbation expansions of the 'inverse Rossby number' ($\sim \Omega L/ V$ where Ω is the mean rotation rate, L and V are the characteristic length and velocity respectively; we shall call it the Coriolis number). The expansions diverge when the Coriolis number is larger than one, which is approximately the case in the solar convection zone, and therefore strictly speaking, the formulation is not applicable. For the same reason, such theories cannot be applied to explain the extreme differential rotation observed on Jupiter and Saturn where the angular velocities form alternately positive and negative latitudinal bands.

Conventional semi-analytical theories emphasize the action of the Coriolis force on the sub-global scales. But actually the effects of the Coriolis force are maximized in the global scale as the Coriolis number increases with the length scale. We therefore proposed a theory which has close ties with conventional semi-analytical theories but emphasizes the global scale interaction of convection and rotation (Chan et al., 1987). This model interprets the differential rotation as the zonal (azimuthal) wind component of a global, axisymmetric mode of convection under the influence of rotation. In conventional theories, the linearized, coupled fluid equations describing the differential rotation are directly driven by latitude-dependent terms in the momentum and/or energy equations, and in most cases, the amount of driving required is excessively large. In our case, instead, we seek an appropriate resonant response to such driving. The response being resonant, the driver can be arbitrarily small and its exact form becomes unimportant. The response is in fact a self-excited convective mode. The pattern of such a mode depends on the balance of the Coriolis force and the buoyance force and therefore the distribution of δV plays a crucial role. This theory can be applied to explain the

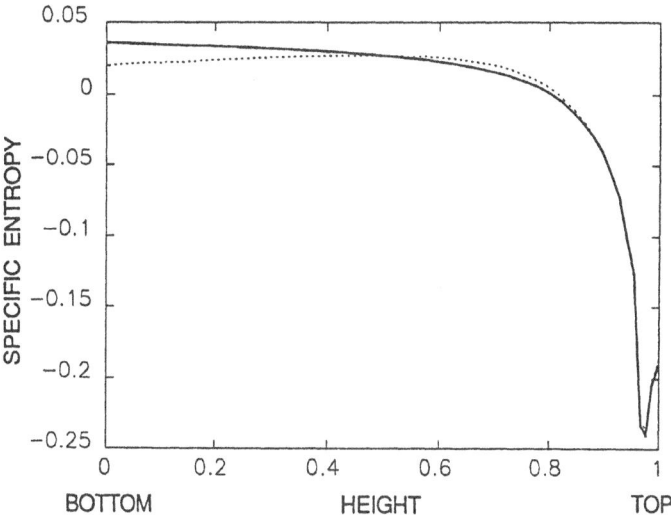

Fig. 1. The effect of rotation on the entropy distribution of a convection zone.

alternating wind bands of Jupiter and Saturn (Mayr et al., 1984; 1991). Both Boussinesq (Chan et al., 1987) and compressible flow (Chan and Mayr, 1991) models have been constructed to describe the solar differential rotation. The compressible model can describe the observed internal solar angular distribution rather well.

Recently, using 3D numerical experiments, Chan and Gigas (1991) found that $\delta\nabla$ can turn slightly negative (subadiabatic) in the lower region of a deep convection zone. Supposing that to be the case, we have tried to construct solar differential rotation models with background structures that followed such behavior. In all the cases, however, the differential rotations could not be made to penetrate significantly into the subadiabatic lower convection zone, in contradiction with observations. A question then arose: Since the Coriolis number in the lower solar convection zone is close to 1, can the rotation alter the sign of $\delta\nabla$ in this region? To answer this question, we performed a numerical experiment on rotating deep convection. The results were striking. The rotation turned the lower convective region from slightly subadiabatic to significantly superadiabatic. Details of this numerical experiment are given in Section 2. The following conclusions are reached: (i) Estimates of $\delta\nabla$ based on the standard mixing length theory are incorrect in the lower part of a deep convection zone. (ii) Strong rotation plays an important role in determining $\delta\nabla$. At the moment, we do not have a theory of $\delta\nabla$ for rotating convection. But we can now assume that it is positive in the whole convection zone and take its distribution as a free parameter in our model. We made a systematic study on the parametric combinations that produce solar-like differential rotation and found a number of possible solutions. The results of this study are summarized in Section 3.

2. A Numerical Experiment of Rotating Convection

Using the numerical code described in Chan and Sofia (1986), we performed a numerical experiment with the following specifications: (i) The fluid is rotating, with the axis of rotation at an angle 45° from the vertical direction. The magnitude of the angular velocity is 0.5 $[(p_t/\rho_t)^{1/2}$ / depth] where the subscript 't' denotes values of variables at the upper boundary. The Rossby number is about 1. (ii) The depth of the convection zone contains 4.5 pressure scale heights. To have better control on the conditions near the top of the convection zone, a stable radiative layer with 1 pressure scale height is attached at the top. (iii) The domain of computation is a 3D rectangular box with an aspect ratio (horizontal width / depth) of 1.5. (iv) The grid consists of 35x35x39 points. (v) The side boundaries are periodic; the upper and lower boundaries are impenetrable and stress-free. (vi) There is a constant input flux from the lower boundary (= 0.25 $[p_t(p_t/\rho_t)^{1/2}])$; the temperature at the upper boundary is fixed.

The effect of rotation on the entropy distribution is illustrated in Fig. 1. The solid curve shows the distribution of the specific entropy for the rotating case; the dashed curve shows the distribution for an identical case without rotation. To overcome the impediment from rotation, the layer adjusts its structure to make δV positive and large enough so that convection prevails.

3. Generation of 'Sun-Like' Differential Rotation

Having found that δV could be positive in the whole convection zone, we returned to the modeling of the solar differential rotation. Using a linearized, axisymmetric version of a recently developed spectral code that solves the compressible fluid equations (Chan et al. 1993), we studied convection zone models with assumed distributions of δV and locked for situations which can produce sun-like differential rotation. The common characteristics of the models are: (i) The domain of computation spans between 0.575 and 0.934 (or 0.99) solar radii. It contains a major portion of the convection zone and the upper part of the stable radiative layer below. (ii) The upper boundary is stress-free but the lower boundary is slip-free (the implication is that magnetic field plays a role in locking up the flow). (iii) The number of vertical grid levels is 52 (or 101). For studying the subcritical responses (see later discussion), the degree of spherical harmonics used is 10. For the supercritical cases which require many iterations for finding the resonances, the degree is 4 (or 6). (iv) The gas can be conveniently approximated as an ideal gas with a ratio of specific heat 5/3. (v) The background distributions of temperature, density, and pressure are held fixed. (vi) Inside the convection zone, the superadiabatic gradient is assumed to vary as $\delta V_t(\rho_t/\rho)^n$ where ρ is the density. δV_t and the power index n are free parameters. The eddy diffusivity κ is assumed to satisfy $f = \kappa \rho T \delta V / H$ where f is the total energy flux and H is the pressure scale height. The eddy Prandtl number Pr, another free parameter, is assume to be independent of depth in the convection zone. (vii) In the stable layer, the radiative conductivity K is taken to satisfy f $= K \, dT/dr$. The viscosity there is arbitrarily set at 10% of the value at the bottom of the convection zone (exact value unessential).

To drive the linearized system, we introduce a perturbation of the form ε f $P_2(\cos\theta)$ in the energy equation, near the bottom of the convection zone (the results are insensitive to the exact location). ε is an amplitude factor, f is the solar flux, P_2 is the normalized

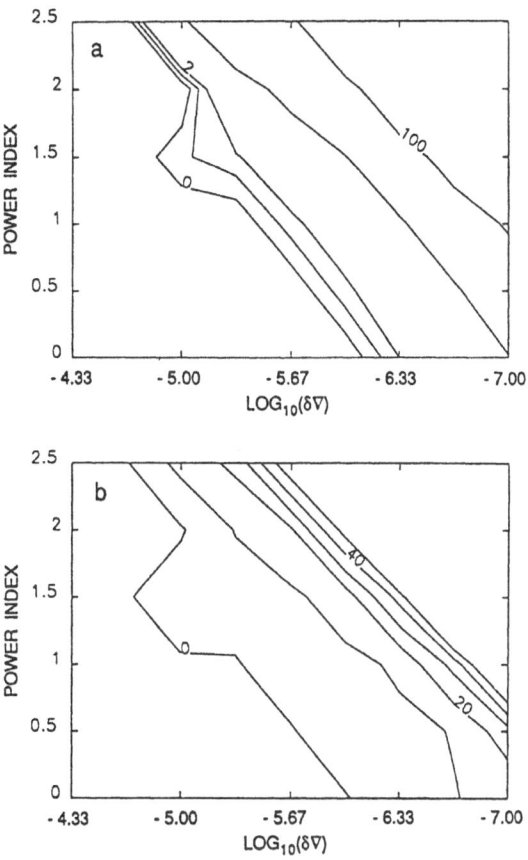

Fig. 2. a. Distribution of the amplitude factor of perturbation in the $(\delta V_t, n)$ plane.
b. Distribution of the maximal surface meridional velocity.

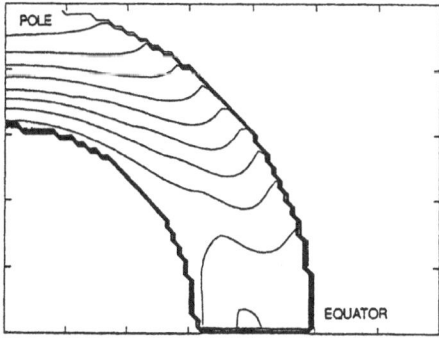

Fig. 3. The angular velocity distribution for a subcritical response with $n = 1$, $Pr = 1$, and
$$\delta V_t = 2.\times10^{-6}.$$

Legendre function of the second degree, and θ is the co-latitude. The physical rationale and the formulation of this procedure are similar to those adopted by the 'latitudinal-dependent heat transport' model. Under these conditions, there exists a large parameter region (see Fig. 2 for the Pr = 1 case; cases with Pr = 0.1 and 10 have similar patterns) which can produce radially oriented isorotation contours in the convection zone (see Fig. 3), similar to those observed. This region however is characterized by large viscosities and diffusivities (small δV_t since f is fixed), corresponding to subcritical effective Rayleigh numbers (no self-excited modes). Over a large portion of this region (towards the right), the pattern of differential rotation as depicted in Fig. 3 is quite robust; it is insensitive to δV_t, n, and Pr, but it relies on the form of the driver $P_2(\cos\theta)$. The magnitude of the driving needed to produce the correct solar amplitude (Fig. 2a) and the size of the surface meridional velocity (Fig. 2b) quickly increase with smaller δV_t and larger diffusivity.

The lines labeled by 0 in Fig. 2a,b approximately delineate the boundary which separates the super- and sub- critical regions. Beyond a thin strip on the right of this critical boundary, the perturbation factor required is unreasonably large ($\varepsilon > 1$; latitudinal variation in flux > f). In the region $1 > \varepsilon > 0$, the angular velocity distributions often show irregularities; sun-like distributions appear only when ε is close to 1; the required size of flux perturbation is at best marginally acceptable.

Resonant responses to the energy driver occur on the left side of the critical boundary at discrete locations of δV_t. These are unstable eigenmodes described by the linearized fluid equations and do not always produce sun-like differential rotation. The angular velocity distributions depend on δV_t, n, and Pr.

A number of solutions with sun-like differential rotation have been found for a variety of combinations of parameters. Table 1 provides some examples (N is the degree of the harmonic expansion). The values of δV_t at which resonances occur are quite insensitive to the location of the upper boundary of the model and the number of spherical harmonics included, and they are close to the critical boundary. They also approximately satisfy an inverse proportion relationship with Pr as derived in Chan et al. (1987; eqn. 21). Since they are self-excited, the resonant modes do not depend on the amplitude and form of the excitation driver. Fig. 4 shows a surface plot of the angular velocity distribution for one such case. The amplitude of the mode is found by matching the surface angular velocity distribution to that observed on the sun. Though the patterns of the zonal flows are simple and similar, these modes possess complicated, multi-cellular meridional flows. The magnitudes and patterns of the meridional circulations are highly dependent on the free parameters. In the case presented by Fig. 4, the maximal meridional velocity at the surface is 28.6 m/s.

In the solar convection zone, the axisymmetric convective modes discussed here are not the first to be excited. Many other modes (especially the small scale ones, e.g., granule scale) grow more readily and at much faster rates. But in the 'mean-field' picture (as identified by Tuominen et al., 1993) adopted by the semi-analytical models and inherited here, these nonlinearly interacting and fully developed sub-global scale or non-axisymmetric modes are treated as part of the background turbulence and their actions are lumped into the mean-field transport coefficients (e.g. eddy diffusion). The lowest degree axisymmetric mode may thus represent the 'first' growing mode in this idealization, and

TABLE I

Examples of δV_t at which resonances with 'sun-like' differential rotation occur.

Pr =	10	5	3
$r_t = 0.934$; N = 4	9.57×10^{-6}	1.76×10^{-5}	2.79×10^{-5}
$r_t = 0.990$; N = 4	1.13×10^{-5}	$(1.75 \times 10^{-5})^*$	2.38×10^{-5}
$r_t = 0.934$; N = 6	9.7×10^{-6}	1.72×10^{-5}	2.62×10^{-5}

* a local maximum in response, not a resonance

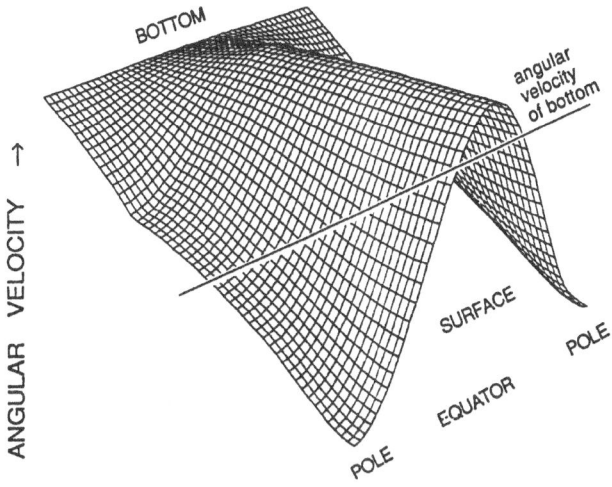

Fig. 4. Angular velocity distribution for the convective mode with Pr = 10, $r_t = 0.990$, N = 4, and $\delta V_t = 1.13 \times 10^{-5}$.

the linear mode may then approximately describe the 'mildly nonlinear' situation. Whether the mean-field simplification is valid and whether a resonant mode of the linearized mean-field equations can indeed mimic the global scale circulation are questions that need to be addressed by future numerical experiments.

Acknowledgement

We are indebted to Drs. Axel Brandenburg, Bernard Durney, and Kenneth Schatten for discussions. KLC thanks NSF (ATM-9108315) for support and the Pittsburgh Supercomputing Center for providing computer time.

References

Belvedere, G., and Paterno, L.: 1977, *Solar Phys.* **5 4**, 289.

Brandenburg, A., Moss, D., Tuominen, I.: 1992, *Astron. Astrophys.* **2 6 5**, 328.

Brown, T. M., and Morrow, C. A.: 1987, *Ap. J. Lett.* **3 1 4**, L21.

Chan, K. L., and Gigas, D.: 1991, *Astrophys. J.* **3 8 9**, L87.

Chan, K. L., and Mayr, H. G.: 1991, in *The Sun and Cool Stars: Activities, Magnetism, and Dynamos*, eds. I. Tuominen, D. Moss, and G. Rudiger, Springer-Verlag, Berlin, 178.

Chan, K. L., and Serizawa, K.: 1991, in *The Sun and Cool Stars: Activities, Magnetism, and Dynamos*, ed. I. Tuominen, D. Moss, and G. Rudiger, Springer-Verlag, Berlin, 15.

Chan, K. L., and Sofia, S.: 1986, *Astrophys. J.* **3 0 7**, 222.

Chan, K. L., Sofia, and Mayr, H. G.: 1987, in *The Internal Solar Angular Velocity*, eds. B. R. Durney and S. Sofia, D. Reidel, Dordrecht, 347.

Chan, K. L., Mayr, H. G., Mengel, J. G., and Harris, I.: 1993, *J. Comput. Phys.*, in press.

Durney, B., and Roxburgh, I. W.: 1971, *Solar Phys.*, **1 6**, 3.

Durney, B., and Spruit, H. C.: 1979, *Astrophys. J.* **2 4 3**, 1067.

Duvall, T. L., Harvey, J. W., and Pomerantz, M. A.: 1986, *Nature* **3 2 1**, 500.

Glatzmaier, G. A.: 1984, *J. Comput. Phys.* **5 5**, 461.

Gilman, P. A.: 1977, *Geophys. Astrophys. Fluid Dyn.* **8**, 93.

Gilman, P. A., and Miller, J.: 1986, *Astrophys. J. Suppl.* **6 1**, 585.

Hickey, J. R., Alton, B. M., Kyle, H. L., and Major, E. R.: 1988, *Adv. Space Res.* **8**, (7) 5.

Kippenhahn, R.: 1963, *Astrophys. J.* **1 3 7**, 664.

Kohler, H.: 1970, *Solar Phys.* **1 3**, 3.

Libbrecht, K. G.: 1989, *Astrophys. J.* **3 3 6**, 1092.

Mayr, H. G., Harris, I., and Chan, K. L.: 1984, *Earth, Moon, and Planets* **3 0**, 245.

Mayr, H. G., Chan, K. L., Harris, I., and Schatten, K.: 1991, *Astrophys. J.* **3 6 7**, 361.

Parker, E. N.: 1955, *Astrophys. J.* **1 2 2**, 293.

Pidatella, R. M., Stix, M., Belvedere, G., and Paterno, L.: 1986, *Astron. Astrophys.* **1 5 6**, 22.

Rhodes, E. J., Cacciani, A., Woodard, M., Tomczyk, S., Korennik, S., Ulrich, R. K.: 1987, in The Internal Solar Angular Velocity, eds. B. Durney and S. Sofia, D. Reidel, Dordrecht, 75.

Rudiger, G.: 1980, *Geophys. Astroshys. Fluid Dyn.* **1 6**, 239.

_____ 1989, *Differential Rotation and Stellar Convection: Sun and Solar Type Stars*, Gordon and Breach, New York.

Schmidt, W.: 1982, *Geophys. Astrophys. Fluid Dyn.* **2 1**, 27.

Tuominen, I., and Rudiger, G.: 1989, *Astron. Astrophys.* **2 1 7**, 217.

Tuominen, I., Brandenburg, A., Moss, D., and Rieutord, M. 1993, *Astron. Astrophys.*, in press.

Wasiutynski, J.: 1946, *Astrophys. Norvegica* **4**.

Weiss, N. O.: 1965, *Observatory* **8 5**, 37.

Willson, R. C., and Hudson, H. S.: 1991, *Nature* **3 5 1**, 42.

VARIATIONS OF THE MAGNETIC FIELDS OF THE SUN
AND THE EARTH IN 7–50 DAY PERIODS

V. P. BOBOVA and N. N. STEPANIAN

Crimean Astrophysical Observatory

334413 p/o Nauchny, Crimea, Ukraine

Abstract. The time variations of solar and terrestrial magnetic fields (background magnetic field, power of the active regions, AE and aa-indices) have been studied. The analysis of these data shows that multiplets of 27, 13.5, 9 and 7 day periods exist in the solar data as in the terrestrial data. The solar multiplets 13.5 and 9 days appear predominantly close to the equatorial zone of the Sun and can plausibly be explained by the presence of active longitudes. The similarity of the variations in period in solar and geophysical data provides evidence that the magnetosphere of the Earth is actually a continuation of the heliosphere. The variations of the terrestrial magnetic field are mainly determined by the solar background magnetic fields in middle heliographic latitudes.

1. Data and Method of Analysis

1.1. SOLAR DATA

1.1.1. Synoptic Maps

The background magnetic fields were analyzed on the basis of Hα synoptic maps published in Solar Geophysical Data. The synoptic maps for 1969–1980 are represented by digits +1 and −1 as a function of the background magnetic field sign with a spatial resolution of $10° \times 10°$ for 10° latitude zones ($\varphi = \pm 50°$) and 36 longitude intervals (L=0°–360°).

1.1.2. Plages

Daily observations of area and intensity of plages in the interval 1969–1980 from Solar Geophysical Data were used to determine plage power, M. This is the sum of the product of the plage brightness times its area for each day the plage was on the disk. Thus, the plage power, M, is proportional to the full energy radiated by the plage during the time it was on the visible hemisphere. The spatial distribution of the background magnetic field polarities and the plage power in all 150 synoptic maps were written in the form of a two-dimensional array of 10×5400 terms. Each one of the 10 lines of the array corresponds to a 10-degree latitude zone. For plages, each term in this line corresponds to the power of the plages located in the central meridian zone ($\pm 5°$). For the background fields, each term in a line labeled +1 or -1 corresponds to the polarity of the background magnetic field at that location. The time interval between two neighboring values is 0.75 days (10° of longitude).

Solar Physics **152**: 291–296, 1994.

© 1994 *Kluwer Academic Publishers.*

1.1.3. Relative Sunspots Numbers

Time sets are represented by daily average relative numbers of sunspots on the visible solar disk—R_z (Wolff numbers). These data are taken from Solar Geophysical Data for the corresponding years.

1.2. GEOMAGNETIC DATA

1.2.1. AE index

This is the global index of the polar electrojet. It characterizes the change of H components of the magnetic fields for the auroral zone (Davis and Sugiura, 1966). Data for the interval 1966–1984 were obtained from NGDC (USA).

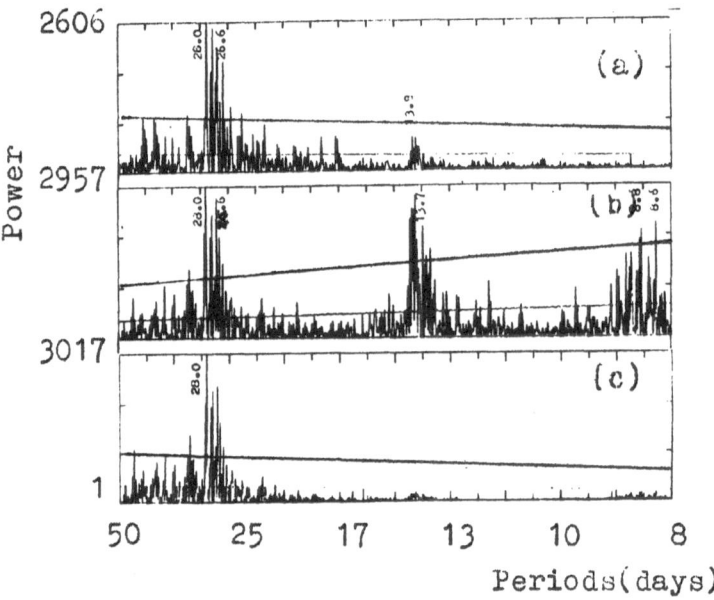

Fig. 1. The power spectra (1969-1980), (a) Wolff numbers, (b) plages ($\sum_\varphi$ M), (c) plages ($\sum_\varphi \sum_\theta$ M).

1.2.2. aa index

Data from two stations at antipodes (Greenwich and Melbourne) were used for constructing the aa index. The average k-indices were transformed into magnetic-field amplitude (Mayaud, 1972). The values of the aa-index were obtained from IDC B2 (Moscow) for 1868–1982.

For a study of the dynamical peculiarities of our time sets, a spectral-harmonic analysis (a superposed epoch with Fourier analysis) was used. Solar and geomagnetic data were analyzed using the same technique. This allows us to obtain power spectra, the most stable periods, and to estimate the full energy of the variations and study its distribution between separate harmonics during different time intervals. We analyzed the main range of 7–50 day periods with a step in frequency of 4.3×10^{-5} days.

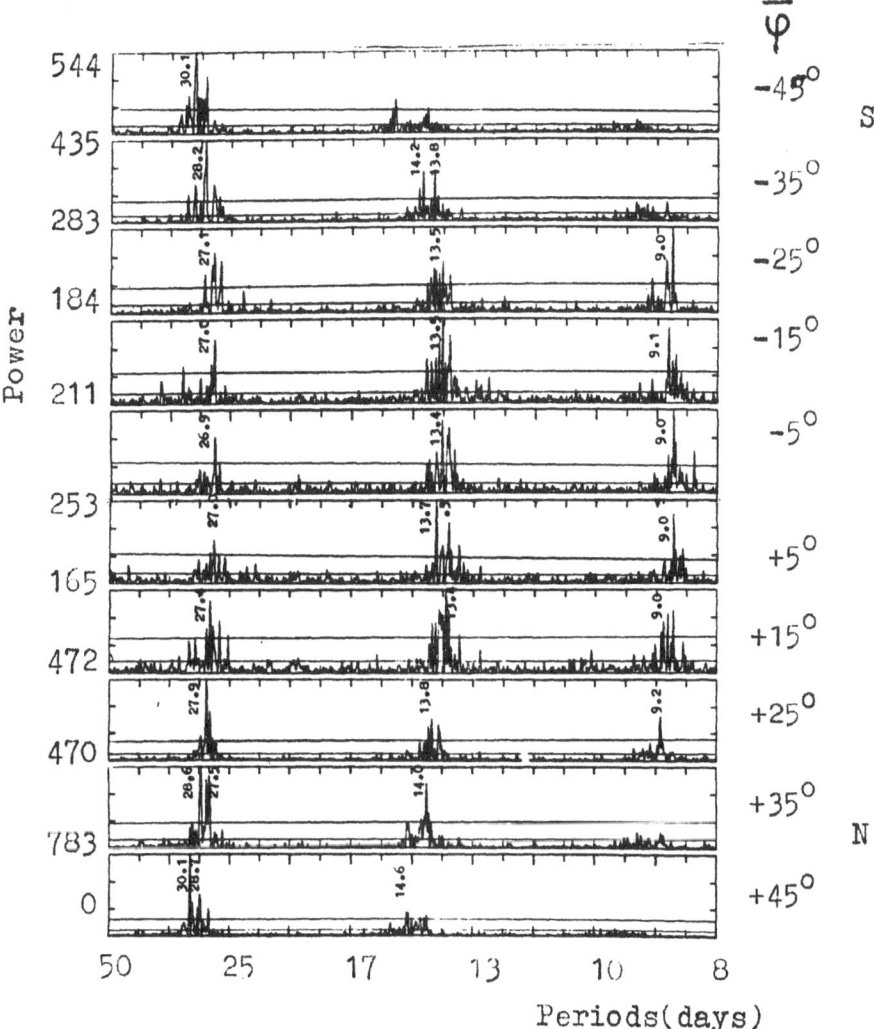

Fig. 2. Power spectra of background magnetic fields for different latitude intervals. Mean latitudes are shown at the right.

2. Results

The Wolff-number power spectrum and the plage power averaged over all latitudes are shown in Figures 1a and 1b respectively. As is seen in Figure 1a, the variations with periods 9 and 13 days are not appreciable. The reason is the following: Wolff numbers include sunspots over the whole disk. If we average the plage power over latitude and longitude, we have the result which is plotted in Figure 1c. Periods 9 and 13 days are absent in this spectrum.

The dependence of the power spectra on the spatial distribution of the solar background fields is shown in Figure 2. One can see that the spectra are different for various latitudes. The 27-day multiplet periods increase with latitude from 26 to 30 days. The structure of the 27-day period is explained by differential rotation. Periods of 9 and 13 days are observed in the equatorial zone, where the solar activity is the highest. The structure of all multiplets changes with time (Figure 3).

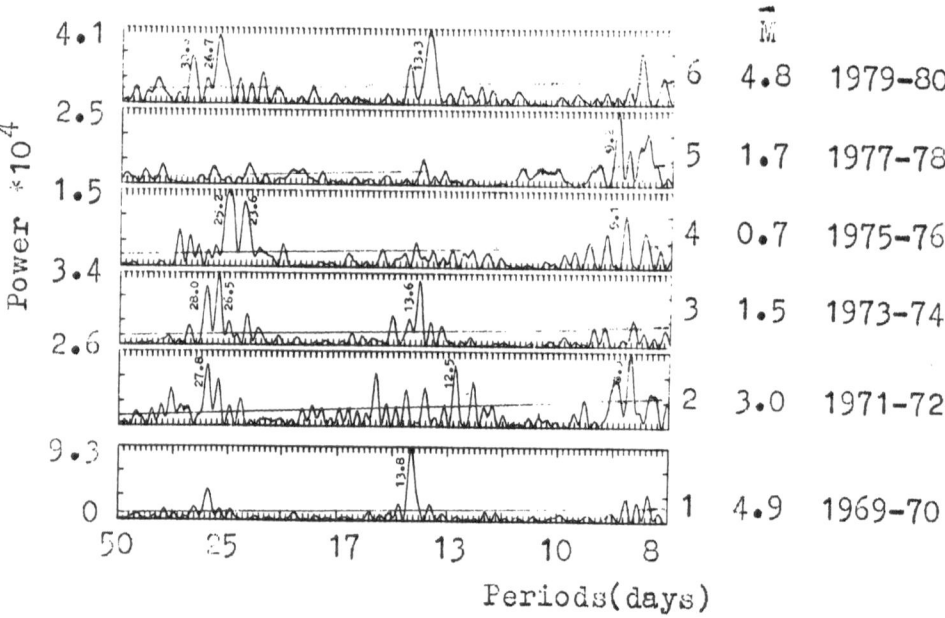

Fig. 3. The power spectra of plages ($\sum$ M) for 25 solar rotations. The mean value (M) is on the right.

Properties of the AE and aa-indices for the same dates were studied in the same way. The spectrum of variation of the geophysical data is similar to the solar spectrum (Figure 4).

The energetic characteristics of different multiplets—spectral power (SP) and the integral energy of the full spectrum (ISP) were studied. Variations of SP and ISP with time are not identical at different latitudes. Differences in SP take place for the most part at the middle latitudes. The smallest differences are observed in the central zone. The largest amplitude of 27-day multiplets of the background field are seen in the middle latitudes where the level of solar activity is low. In the central zone, where the level of activity is high, the spectrum of the background field is closer to the plage spectrum. The power of the different multiplets is almost equal ($SP_{27}:SP_{13}:SP_9=1:1:1$).

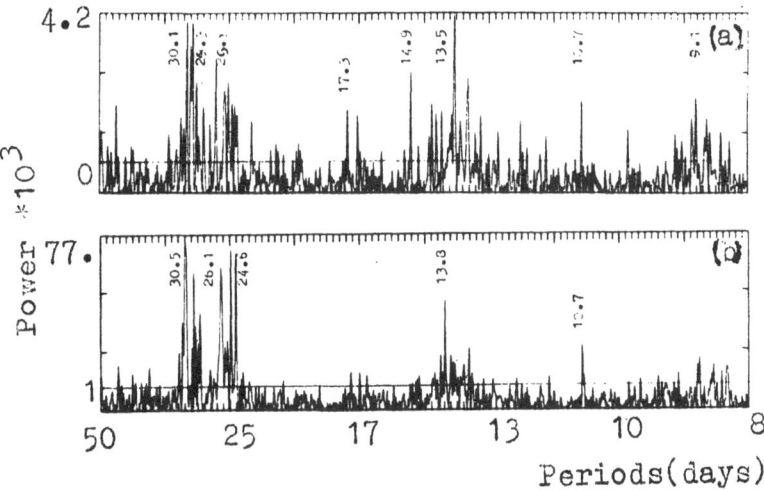

Fig. 4. The power spectra of the geomagnetic index for the interval 1969–80. (a) aa index, (b) AE index.

The ratios of the spectral powers of these multiplets for geophysical data are the same as for the solar background field in middle latitudes ($SP_{27}:SP_{13}:SP_9=3:2:1$).

Analysis of these results shows that the terrestrial magnetic field variations are determined by the solar background magnetic field in middle latitudes. This conclusion is confirmed by the connections found between ISP(aa) and solar activity for growing and decaying stages of the 11 year cycle (Figure 5).

3. Conclusions

We make the following conclusions:

1) The spectra of solar and geomagnetic data in the range of 7–50 day periods are characterized by a number of multiplets near 27, 13.5, 9, and

7 days. Their structure and properties depend on time, solar latitude and longitude.

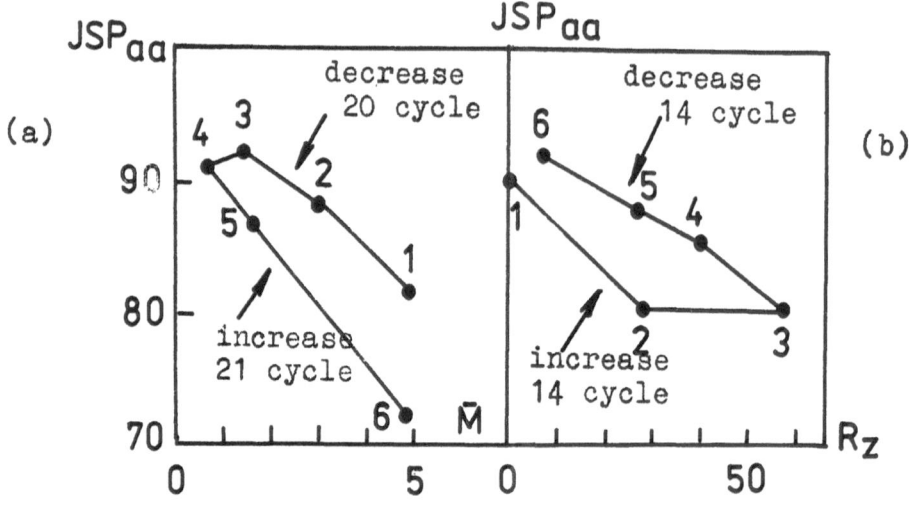

Fig. 5. The change of ISC_{aa} resulting from (a) plage activity, and (b) Wolff number for cycle 14.

2) The 27-day multiplets are caused by solar background magnetic field structures and their differential rotation.

3) The value of ISP for the background fields increases with increasing latitude.

4) The 13-, 9-, and 7-day periods are caused by the spatial distribution of solar activity.

5) Variations of the geomagnetic field are caused mainly by the solar background fields in middle latitudes.

6) An increase of solar activity causes a decrease of the integrated spectral energy, ISP, and a decrease in the stability of the multiplet structure.

References

Davis J. N., and Sugiura M.: 1966, *J. Geophysical Res.* **71**, 785.

Mayaud P. N.: 1972, *J. Geophysic. Res.* **77**, 6870.

Solar Geophysical Data: p. 1, *WDC-A, NOAA, E/GC2*, Boulder, CO, 80303, USA 1969–1980.

DISTINCTION BETWEEN THE CLIMATIC EFFECTS OF THE SOLAR CORPUSCULAR AND ELECTROMAGNETIC RADIATION

T. BARANYI and A. LUDMÁNY

Heliophysical Observatory of the Hungarian Academy of Sciences
H-4010 Debrecen P.O.Box 30. Hungary

Abstract. We study the possibilities of the separation of solar electromagnetic and corpuscular impacts on the terrestrial lower atmosphere by examining their characteristic differences. We focus on the behaviour of the solar-meteorological correlation with respect to characteristic magnetic properties. Examples are given that the solar meteorological correlation - the efficiency of the solar impact - depends on the Sun-Earth attitude, the polarity of the solar main dipole field (and IMF) and the type of the geomagnetic events. This can explain the virtual disappearance or reversal of certain solar-meteorological effects.

1. Introduction

The majority of the solar-meteorological literature uses the tacit assumption that the variations of the meteorological processes are the results of the variations in the solar radiation flux because the other possible energy source, the corpuscular flux is 6-7 orders of magnitude smaller than the radiative one. However, as is well known, most of the reported regularities and correlations may disappear or alter their signs apparently without detectable reasons.

We think that it is a simplification when the intensity is the only measure of the input, as in the electomagnetic case. The impact of the corpuscular flux can be confused with that of the radiative flux in certain periods, because they have more or less parallel trends. However, the corpuscular radiation has additional properties: strong anisotropy and the varying spatial features of the interplanetary magnetic field (IMF), so these conditions should be taken into account and perhaps their variations could explain some previously unexplained results. Moreover, the relatively small but highly anisotropic corpuscular impact might be more efficient in generating atmospheric inhomogeneities and circulations than the electromagnetic radiation.

We study not only the simple correlations between geomagnetic activity (induced overwhelmingly by corpuscular streams) and surface temperature but also its dependence on additional conditions such as the Earth's position in the IMF, polarity of the IMF and the nature (temporal run) of the particle events. We refer partly also to results of former papers: Baranyi and Ludmány (1992) (Paper I) and Baranyi et al. (1993) (Paper II). The aa-index of the geomagnetic activity was used (Mayaud, 1972) along with

Solar Physics **152**: 297–302, 1994.

surface temperatures of 22 European stations on a hundred-year period (see detailed list in Paper II).

2. Dependence on the Earth's Position

Twice a year the Earth gets into the "right" positions, around the equinoxes, enabling higher geomagnetic activity levels than around the solstices; this causes the well-known semiannual wave of geomagnetic activity (Crooker and Siscoe, 1986). Similar semiannual fluctuation can be observed in the correlations between solar and meteorological parameters measured in Budapest, Hungary (Paper I). We computed correlation values for monthly mean aa-index versus monthly mean temperature and monthly total precipitation for data sets of each separate month of the year, i.e. for 119 january values, 119 february values and so on. The distribution of these correlation values shows a similar property to the geomagnetic activity: extremes around the equinoxes. The extremes are positive for temperature and negative for precipitation. This can be interpreted also in terms of the varying terrestrial orientation: the equinox position is more efficient in transmitting the solar corpuscular effects to the lower atmosphere than is the solstice.

Surprisingly, the Wolf-number - precipitation correlation distribution has also a semiannual character, which can be explained only by the above mentioned aspect: the radiation variability obviously cannot cause a semiannual fluctuation but the corpuscular variability certainly can, and as the different factors of solar activity vary more or less similarly, the precipitation and Wolf-numbers may have a virtual interconnection. This is an indirect artifact demonstrating the need of a corpuscular electromagnetic distinction.

3. Dependence on the IMF-Polarity

For the sake of completeness we refer to an even more characteristic result which has recently been submitted for publication (Paper II). We have studied the temperature data of 22 European stations. The semiannual fluctuation cannot be pointed out at all stations, but if we separate two subdivisions of the whole interval studied according to the polarities of the main solar magnetic dipole field (Makarov and Sivaraman, 1986) which may be considered as the IMF-polarity (Rosenberg and Coleman 1969), then the semiannual fluctuation works by parallel solar and terrestrial magnetic fields, and it is missing in the antiparallel intervals. This is not true at three seaside-towns (Genova, Lisbon, Trondheim) but it is fairly remarkable at the other 19 stations plotted in Figure 1. Further details are in Paper II. We conclude that the IMF-polarity plays an important role in these relations, this result is consistent with those of Wilcox et al. (1976) and Rostoker and Sharma (1980).

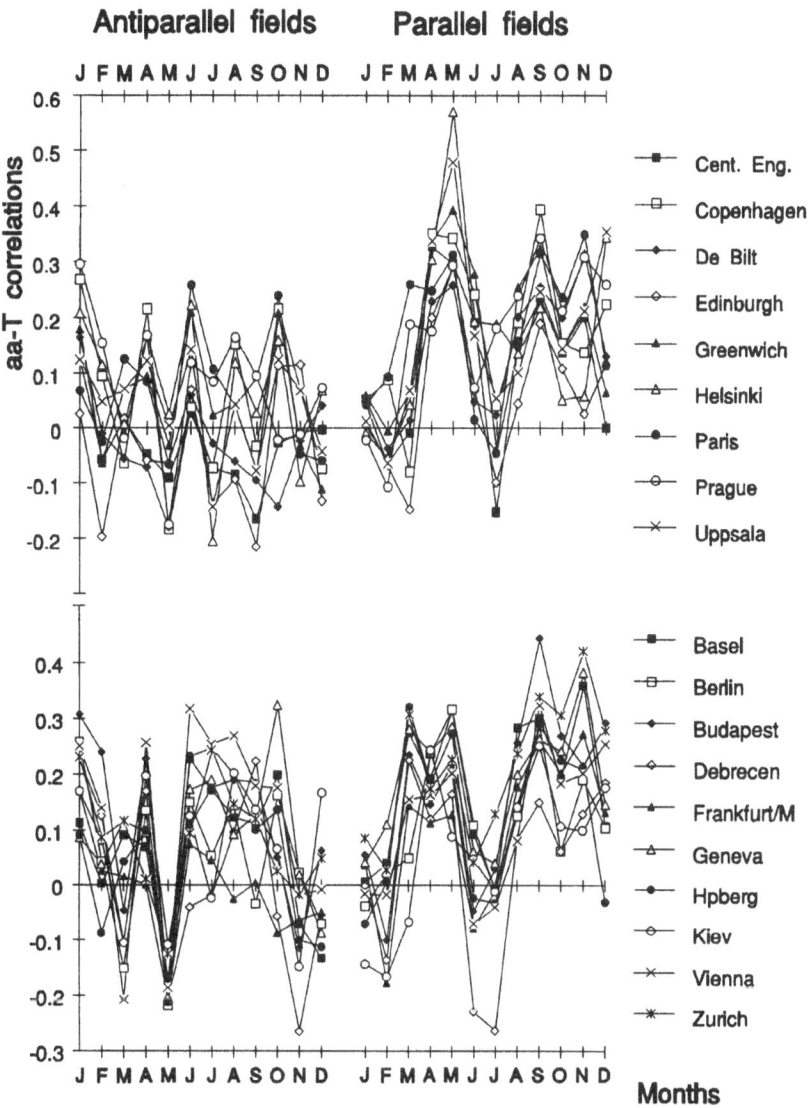

Fig. 1. Annual distributions of the temperature-aa index correlations in the years of antiparallel and parallel orientations of solar and terrestrial magnetic dipole fields at 19 European stations. For a given station and orientation the January value means the correlation between monthly mean temperature and monthly mean aa-index values of all Januaries involved, and so on.

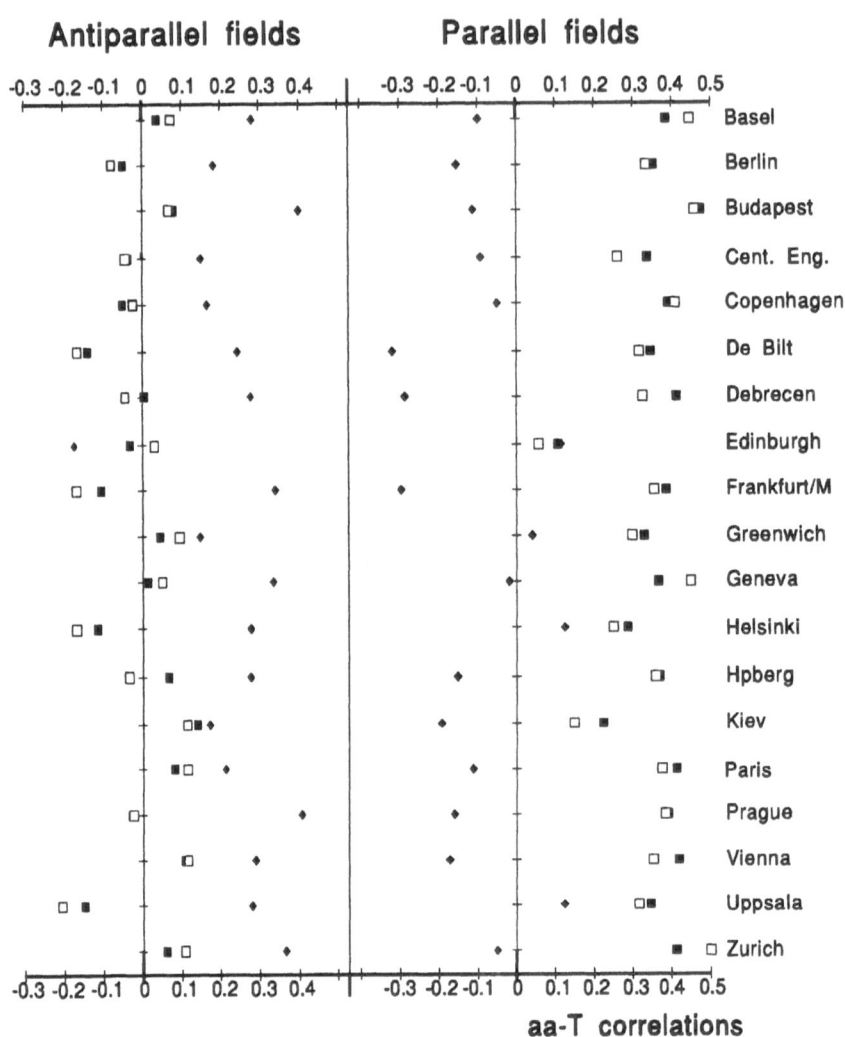

Fig. 2. Dependence of the temperature - aa-index correlation on the orientation of the solar vs. terrestrial magnetic dipole fields as well as on the type of the corpuscular impact (shock, recurrent or fluctuating) for 19 European stations. The temperature data are annual means, the aa-index data are annual sums of the given type.

4. Dependence on the Type of Disturbance

The most pronounced and complex effect of the corpuscular radiation revealed in our studies so far is the clear dependence on the type of the geomagnetic disturbance. These disturbances may be caused by different impacts
of solar origin. (1) The recurrent geomagnetic activity arises on account of
the streams from the polar coronal holes, (2) the shock activity is caused by
active regions and (3) the fluctuating activity is released by moderate velocity wind sources. This distinction is extensively studied by Legrand and
Simon (1985) and Simon and Legrand (1986, 1987), they also compiled classification lists of these events. On the basis of their classification scheme, a
remarkable regularity can be revealed by using the annual sums of these activity types as different aa-activity indices (Figure 2): the correlations of the
temperature with the shock, fluctuating and recurrent disturbances show
opposite behaviour in the periods of parallel and antiparallel orientations
of the solar and terrestrial magnetic fields in most cases. The separation is
clearer in the parallel case like the semiannual fluctuation.

As the shock and fluctuating activity originate from the solar equatorial
belt, it is not surprising that their correlations with the temperature behave
similarly to each other and reverse to that of the recurrent disturbances
coming from the polar region. But it is not quite obvious that they change
their roles after the IMF polarity reversal, this may refer to an important
feature that is not yet clear.

5. Discussion

The figures imply a rather complicated process, a complex spatial and temporal behaviour; they reflect some features of its peculiar geometry: different
solar locations of different disturbance types, their different temporal runs,
varying position of the Earth with respect to the IMF-environment and
alternating polarity of the IMF. All these factors play some roles in the
solar-meteorological effects.

We think that the disappearances of many relationships might be related
to these phenomena. Any virtual relationship can be observed with, say, the
Wolf numbers and after, for instance, the IMF-polarity reversal it disappears
unexpectedly, simply because we did not observe the relevant factors and the
really significant process. In addition, if we mix the inversely working periods
or factors, then the effect can be smeared out which is also demonstrated
by the results of Labitzke and Van Loon (1988).

We can draw the conclusion that the solar corpuscular radiation is probably concurrent with the electromagnetic influence; it may even prevail in
spite of its comparative weakness, at least on a shorter range. This can be
the case even on a cycle-long interval as was shown in Paper I. On longer

time scales the radiative flux is surely determinant.

6. Acknowledgements

We are deeply indebted to J. P. Legrand who kindly put their classification scheme at our disposal. The work was partly supported by the Hungarian Foundation for Scientific Researches under grants Nos. OTKA-F4142/1992 as well as by the Foundation for Hungarian Science.

References

Baranyi, T. and Ludmány A.: 1992, *J. Geophys. Res.* **97**, 14,923 (Paper I.).
Baranyi, T., Ludmány, A. and Terdik, Gy.: 1993, submitted to *J. Geophys. Res.* (Paper II.).
Crooker, N.U. and Siscoe, G.L.: 1986, in P.A. Sturrock (ed.), 'Physics of the Sun' Vol III., D.Reidel, Dordrecht, p. 193.
Labitzke, K. and Van Loon, H.: 1988, *J. Atmos. Terr. Phys.* **50**, 197.
Legrand, J.P. and Simon, P.A.: 1985, *Astron. Astrophys.* **152**, 199.
Makarov, V.I. and Sivaraman, K.R.: 1986, *Bull. Astr. Soc. India* **14**, 163.
Mayaud, P.N.: 1972, *J. Geophys. Res.* **77**, 6870.
Rosenberg, R.L. and Coleman, P.J.Jr.: 1969, *J. Geophys. Res.* **74**, 5611.
Rostoker, G. and Sharma, R.P.: 1980, *Can. J. Phys.* **58**, 255.
Simon, P.A. and Legrand J.P.: 1986, *Astron. Astrophys.* **155**, 227.
Simon, P.A. and Legrand J.P.: 1987, *Astron. Astrophys.* **182**, 329.
Wilcox, J.M., Svalgaard, L. and Scherrer, P.H.: 1976, *J. Atmos. Sci.* **33**, 1113.

EUDOSSO: A SPACE PROJECT FOR SOLAR OSCILLATIONS AND LONG TERM VARIABILITY RELEVANT TO CLIMATIC CHANGES

L. PATERNÒ
Istituto di Astronomia dell'Università, Città Universitaria, 95125 Catania, Italy

S. SOFIA
Dept. of Astronomy, Yale University, P.O. Box 6666, New Haven, CT 06511, USA

ABSTRACT: The basic design and the scientific aims of the EUDOSSO project are described. EUDOSSO is a space instrument conceived for achieving very high accuracy and long term stability in measuring solar diameter time variations. Therefore it is suitable for measuring long period g-modes, oblateness and long-term diameter variations possibly relevant to climatic changes.

1. INTRODUCTION

The EUDOSSO experiment has been conceived for accurate and stable measurements of the solar diameter time variations, along different directions. This experiment can accomplish three important scientific objectives, which cannot be reached by means of any ground-based experiment or other space experiments presently under development.

The scientific objectives are:

1) measurement of the long-period, low-degree and high-order g-mode spectrum, whose knowledge is important for a deep understanding of the structure and dynamics of the Sun's core;

2) measurement of the solar oblateness, related to the gravitational quadrupole moment;

3) measurement of the long term solar diameter variations, as related to total luminosity changes of relevance for terrestrial climate.

The diameter variations produced by the phenomena we intend to study are of the order of 0.01 arc-s, with time-scales ranging from hours to years. The characteristics of the proposed instrument allows us to achieve an accuracy better than 0.003 arc-s per measurement with a stability better than 0.003 arc-s/year.

The instrument was selected by ESA to fly on board the EURECA-3 platform, one of the Columbus precursor flights, in view of the Columbus utilization. Presently, since EURECA reflights have been hibernated due to the lack of funds, the instrument has been proposed for ESA small missions M3 as a part of the experiment GSE (Global Solar Explorer) which also includes the active cavity radiometer VIRGO, scheduled to fly on SOHO, and MGS, an EUV solar imager already selected by ESA for EURECA-3.

Solar Physics **152**: 303–308, 1994.
© 1994 *Kluwer Academic Publishers.*

A phase-A study of EUDOSSO is supported by the Italian Space Agency (ASI), which is also committed to finance the project to completion after the final approval by ESA.

2. SCIENTIFIC COOPERATION

The experiment is carried out through a collaboration among several institutions (Institute of Astronomy of the Catania University, Catania Astrophysical Observatory, Dept. of Astronomy and Space Science of the Florence University, Arcetri Astrophysical Observatory, Center for Aerospace Research of the Rome La Sapienza University, Dept. of Physics of Rome Tor Vergata University, Dept. of Physics of L'Aquila University, Dept. of Astronomy Yale University, USA, NASA - Goddard Space Flight Center, USA, High Altitude Observatory, USA, Mullard Space Science Laboratory, UK, CNRS - Service d'Aeronomie, France) for a total number of 30 scientists and 12 technicians.

3. INSTRUMENT HERITAGE

The EUDOSSO instrument is an upgraded, space-qualified version of the Solar Disk Sextant (SDS) developed at Goddard Space Flight Center (Chiu 1984; Chiu et al. 1984; Sofia et al. 1984), which flew on board stratospheric balloons in 1988, 1990 and 1992 (Sofia et al. 1990, 1991; Maier et al. 1992).

The results of the first two flights are discussed in detail in Sofia et al. (1990), Sofia et al. (1991) and Maier et al. (1992), while the results of the third flight are, at the present, being analysed in detail, and they will be matter of a future paper.

The balloon flights have clearly demonstrated that the space experiment is feasible and that the accuracy required for helioseismological, oblateness and long-term variability investigations is easily attainable.

The time series of data carried out during the balloon flight of 11 October 1990 have clearly detected the tiny effect of the increase of the apparent size of the solar diameter as the Earth approaches perihelion. The least-squared fitted data, during three 20 min measurement time series at fixed telescope angles, gave a slope of 0.0067 m arc-s/s consistent with the expected value of the perihelion.approach of 0.0065 m arc-s/s. The rms error per measurement was in this case about 15 m arc-s.

A new wedge, built by molecular contact, was used for the flight on 30 September 1992. The results show a drastic reduction of the rms error per measurement to a value of the order of 6 m arc-s. A complete rotation of the instrument about its optical axis during 33 min of continuous measurements does not show distortions and drifts as in the previous flight. The oblateness signal deduced from data, once cleaned by the differential atmospheric refraction still detectable at 5 mbars altitude, is $(8.6 \pm 0.7) \times 10^{-6}$.

Early ground based oblateness measurements gave values ranging from $(5.0 \pm 0.7) \times 10^{-5}$ (Dicke and Goldenberg 1967) to $(9.6 \pm 6.5) \times 10^{-6}$ (Hill and Stebbins 1975). However, more recently, Dicke et al. (1986) have produced results differing from the early ones in ways the authors could not explain, unless assuming the oblateness varied with solar cycle.

4. LONG TERM DIAMETER VARIATIONS

The scientific objectives of the EUDOSSO project concerning g-mode helioseismology and oblateness measurement implications have already been discussed in detail elsewhere (Paternò 1993). Here we want to stress the importance of the third aspect of the EUDOSSO measurement capabilities, which concerns the long-term solar variability as a possible input for terrestrial climatic changes (Sofia 1993). This is an important point which involves the study of the Sun during the phases of non-thermal equilibrium and requires a fresh approach to the theory of solar structure and evolution.

The measurement of the solar diameter variations ΔR for a period comparable with that of one solar cycle is important for understanding the mechanisms which control the solar luminosity during the phases of non-thermal equilibrium.

For this purpose to be fulfilled, it is necessary to have also simultaneous measurements of solar total irradiance variations, ΔH.

The total irradiance outside the Earth's atmosphere is a measure of the Sun's total luminosity, L, once all surface modulations (e.g. spots, plages, faculae, network, ...Schatten et al. 1985) have been accounted for.

The measurement of the function $W(t) = (\Delta R/R)/(\Delta L/L)$ is not only important for astrophysics, but also for the possibility of constructing a predictive scheme for terrestrial climatic variations induced by the solar component.

The solar radius variations ΔR are necessarily associated to the expansion or contraction of some layers. This implies that work is done by the layers or on the layers, and consequently the gravitational potential energy changes, inducing luminosity variations ΔL.

In standard models of the Sun, the effects produced by rotation, magnetic field and mass loss are neglected and a stationary theory of convection, the mixing-length, is used. The present Sun is obtained by evolving an initially homogeneous Sun through thermal equilibrium phases. In these conditions, ΔL is simply related to ΔR through the virial theorem and it is not possible to vary a global parameter independently of the others. The changes in global parameters occur on the nuclear time-scale ($\simeq 10^{10}$ ys), much longer than the thermal time-scale ($\simeq 10^7$ ys).

In order to understand how ΔL is related to ΔR, it is necessary to consider the response of the highly non-linear differential equation system, which describes the stellar structure in non-thermal equilibrium, for times long with respect to the dynamical time-scale ($\simeq 1$ h) and short with respect to the thermal time-scale. The response should depend on the nature and properties of the particular physical process which produces the changes.

It is thus necessary to construct more realistic solar models which include, in particular, the effects of rotation, magnetic field and a non-stationary treatment of convection.

Work in this direction has already been developed and is now being improved by research groups participating in this experiment (Endal et al. 1985; Chan and Sofia 1989; Fox et al. 1991). The results of this research can be summarized in three main points:

1) the luminosity variations depend in a dominant way on the surface temperature variations, rather than on the radius variations;

2) the relationship between ΔL and ΔR depends on the particular mechanism which produces the changes and on the depth at which the mechanism operates;

3) the process appearing to be the most serious candidate for producing significant radius variations (tenths of arc-s in time-scales of years) is a variable, strong magnetic field located below the base of the convection zone, in agreement with the most recent dynamo theories of the solar cycle.

The measurement of $W(t)$ would permit to verify the validity of the models and obtain suitable information for elaborating new, more accurate models.

The solar total irradiance detectors ACRIM, operating on SMM, ERB, operating on NIMBUS 7, and ERBS and NOAA-9 have revealed ΔH variations of about 0.1 - 0.2 %, which, when extrapolated to time-scales of some tens of years, could constitute a significant component of terrestial climatic changes. Moreover, recent researches have shown the existence of positive correlations between solar parameters and climatic data. Therefore the suspected link between climate and solar variations has shifted from the realm of speculation to what must now be considered worthy of a serious research effort.

Besides the purely scientific interest, the possibility of establishing how much climatic changes depend on solar variations and predicting which changes are to be induced in the future by solar variability is an important question which involves the world economy.

The characteristic time-scales of the climatic variations range from some tens to some hundreds of years and therefore they are included in the non-thermal equilibrium scales of solar variations.

In order to evaluate the importance of the effects produced by long-term solar variability on climatic changes, one should know how the solar irradiance varied in the past. Since direct and reliable measurements of ΔH do not exist for the past, and the correlations between solar parameters and climatic data are uncertain, the problem of the Sun's long-term influence on the climate is still a controversial matter (Roberts 1989). On the other hand, the current accurate measurements of ΔH cover too short a period to infer a long-term Sun-climate relationship and, even more, to construct a forecast scheme based on the observations alone.

The method we propose here to tackle the problem is based on the measurement of $W(t)$ [$\Delta L = f(\Delta H)$ must be supplied by simultaneous measurements of ΔH] for a period of a solar cycle or possibly longer, and the use of realistic models of the Sun, which indicate how the possible physical processes, operating over the time-scales of interest, modify radius and luminosity.

For this semi-empirical approach to be meaningful in terms of forecasting schemes, it needs to be verified in the past, over climatic time-scales.

This is possible because reliable ΔR data (0.1 %) concerning the last 270 years are available.

These data are based on Mercury and Moon transits on the solar disk, the latter during the total eclipses, and therefore they are free from errors introduced by the presence of the Earth's atmosphere.

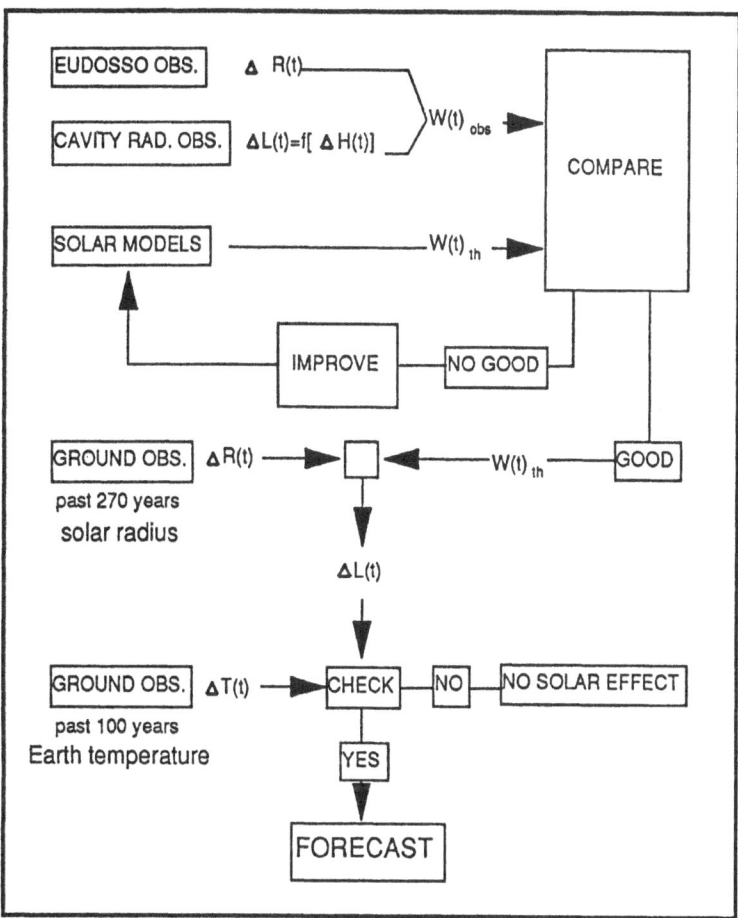

Figure 1: Forecasting scheme for climatic changes caused by solar luminosity secular changes.

The results indicate that there is no evidence of a significant trend in radius increase or decrease, but evidence of an oscillation with an 80.4 y period and 0.25

arc-s amplitude, which shows extreme variations of about 1 arc-s (Parkinson et al. 1988).

This fact poses quite tight bounds on ΔR in the past 270 ys and consequently on ΔL. The effect of ΔR on ΔL can thus be verified in the past by means of the present solar models and $W(t)$ measurements during the space mission, for a sufficiently long period to ascertain the influence of ΔL on climatic changes, especially in the last one hundred years when temperatures at the Earth's surface were systematically and accurately recorded. In Figure 1 we show the flow-chart of such a predictive scheme for climatic changes.

If non-thermal equilibrium solar models are calibrated so as to reproduce the observations during the mission period and the application to the past ΔR data gives results consistent with the recorded climatic changes, climate forecasting may then be possible.

REFERENCES

Chan, K. L., and Sofia, S.: 1989, *Astrophys. J.* **336**, 1022.

Chiu, H. Y. 1984, *Applied Optics* **23**, 1226.

Chiu, H. Y., Maier, E., Schatten, K. H., and Sofia, S.: 1984, *Applied Optics* **23**, 1230.

Dicke, R.H., and Goldenberg, H.M.: 1967, *Phys. Rev. Lett.* **18**, 313.

Dicke, R.H., Kuhn, J. R., and Libbrecht, K.G.: 1986, *Astrophys. J.* **311**, 1025.

Endal, A. S., Sofia, S., and Twigg, L. W.: 1985, *Astrophys. J.* **290**, 748.

Fox, P., Sofia, S., and Chan, K. L.: 1991, *Solar Phys.* **135**, 15.

Hill, H.A., and Stebbins, R.T.: 1975, *Astrophys. J.* **200**, 748.

Maier, E., Twigg, L.W., and Sofia, S.: 1992, *Astrophys. J.* **389**, 447.

Parkinson, J. H., Stephenson, F. R., and Morrison, L. V.: 1988, in *Secular Solar and Geomagnetic Variations in the Last 10,000 Years*, p. 203.

Paternò, L.: 1993, in *Space Projects to Probe the Internal Structure and Magnetic Activity of the Sun and Sun-like Stars*, eds. G. Belvedere and M. Rodonò, *Mem. Soc. Astron. Ital.* **64**, 195.

Roberts, L.: 1989, *Science* **246**, 992.

Schatten, K. H., Miller, N., Sofia, S., Endal, A.S., Chapman, G., and Hickey, J.: 1985, *Astrophys. J.* **294**, 689.

Sofia, S.: 1993, in *Space Projects to Probe the Internal Structure and Magnetic Activity of the Sun and Sun-like Stars*, eds. G. Belvedere and M. Rodonò, *Mem. Soc. Astron. Ital.* **64**, 207.

Sofia, S., Maier, E., and Twigg, L.W.: 1990, in *Helioseismology from Space*, Proc. Scientific Meeting ME.7, The Hague, NL, p. 121.

Sofia, S., Maier, E., and Twigg, L.W.: 1991, *Adv. Space Res.* **11**, 123.

Sofia, S., Chiu, H. Y., Maier, E., Schatten, K. H., Minott, P., and Endal, A. S.: 1984, *Applied Optics* **23**, 1235.

SOLAR MAGNETIC AND BOLOMETRIC CYCLES
RECORDED IN SEA SEDIMENTS

G.Cini Castagnoli, G.Bonino, C.Taricco
Istituto di Cosmogeofisica del C.N.R., C.so Fiume 4, 10133 Torino, Italy
and
Istituto di Fisica Generale dell'Universita', Via P.Giuria 1, 10125 Torino, Italy

Abstract. The total carbonate and thermoluminescence (TL) profiles of the GT89-3 Ionian sea sediment core have been measured in the upper 200 cm of the core spanning the last 3100 years in order to test the presence of the Gleissberg (80-90 yr) cycle in the two different time series recorded in the same archive. Two different sampling intervals respectively of 2.5 mm and 2 mm have been chosen for the measurements in order to obtain results independent from sampling effects in the time series. We have revealed the Gleissberg cycle at 83 and 92 yr in both records.

Introduction

In a previous paper (Cini Castagnoli et al.,1992), we have pointed out the presence of the Gleissberg solar cycle in the $CaCO_3$ time series obtained from the GT90-3 Ionian sea core, using power spectra analysis, by both MEM and Fourier methods. We analyze here the profiles of the carbonate content and the thermoluminescence signal in another core GT89-3 (latitude 39°45'43"N, longitude 17°53'55"E), taken in the same area. The carbonate has been measured at depth intervals of 2.5 mm (Cini Castagnoli et al., 1990a) and the thermoluminescence at intervals of 2 mm. To transform the depth profiles of the cores into well-defined time series, it is necessary to carefully determine the sedimentation rate and to verify whether it remained constant over the time interval spanned by the cores. A detailed evaluation of the dating of the Ionian sea cores, performed by both radiometric and tephroanalysis methods, is reported elsewhere (Bonino et al., 1993). The results of the dating procedure allow for the accurate evaluation of the sedimentation rate, wich is S = (0.0646±0.0007) cm/yr. Therefore the time intervals corresponding to 2.5 mm and 2 mm are respectively 3.87 yr and 3.096 yr, with an error of about 1%. In this paper, the $CaCO_3$ and TL time series of the GT89-3 core are analysed by the cyclogram method and by the superposition of epochs.

Experimental procedure

The total $CaCO_3$ content of each sample of thickness 2.5 mm of the GT89-3 core was determined by titration with EDTA (ethilendiaminetetracetic acid). The material of each sample was dried at 110°C overnight and the $CaCO_3$ from the powder (typically 100 mg) was leached by boiling 2 or 3 times with 50 ml of 2% acetic acid. Titration was performed on three aliquots of the solution and frequent titration of standard solutions was carried out to ensure reproducibility. The triplicate analyses agreed within 0.3% (Cini Castagnoli et al., 1990b).

Here we study the TL profile of the GT89-3 core, sampled at regular intervals of 2 mm, for a total of 1000 samples, taken from the upper part (200 cm) of the core. The preparation of the samples was done in red light, treating the still wet material (~1 g) by successive washing in NaOH, water and acetone, and using a centrifuge, in such a way as

Solar Physics 152: 309–312, 1994.

to preserve the original composition of the polyminerals contained in the sediment (procedure adopted previously in Cini Castagnoli et al., 1990c). After drying in oven at 40°C overnight, the powder was gently sieved and the fraction <44 μm was weighted in samples of 15 mg for the TL measurement. Glow curve measurements were obtained at a heating rate of 5 °C/s. We consider the average TL signal provided by the glow curves at five different temperatures in the plateau, namely at 300, 320, 340, 360, 380 °C.

Data analysis and discussion of results

The $CaCO_3$ and the TL profiles of the GT89-3 core were transformed in two time series as mentioned in the introduction. The two time series, consisting respectively in 801 and 1000 data points, covering the time interval of 3096 yr from the top of the core dated 1979 AD, are analysed by the method of cyclograms (Galli, 1988) and the method of superposition of epochs. The cyclogram method allows a time series to be analysed interval-wise by plotting the running average of the complex amplitudes of the Fourier transform for a given periodicity. A window is moved along the time series, continuously testing for the chosen periodicity. When the chosen periodicity is present and correctly determined, the cyclogram tends to a straight line. If no regular oscillation with periodicity around the chosen test period is present in the data, then the cyclogram has the appearance of a random walk on the plane of the complex Fourier amplitudes. Thus computing several cyclograms for different test periods around a periodicity of interest allows a) for its careful determination and b) for testing its stability along the time series. The cyclograms were computed using test periods between 80 and 100 yr, at yearly steps. In this interval we found straight cyclograms yielding a specific period only at $\tau=92$ yr.

In Figure 1 we show the cyclograms of the $CaCO_3$ and TL time series for the test period $\tau=92$ yr, with a window of $T=\tau$. Each cyclogram is clearly straight and shows that 33 independent vectors (whose extrema are indicated by squares) are almost aligned from 1117 BC to 1979 AD. The average directions of the two cyclograms are about in quadrature over the whole time interval. In the test at yearly steps, we have found an indication of stretching of the cyclograms also for $\tau=83$ yr.

To further confirm these results, the periodicities in the $CaCO_3$ and TL time series have been investigated by the technique of superposition of epochs. The two time series are partitioned into consecutive segments of fixed length T, wich are then superposed on a single interval. In this way, all the periodicities in the time series are averaged to about zero except those with period T/n, where n is an integer number. The basic interval is divided in 5 subdivisions and the mean and the standard deviation of the experimental data falling in each subdivision are computed. In Figures 2a,b are shown the results of this method at T=92 yr for the $CaCO_3$ and TL time series respectively; in Figures 3a,b are shown the results for T=83 yr. The least-square fit of the superposed data to a sinusoidal wave with period T is evaluated and the amplitude (A) and phase (φ) of the best-fit wave are determined together with its correlation coefficient (r^2) to each of the two superposed time series. The values of A, r^2 and φ (with respect to 247.4 AD) are listed in the captions of the Figures 2 and 3. In these figures, in order to better visualize the waves, both the superposed data and the least-square-fit sinusoid have been repeated twice. The error bars (at 1 sigma) appear to be about half of the magnitude of the periodic variations. The high correlation coefficients show the presence of the two specific periodicities at T=92 yr and T=83 yr.

In conclusion, these results indicate that i) the Gleissberg cycle is present in both the records ($CaCO_3$ and TL) of Ionian sea sediments and ii) if the TL signal is forced by the solar activity, as previously proposed (Cini Castagnoli et al., 1990c), the precipitation of the carbonate may be forced by the bolometric solar output coupled to the sunspot activity. This may support the hypothesis of the influence of the solar activity on the climate on century scale time variations.

Acknowledgements
We thank Prof.Carlo Castagnoli for helpful discussions. The technical assistance of Mr.A.Romero is gratefully acknowledged.

References
Bonino, G., Cini Castagnoli, G., Callegari, E., Zhu, G.M.: 1993, Nuovo Cimento C 16, 155.
Cini Castagnoli, G., Bonino, G., Caprioglio, F., Provenzale, A., Serio, M., Zhu, G.M.: 1990a, Geophys. Res. Lett.17, 1937.
Cini Castagnoli, G., Bonino, G., Caprioglio, F., Provenzale, A., Serio, M., Bhandari, N.: 1990b, Geophys. Res. Lett.17, 1545.
Cini Castagnoli, G., Bonino, G., Provenzale, A., Serio, M.: 1990c, Solar Phys.127, 357.
Cini Castagnoli, G., Bonino, G., Serio, M., Sonett, C.P.: 1992, Radiocarbon 34, 798.
Galli, M.: 1988, Solar-Terrestrial Relationships and the Earth Environment in the Last Millennia, Proceedings of the International School of Physics "E.Fermi", Course XCV (1985), edited by G.Cini Castagnoli (Elsevier, Amsterdam, 1988), 246.

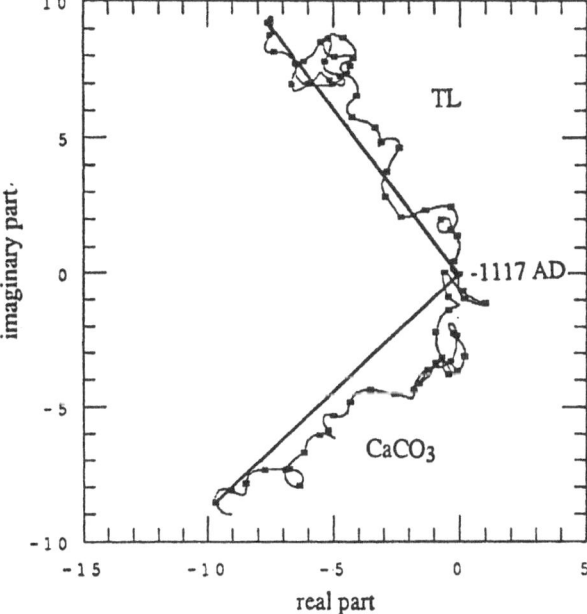

Fig. 1: Cyclograms of the $CaCO_3$ and TL time series of the GT89-3 sea core for a test period τ and a window T, $\tau=T=92$ yr.

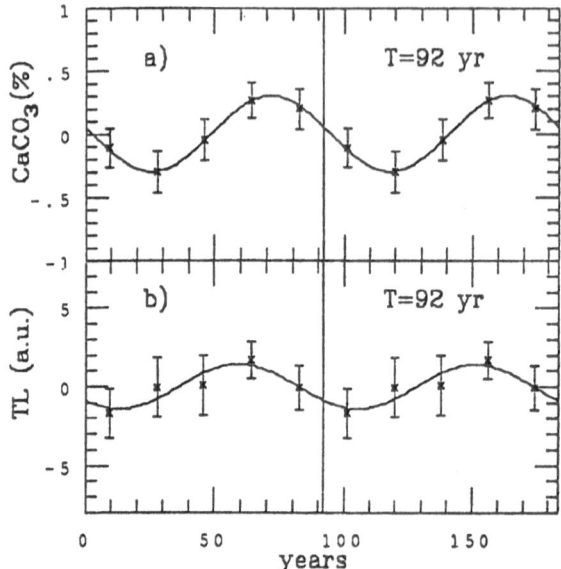

Fig. 2a,b: Superposed data and least-square-fit sinusoid at T=92 yr for the CaCO₃, a), and
for the TL, b), time series, as given by the method of superposition of epochs. a) A=0.3,
φ=2.95, r²=0.99; b) A=1.4, φ=3.80, r²=0.92. The phases are referred to 247.4 AD.

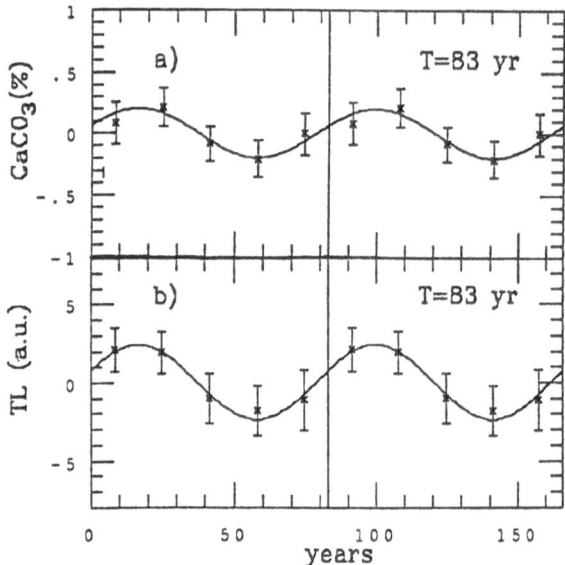

Fig. 3a,b: Superposed data and least-square-fit sinusoid at T=83 yr for the CaCO₃, a), and
for the TL, b), time series, as given by the method of superposition of epochs. a) A=0.2,
φ=0.31, r²=0.97; b) A=2.4, φ=0.31, r²=0.99. The phases are referred to 247.4 AD.

LONG-TERM PERSISTENCE OF SOLAR ACTIVITY

ALEXANDER RUZMAIKIN

Department of Physics and Astronomy, California State University Northridge, 18111 Nordhoff Str., Northridge, CA 91330, U.S.A.

and

JOAN FEYNMAN and PAUL ROBINSON

Jet Propulsion Laboratory, California Institute of Technology, 4800 Oak Grove Dr., Pasadena, CA 91109, U.S.A.

Abstract. The solar irradiance has been found to change by 0.1% over the recent solar cycle. A change of irradiance of about 0.5% is required to effect the Earth's climate. How frequently can a variation of this size be expected?

We examine the question of the persistence of non-periodic variations in solar activity. The Hürst exponent, which characterizes the persistence of a time series (Mandelbrot and Wallis, 1969), is evaluated for the series of ^{14}C data for the time interval from about 6000 BC to 1950 AD (Stuiver and Pearson, 1986). We find a constant Hürst exponent, suggesting that solar activity in the frequency range of from 100 to 3000 years includes an important continuum component in addition to the well-known periodic variations. The value we calculate, $H \approx 0.8$, is significantly larger than the value of 0.5 that would correspond to variations produced by a white-noise process. This value is in good agreement with the results for the monthly sunspot data reported elsewhere, indicating that the physics that produces the continuum is a correlated random process (Ruzmaikin et al., 1992), and that it is the same type of process over a wide range of time interval lengths.

We conclude that the time period over which an irradiance change of 0.5% can be expected to occur is significantly shorter than that which would be expected for variations produced by a white-noise process.

The full paper has been submitted to Solar Physics. Part of the research decribed here was carried out by JPL, Caltech under a contract with NASA.

References

Mandelbrot, B., and Wallis, J.: 1969, *Water Resources Research* **5**, 321.
Ruzmaikin, A. A., Feynman, J., and Kosacheva, V. P.: 1992, in Karen L. Harvey (ed.), *The Solar Cycle*, ASP Conference Series **27**, 547.
Stuiver, M., and Pearson, G. W.: 1986, *Radiocarbon* **28**(2B), 805.

STELLAR IRRADIANCE VARIATIONS DUE TO SURFACE TEMPERATURE INHOMOGENEITIES

K. G. STRASSMEIER

Institute for Astronomy, University of Vienna,
Türkenschanzstraße 17, A-1180 Wien, AUSTRIA

ABSTRACT. Observations of rotational modulation of continuum brightness and photospheric and chromospheric spectral-line profiles of late-type stars indicate the presence of very inhomogeneous surface temperature distributions. We present three stellar examples (VY Ari, HR 7275, HU Vir) where time-series photometry is used to trace the evolution of spotted regions. Simultaneous spectroscopy and Doppler imaging for one of the three stars (HU Virgo, Fig. 1) makes it possible to compute the temperature distribution of the photosphere and the relative intensity distribution of parts of the chromosphere (from Ca II K and Hα line profiles). The combination of time-series spot modeling and Doppler imaging enabled us to determine the *sign* and amount of differential surface rotation on HU Vir. We found a big, cool polar spot (see figure below) and a differential (surface) rotation law

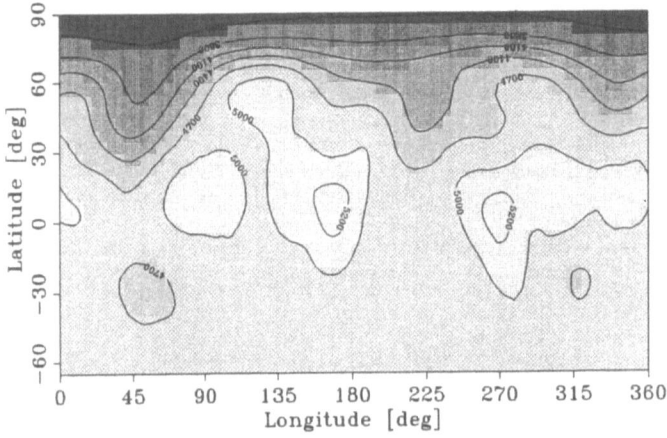

Fig. 1

where higher-latitude regions rotate faster than lower-latitude regions (opposite to what we see on the Sun). Currently, this ensemble of techniques – time-series photometry and photospheric and chromospheric Doppler imaging – is only applicable to stars overactive by approximately a factor of 100 as compared to the active Sun, e.g. the evolved components in RS CVn-type binaries and some rapidly-rotating, single, pre-main sequence stars or giant stars. Stellar rotation is a fundamental parameter for (magnetic) activity. Starspots, or any other surface inhomogeneities, allow one to derive very precise stellar rotation rates and, if coupled with seismological observations of solar-type stars, could provide information on the internal angular momentum distribution in overactive late-type stars.

To be published in Astronomy & Astrophysics.

Solar Physics **152**: 314, 1994.

Voyager EUV Solar Spectra During 1981-1993

Giuliana de Toma[1], R. J. Vervack, Jr.[2], B. R. Sandel[2], and R. Stalio[1]

[1] Dipartimento di Astronomia, Università di Trieste, 34131 Trieste, and CARSO, Area di Ricerca, 34012 Trieste, Italy

[2] Lunar and Planetary Laboratory, University of Arizona, Tucson, AZ 85721

EUV solar observations have been obtained with the Voyager 1 and Voyager 2 Ultraviolet Spectrometers since 1981. The two instruments, which are now at a distance of more than 40 AU, cover the wavelength range between 500 and 1700 Å at a resolution of about 30 Å. Stellar observations have demonstrated their photometric stability, and both instruments are operating correctly and capable of solar observations for the next several years. This long term data set represents the longest monitoring of the sun in this spectral region performed by the same instrument.

Our present work focuses on the calibration and analysis of the 500-1250 Å wavelength region with particular interest in H Ly α, which for the first time is observed far from the Earth's atmospheric contamination. EUV solar variability on long (11-year cycle) and short (27-day) time scales is examined. H Ly α and the H continuum (600-912 Å) are shown to vary in time with very high correlation. Our recent results also indicate that the ratio between the H continuum and the emission lines in the region of 970-1040 Å may differ from the presently accepted value. Our data, compared with other reference spectra, show a higher H continuum. This feature is consistent with the spectrum of sunlight reflected by Saturn's rings that was recorded by the Voyager 2 UV spectrometer in 1981 with the detector operating in a different mode.

Solar Physics **152**: 315, 1994.

ON P-MODE OSCILLATIONS IN STARS
FROM 1 $M_\odot$ TO 2 $M_\odot$

N. AUDARD and J. PROVOST

Département Cassini, Observatoire de la Côte d'Azur BP 229,
F-06304 Nice Cedex 4, France

Abstract. The structure of stars more massive than about $1.2M_\odot$ is characterized by a convective core. We have studied the evolution with age and mass of acoustic frequencies of high radial order n and low degree ℓ for models of stars of 1, 1.5 and 2 $M_\odot$. Using a polynomial approximation for the frequency, the p-mode spectrum can be characterized by derived global asteroseismic coefficients, i.e. the mean separation $\nu_0 \sim \nu_{n,\ell} - \nu_{n-1,\ell}$ and the small frequency separation $\Delta\nu_{0,2} \sim \nu_{n,\ell=0} - \nu_{n-1,\ell=2}$. The diagram $(\nu_0, \Delta\nu_{0,2}/\nu_0)$ plotted along the evolutionary tracks would help to separate the effects of age and mass. We study the sensitivity of these coefficients and other observable quantities, like the radius and luminosity, to stellar parameters in the vicinity of 1 $M_\odot$ and 2 $M_\odot$; this sensitivity substantially depends on the stellar mass and must be taken into account for asteroseismic calibration of stellar clusters. Considering finally some rapid variations of the internal structure, we show that the second frequency difference $\delta_2\nu = \nu_{n,\ell} - 2\nu_{n-1,\ell} + \nu_{n-2,\ell}$ exhibits an oscillatory behaviour well related to the rapid variation of the adiabatic exponent γ in the HeII ionization zone.

A more complete discussion is given in Audard N, Provost J, 'Seismological properties of intermediate-mass stars', A&A, 1993, in press.

References

Audard N., Provost J.: 1993, 'About seismological properties of intermediate-mass stars' in Weiss W. W., Baglin A., ed(s)., *Inside the stars*, ASP Conf.Ser., page 544

FACULAR EXCESS RADIATION AND THE ENERGY
BALANCE OF SOLAR ACTIVE REGIONS

M. STEINEGGER, H. HAUPT

Institut für Astronomie, Graz, Austria

and

P.N. BRANDT, W. SCHMIDT

Kiepenheuer-Institut für Sonnenphysik, Freiburg, Germany

Abstract. Plage areas and intensities derived from CaII K spectroheliograms are used as a proxy for the facular irradiance excess of solar active regions for the period 19 August to 4 September 1980. Using a calibration method proposed by Vršnak et al. (1991), the "photospheric facular index" (PFI) with constant facular contrast $C_p = 0.018$ is replaced by a variable C_p, depending on the plage brightness. A significant increase of C_p from 0.015 to $\simeq 0.025$ is found for plage areas varying from a few to approx. $6 \cdot 10^3$ millionths hemispheres.

Combining the facular irradiance excess with sunspot deficits (as determined for the same period by Steinegger et al. 1990) yields good agreement with the irradiance variations measured by ACRIM I, using a center-to-limb variation of C_p according to Chapman and Meyer (1986). The ratio of facular excess to sunspot deficit (integrated over solid angle 2π) decreases from values of 1.5 to 2 for regions with sunspot areas below 100 millionths hemispheres to $\simeq 0.2$ for sunspots of areas > 1000 millionths hemispheres.

Key words: Solar Constant – Solar Irradiance – Photometry of Active Regions – Solar Activity

THE 73-DAY PERIODICITY OF THE FLARE INDEX DURING THE CURRENT SOLAR CYCLE 22

ATİLA ÖZGÜÇ and TAMER ATAÇ
Kandilli Observatory, Boğaziçi University, Çengelköy,
81220 Istanbul, Turkey

Abstract. The flare index of the current solar cycle 22 is analysed to detect intermediate-term periodicities from Sep.1, 1986 to Dec. 31, 1991. Power spectral analysis of the time series of solar flare index data reveals a periodicity around 73 and 53 days. We find that a periodicity of 73 days was in operation from 1988 November to the end of 1991 December. We also find that when the 73-day periodicity or the 154-day periodicity is in operation, the flare index is well correlated with the relative sunspot numbers. As a conclusion, we do not expect to see a resumption of the 154-day or 73-day periodicity, but we do expect only one of the periodicity near the integral multiplies of $25\overset{d}{.}8$ in the next solar cycles.

Acknowledgement

Travel support for the poster presentation of this work was provided by Boğaziçi University Research Fund. The complete version of this paper to be published in Solar Physics.

Phase differences between irradiance and velocity low degree solar acoustic modes revisited

A. Jiménez

Instituto de Astrofísica de Canarias. E-38200.La Laguna . Tenerife.

Abstract:

Since1984,simultaneous observations of irradiance and velocity solar acoustic modes, have been carried out by several authors in order to measure the phase difference between irradiance and velocity modes. Following the earliest observations with a stratospheric balloon (Frolich and van Der Raay,1984), a two ground-based stations (Tenerife and Baja California) were established (Jimenez et al,1990) obtaining coherence results in the frequency range from 2.5 mHz to 4.3 mHz. These phase differences between irradiance and velocity solar acoustic modes are interpreted in terms of the non-adiabatic behaviour of the solar atmosphere. In 1988 the IPHIR (Frolich et al,1988) instrument flown on the PHOBOS-2 mission to Mars and measured the solar irradiance during 150 consecutive days.The best velocity observations obtained in Tenerife for this period were compared with IPHIR data to compute the phase differences (Schrijver et al,1991). The final conclusion is that good agreement is attained between "space", "quasi-space" and "ground observations" which yield a phase difference of about -125 degrees in the frequency range 2.5mHz to 4.2 mHz, with a slight increase suggested by the data running up to 4.6 mHz.

The entire scientific community looks forward to SOHO data , which will increase a great deal the quality of Helioseismology results. Concerning phase differences and gains between intensity and velocity solar oscillations, we will have more raw continous data (GOLF,VIRGO and MDI) which will improve the accuracy of the results, and we will probably be able to extend the results to low and high frequencies and study "changes " with time. Other important projects before the lunch of SOHO are the SOVA experiment (EURECA satellite), which is now in the space and DIFOS experiment on board the CORONAS satellite which will be lunched in autumm 1993. These irradiance data together with ground-based velocity data will offer a good opportunity to improve current results.

References.

Frolich, C., Bonnet, R.M., Bruns, A.V., Delaboudiniere, J.P., Domingo, V., Kotov, V.A.,Kollath, Z., Rashkovsky, D.N., Toutain, T., Vial, J.C., Wehrli, Ch.: 1988, in: *Proc. Symp. Seismology of the Sun and Sun-like Stars*. Noordwijk,ESA SP- 286, p.359.
Frohlich, C., and van der Raay, H.B.: 1984, in:*The Hydromagnetics of the Sun*. Noordwijk, ESA SP-220, p.17.
Jiménez, A., Álvarez, M., Andersen, N.B., Domingo, V., Jones, A., Pallé, P.L., and Roca Cortés, T.:1990, *Solar Phys*, **126**,1
Schrijver, C.J., Jiménez, A.,and Dappen ,W.: 1991, *Astron.Astrophys*. **251**, 655.